Gary Gilleskie, Charles Rutter, Becky McCuen
Biopharmaceutical Manufacturing

Also of Interest

Flow Chemistry, Volume 1 & 2.
Fundamentals and Applications
2nd Edition
Ferenc Darvas, György Dormán, Volker Hessel, Steven V. Ley (Eds.), 2021
Set-ISBN 978-3-11-073679-3

Catalysis for Fine Chemicals
2nd Edition
Werner Bonrath, Jonathan Medlock, Marc-André Müller, Jan Schütz, 2024
ISBN 978-3-11-109609-4
e-ISBN 978-3-11-110267-2

Solubility in Pharmaceutical Chemistry
Christoph Saal, Anita Nair (Eds.), 2020
ISBN 978-3-11-054513-5
e-ISBN 978-3-11-055983-5

Downstream Processing in Biotechnology
Venko N. Beschkov, Dragomir Yankov (Eds.), 2021
ISBN 978-3-11-057395-4
e-ISBN 978-3-11-057411-1

Gary Gilleskie, Charles Rutter, Becky McCuen

Biopharmaceutical Manufacturing

—

Principles, Processes, and Practices

2nd Edition

DE GRUYTER

Authors
Dr. Gary Gilleskie
Golden LEAF Biomanufacturing
Training and Education Center
Suite 195
850 Oval Drive
Raleigh, NC 27606
USA
glgilles@ncsu.edu

Dr. Charles Rutter
Golden LEAF Biomanufacturing
Training and Education Center
NC State University
850 Oval Drive, Campus Box 7928
Raleigh, NC 27695-7928
USA
cdrutter@ncsu.edu

Dr. Becky McCuen
KBI BioPharma, Inc.
1101 Hamlin Rd
Durham, NC 27704
USA
bmccuen@kbibiopharma.com

ISBN 978-3-11-111206-0
e-ISBN (PDF) 978-3-11-111245-9
e-ISBN (EPUB) 978-3-11-111396-8

Library of Congress Control Number: 2025931503

Bibliographic information published by the Deutsche Nationalbibliothek
The Deutsche Nationalbibliothek lists this publication in the Deutsche Nationalbibliografie;
detailed bibliographic data are available on the internet at http://dnb.dnb.de.

www.degruyter.com
Questions about General Product Safety Regulation:
productsafety@degruyterbrill.com

Preface to the second edition

The preface to the first edition of this book conveyed the key motivation for writing it: to provide concise, comprehensive coverage of the foundations of biopharmaceutical production, inspired by the dynamic learning environment of the Biomanufacturing Training and Education Center at NC State University. This aim and inspiration are as true now as in 2021 when the first edition of this book was published.

So why a second edition and why now?

The biopharmaceutical industry continues to experience unprecedented growth, driving transformative advancements in human health. Innovations in biopharmaceuticals not only fuel economic growth but also profoundly impact millions of lives worldwide, offering hope and improved quality of life. While estimates vary, one source projects that the global biopharmaceutical market will grow from approximately US$511 billion in 2023 to US$1,375 billion by 2033.* The list of diseases preventable or treatable by biopharmaceuticals is long and expanding, now bolstered by cutting-edge modalities such as mRNA vaccines, gene therapies, and cell therapies. Among these breakthroughs is the first United States Food and Drug Administration-approved therapy utilizing CRISPR (clustered regularly interspaced short palindromic repeats) gene-editing technology – Casgevy®, now available on the market to treat sickle cell disease. Ensuring that their impact continues and grows requires a workforce that is highly educated and skilled in the complex processes and regulations required of biopharmaceutical production. Now, more than ever, resources like this book must remain clear, current, and comprehensive to meet the growing demands of the industry.

As with the first edition, the second edition offers a unique blend of theoretical concepts and practical applications, making it an invaluable resource for both students and professionals alike. While the topics of chapters remain unchanged, this edition includes several updates and enhancements, including

- New examples of impactful biopharmaceuticals like Ozempic®, a treatment for type 2 diabetes, and Comirnaty®, a vaccine against COVID-19.
- Updated content on rapidly evolving technologies, such as single-use systems and continuous processing, to reflect the latest advancements.
- Refined explanations of a number of key concepts to enhance clarity and ensure the book remains an engaging resource for students and professionals.
- New review questions added to Chapter 3 and updates to review questions in most other chapters.

*Biologics Market Size, Share & Trends Analysis Report By Source (Microbial, Mammalian), By Product (MABs, Recombinant Proteins, Antisense & RNAi), By Disease Category, By Manufacturing, By Region, and Segment – Global Industry Analysis, Size, Share, Growth, Trends, Regional Outlook, and Forecast 2024–2033. 2023. Nova One Advisor. (Accessed August 13, 2024, at https://www.novaoneadvisor.com/report/biologics-market#:~:text=Report%20Description,11.8%25%20during%20the%20forecast%20period).

https://doi.org/10.1515/9783111112459-202

These revisions ensure that the book continues to serve as an engaging, up-to-date resource for learning and reference.

In the first edition, we acknowledged a number of colleagues and friends who helped in many ways to bring the book together. Their foundational contributions remain vital to this second edition, and so our appreciation bears mentioning again.

Lauren Lancaster, Ben Lyons, Michele Ray-Davis, Arjun Shastry, and John van Zanten provided data and process-related information that helped in creating real-world examples and problems. Mark Burdick, Sahr James, Mark Pergerson, Jennifer Sasser, and John Taylor supported the creation of a number of figures. Donna Gilleskie, Matt Gilleskie, Erik Henry, Russ McCuen, Baley Reeves, and Caroline Smith-Moore all read through early versions of various chapters and provided helpful feedback that led to continuous improvement in content and clarity during the writing process. John Amara, Suzanne Bellemore, Akshat Gupta, Thomas Parker, and Jonathan Steen from MilliporeSigma provided images and reviewed filtration chapters, greatly enhancing their quality.

Brian Herring worked tirelessly to create most of the figures that are included in the book, and did the same to update figures for this second edition. And Patty Brown undertook many editing duties.

We also want to express special thanks to those whose contributions enriched this second edition. Our appreciation extends to Jennifer Pancorbo for her expertise in vaccines; Sara Siegel and Kurt Selle for their contributions in microbiology and molecular biology; Lucia Clontz for her insights into microbial contamination control; and Danny Schmitt for his expertise in lyophilization. Their contributions have made this edition stronger and more valuable.

We are deeply grateful for the collective contributions of these individuals, without whom this second edition would not have been possible.

On behalf of the authors,
Gary Gilleskie

Contents

Chapter 9
Formulation operations: ultrafiltration —— 256

Chapter 10
Summary and trends in biopharmaceutical processing —— 296

Chapter 1
An introduction to biopharmaceutical products

Because this book is focused on the processes and methods for manufacturing bio-pharmaceuticals, a logical starting point is an in-depth discussion of the output of these processes – biopharmaceutical drug products – to establish a clear understanding of what exactly is being produced. As you will see, biopharmaceuticals are a class of medicines that is among the most effective and biggest selling in the world. Their active ingredients are complex biological molecules and components (e.g., proteins, viruses, and cells) that are used to treat, prevent, and, in some cases, cure illnesses that medicines produced by chemical synthesis alone often cannot. Their development has had significant societal impact and marks one of the great achievements of modern science.

This chapter provides an overview of biopharmaceuticals by addressing the following questions:

– What are biopharmaceuticals, how are they different from more conventional small-molecule drugs such as aspirin, and why is it worth making the distinction?
– What are some examples of biopharmaceutical products and the diseases they treat?
– What exactly are the components of biopharmaceutical drug products, and how are these products administered to patients?
– What quality attributes are critical for biopharmaceutical drug products?

1.1 Definitions and background

Let's start by answering the question, "What is a biopharmaceutical?"

1.1.1 Biopharmaceuticals

The term biopharmaceutical has a number of different definitions that depend on whom you ask. In this book, we take a broad view of biopharmaceuticals and define them as medicines inherently biological in nature and manufactured by or from organisms, including cells from living organisms [1]. Examples of such organisms or cells include humans and human cells, animals and animal cells, plants and plant cells, and microorganisms.

An example of a biopharmaceutical that most are familiar with is insulin, which is used to treat diabetes. Insulin is a protein hormone produced by the pancreas that helps to regulate the metabolism of carbohydrates by keeping one's blood sugar from getting too high or too low. Diabetes is a disease that occurs when the body does not

https://doi.org/10.1515/9783111112459-001

make enough insulin (Type 1) or does not use it properly (Type 2), resulting in high blood glucose levels. If glucose levels stay elevated for an extended period, complications such as heart disease, stroke, high blood pressure, neuropathy (nerve damage), kidney disease, vision loss, and others may arise [2]. Treatment for type 1 diabetes requires that a patient takes insulin to make up for the insulin the body does not produce. Treatment for type 2 diabetes may or may not involve taking insulin.

So how is insulin for therapeutic use made? From its first use in the 1920s until 1982, all insulin was animal sourced. Specifically, insulin was extracted from pig and cattle pancreases and injected into human patients. Insulin produced in this way saved millions of lives but produced adverse reactions in many human patients. In addition, only three days' worth of insulin for a diabetic patient could be prepared from a single pig pancreas [3]. Then in October 1982, the U.S. Food and Drug Administration (FDA) approved an insulin product made in a very different way: Humulin®, developed by Genentech and marketed by Eli Lilly. Humulin® became the first insulin analog produced through recombinant deoxyribonucleic acid (DNA) technology – that is, through a type of genetic engineering – and, notably, the first recombinant medicine approved for use in the United States [287]. Recombinant technology is discussed in more detail in the next section. In the case of Humulin®, which is still on the market, the bacteria *E. coli* is manipulated to produce human insulin through the addition of a gene (a segment of DNA that codes for a certain protein) for human insulin [4, 287]. Most insulin (or its analogs) in use today is produced through recombinant DNA technology and produced in either *E. coli* or yeast, such as *Saccharomyces cerevisiae*. Producing insulin through genetic engineering allows for greater control over production – and results in fewer adverse reactions – compared to using pig and cattle pancreases as the insulin source.

Many insulins are now on the market, representing protein therapeutics produced through recombinant DNA technology. Many more examples are discussed throughout this chapter, and recombinant DNA will be described in more detail shortly.

It is worth noting that some writers limit their definition of biopharmaceuticals to only those medicines that have been produced using recombinant DNA technology. While genetic engineering has opened up a world of possibilities for production of biomolecules, the definition of biopharmaceutical used in this book encompasses all medicines made by or from living organisms; this includes medicines produced by nonrecombinant methods. Numerous examples exist, but let's consider one with which you are likely familiar: flu vaccine. Many types of flu vaccine are on the market, and most are produced using nonrecombinant sources. Vaccines aim to prevent diseases, in contrast to therapeutics such as insulin, which treat existing diseases. Vaccines often consist of live attenuated or inactivated forms of the disease-causing virus and stimulate the body's immune system to build defenses without causing illness. Consider the specific example of Flucelvax® Quadrivalent, a seasonal flu vaccine made from four different influenza virus strains. Each virus strain is grown in Madin-Darby Canine Kidney (MDCK) cells [5] without the use of recombinant DNA technology. Because the different virus strains are propagated in MDCK cells, Flucelvax® Quadrivalent fits our definition of biopharmaceutical. After propagation, each virus strain is inactivated, disrupted, and purified. The four virus strains are then pooled to produce the vaccine.

We take a broad definition of biopharmaceuticals in this book because, whether or not recombinant production sources are used, many similarities exist in manufacturing operations and regulatory expectations for products produced by or from living organisms. It is worth noting an exception to our definition of biopharmaceuticals. Medicines based on small molecules that can be produced from cells, such as antibiotics, are generally not classified as biopharmaceuticals, even though they seem to fit our definition, because (1) the active molecule is much smaller than a protein therapeutic or the antigen that makes a vaccine, for example, and (2) the process for production, particularly after the culture step, can be different from processes used to produce large molecules such as proteins and larger viruses.

1.1.2 Recombinant medicines explained

Let's start this discussion by considering DNA. DNA is a macromolecule that consists of four chemical bases and exists within cells. DNA provides the code – via transcription to ribonucleic acid (RNA) and translation of RNA to protein – for protein production in living organisms. Recombinant medicines are those produced by recombinant DNA technology, a type of genetic engineering. Recombinant DNA is DNA that has been combined from multiple sources. Recombinant DNA inserted into a cell allows the cell to produce a protein that it normally would not. By manipulating DNA in cells such as *E. coli*, the genetically modified cells can be grown and used as a kind of living factory to produce a product, such as a protein like insulin, that it would not naturally produce. (*E. coli* is a bacterium and most strains are harmless, despite the bad press the pathogenic strains give it.) This concept is particularly useful for medicines based on large and complex molecules because chemical synthesis of these molecules can be cost prohibitive or just not possible with current technology.

1.1.3 Proteins: key ingredients in most biopharmaceutical products

The majority of biopharmaceuticals are either proteins – such as insulin, or protein based – such as flu vaccines made from influenza virus, which has significant protein content. Given their prevalence, let's consider proteins in more depth.

Proteins are macromolecules composed of amino acids. In humans, 20 amino acids are used in protein synthesis. As shown in Figure 1.1, an amino acid consists of an α carbon in the center, to which is attached an amine group, a carboxyl group, a hydrogen atom, and side chain designated as "R," that is different for each amino acid.

Amine group   Carboxyl group

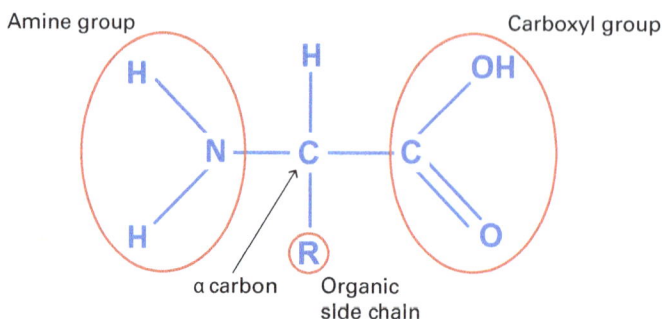

Figure 1.1: Chemical structure of an amino acid, the building block for proteins. Note that at neutral pH, the carboxyl group is deprotonated and carries a negative charge, while the amine group is protonated and carries a positive charge. Image © NC State University; reprinted with permission.

Figure 1.2 shows R groups for the 20 different amino acids. R groups have different properties – for example, some are positively charged, some negatively charged, some are polar (soluble in water) but not charged, and some are nonpolar (limited solubility in water). Differences in R groups among amino acids mean different proteins have different properties, such as charge and hydrophobicity. These properties can be exploited in the design of a manufacturing process, particularly for the separation of product from impurities. This topic is discussed in more detail in Chapter 8.

To form proteins, amino acids combine through the reaction of the amine group of one amino acid with the carboxyl group of another. The reaction is referred to as a condensation reaction because water is formed as a product. Within cells, condensation reactions are catalyzed by ribosomes. The resulting bond is referred to as a peptide bond. As more amino acids react, the polymer becomes larger and a protein is formed. Proteins have a characteristic three dimensional structure. To describe protein structure, biochemists typically refer to four different levels – primary, secondary, tertiary, and quaternary – each defined below and illustrated in Figure 1.3, using insulin as an example:

- **Primary structure** refers simply to the sequence of amino acids in the protein.
- **Secondary structure** refers to regularly repeated local structures created by hydrogen bonds on nearby amino acids. The result of this hydrogen bonding is the formation of alpha helices and beta strands and sheets.
- **Tertiary structure** refers to the overall three-dimensional structure of a single protein molecule.
- **Quaternary structure** is the structure that results from the interaction of more than one protein molecule.

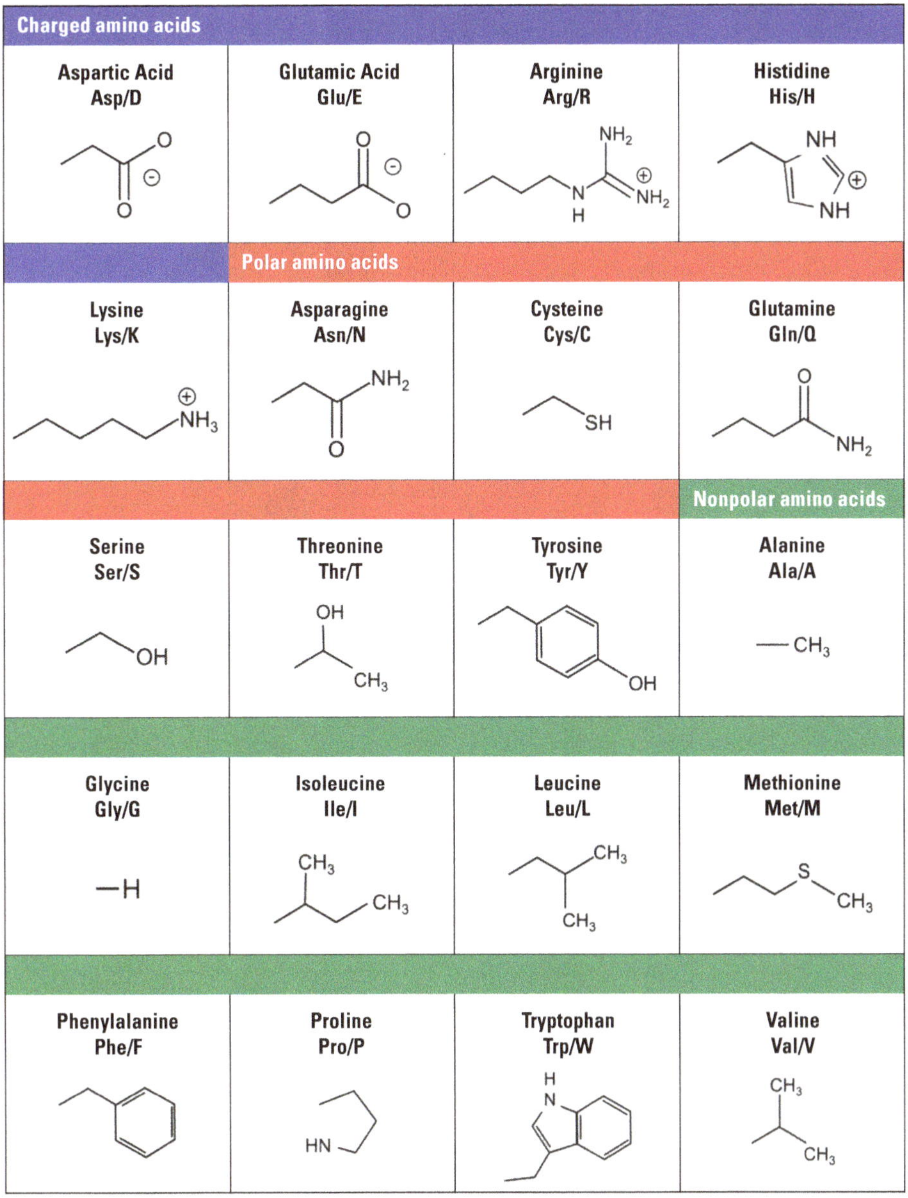

Charged amino acids			
Aspartic Acid Asp/D	**Glutamic Acid** Glu/E	**Arginine** Arg/R	**Histidine** His/H
Polar amino acids			
Lysine Lys/K	**Asparagine** Asn/N	**Cysteine** Cys/C	**Glutamine** Gln/Q
		Nonpolar amino acids	
Serine Ser/S	**Threonine** Thr/T	**Tyrosine** Tyr/Y	**Alanine** Ala/A
Glycine Gly/G	**Isoleucine** Ile/I	**Leucine** Leu/L	**Methionine** Met/M
Phenylalanine Phe/F	**Proline** Pro/P	**Tryptophan** Trp/W	**Valine** Val/V

Figure 1.2: Side chains of the 20 amino acids that make up proteins in humans. Image © NC State University; reprinted with permission. Note that aspartic acid and glutamic acid are shown in their anionic form.

(a)

(b)

(c)

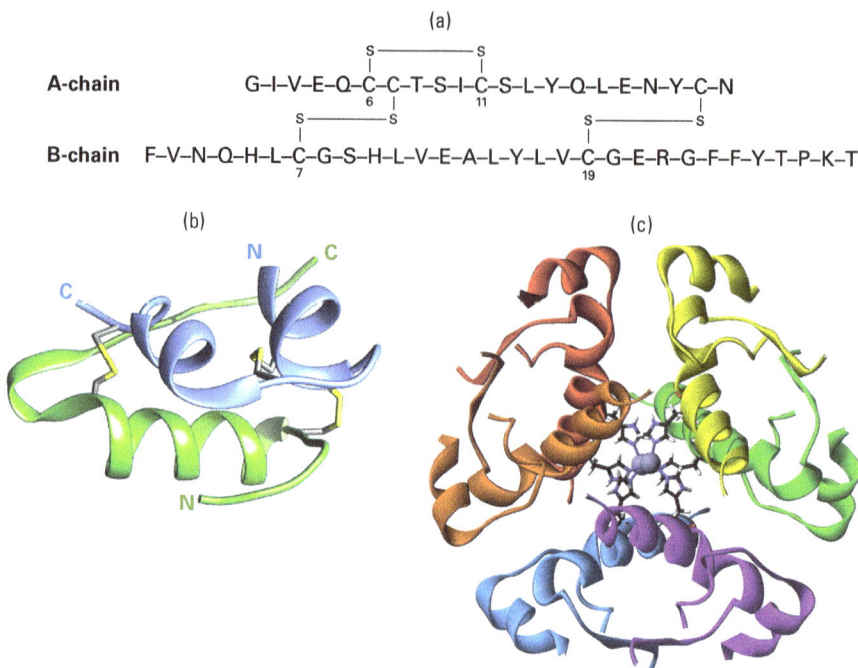

Figure 1.3: The primary, secondary, tertiary, and quaternary structure of human insulin. (a) is the primary structure and shows the 51 amino acids that make up insulin along with its three disulfide bridges. (b) is a ribbon diagram illustrating the secondary and tertiary structure, with the A chain (blue) forming two short helices, and the B chain (green) a helix and a strand structure. (c) is the quaternary structure, a hexameric form of insulin made up of three insulin dimers. Insulin is stored within the cells as hexamer. (Note that each of the six insulin monomers is shown in a different color.) Image (a) © NC State University; reprinted with permission. Image (b) by PDB-101 (PDB101.rcsb.org) at https://pdb101.rcsb.org/global-health/diabe tes-mellitus/drugs/insulin/insulin, used under CC-BY-4.0 license (https://creativecommons.org/licenses/by/4.0/legalcode). Image (c) Benjah-bmm27, Public domain, via Wikimedia Commons.

Three-dimensional structure is relevant to biopharmaceutical manufacturing because protein structure is critical to its function. If that structure is disrupted during processing, it is likely that the protein will not function as intended. Loss of secondary, tertiary, and/or quaternary structure is referred to as denaturation.

To avoid denaturing a biopharmaceutical product, it is important to know the causes for denaturation. A short list is given below:

– Heat. Heat disrupts weak bonds (e.g., hydrogen bonds) that maintain a protein's three-dimensional structure. For example, denaturation of immunoglobulin G (IgG) becomes irreversible at above 65 °C [6]; however, denaturation temperatures vary depending on the specific protein and its environment.
– High or low pH. Acidic and basic conditions impact the charge of amino acids that make up the protein. A change in charge of the amino acid side chains can

disrupt hydrogen bonding and salt bridges, both of which impact the three-dimensional structure of the protein molecule.
- Shear. Shear can be a cause of denaturation, particular when a gas/liquid interface is present [7].
- Certain chemical agents. Chaotropic agents destabilize proteins by disrupting hydrogen bonding between water molecules that surround a protein in solution or by directly interacting with the protein [8]. Examples include ethanol, guanidine hydrochloride, and urea.

It is worth noting that the terms *protein* and *peptide* are sometimes used interchangeably; however, a generally accepted distinction is that peptides are made up of shorter chains of amino acids than proteins. The U.S. FDA specifically defines a peptide as containing 40 or fewer amino acids [9]. One of the biggest-selling drugs worldwide, Ozempic® [11], is a peptide under the definition of the U.S. FDA. It is a glucagon-like peptide 1 (GLP-1) receptor agonist produced in a yeast and approved for the treatment of type 2 diabetes. It is administered subcutaneously as a sterile aqueous solution [288].

1.1.4 A final comment regarding terminology

In these first few sections, we have introduced several important definitions and clarified important terms. Before continuing, it's important to address one more relevant term.

The term biological product is one that is commonly used in a regulatory context. In this text, the terms *biopharmaceutical* and *biological product* are used synonymously. According to the U.S. FDA, biological products "can be composed of sugars, proteins, or nucleic acids or complex combinations of these substances, or may be living entities such as cells and tissues. Biologics are isolated from a variety of natural sources – human, animal, or microorganism – and may be produced by biotechnology methods and other cutting-edge technologies." Further, "biological products include a range of products such as vaccines, blood and blood components, allergenics, somatic cells, gene therapy, tissues, and recombinant therapeutic proteins" [10]. This definition and the examples provided align with the definition of biopharmaceutical used in this book.

1.2 Biopharmaceuticals versus conventional small-molecule pharmaceuticals

Many of the most commonly used medicines in the world are made via chemical synthesis (think aspirin) rather than through biological methods. However, if we relied solely on chemical synthesis for pharmaceutical production, the medicines with large, relatively complex active ingredients such as insulin or flu vaccine, would likely not exist. As the size of these active components (e.g., proteins) become larger and more complex, chemical synthesis becomes more challenging, particularly at the scales required for commercial production, or even impossible as mentioned previously.

In this section, we take a deeper look at biopharmaceuticals by comparing them to the more familiar small-molecule pharmaceuticals that are produced via chemical synthesis. A number of important differences exist, as described below.

1. *Biopharmaceuticals are large, complex molecules, viruses, or even cells*, while more traditional pharmaceuticals are small molecules with well-defined structures. Consider an example of each. Aspirin, shown in Figure 1.4, which is produced by chemical synthesis, is one of the most widely used drugs in the world. It provides relief from pain, fever, and inflammation caused by a variety of conditions. Humira®, a biopharmaceutical, was the largest-selling (based on revenue) drug in the world from 2012 to 2020 [11]. It is a human immunoglobulin G of subclass 1 (IgG1) monoclonal antibody, with a structure similar to that shown in Figure 1.4, and is used to treat rheumatoid arthritis, among a number of autoimmune diseases. It has a molecular weight of approximately 148,000 Da [12] and contains over 20,000 atoms; in comparison, the molecular weight of aspirin is only 180 Da and consists of 21 atoms [13]. Table 1.1 shows a comparison of sizes of some well-known small-molecule drugs with biopharmaceuticals.

2. *Biopharmaceuticals are produced using relatively complex processes that are biological in nature.* Small-molecule pharmaceuticals like aspirin are most often produced by chemical synthesis. We have already discussed that the biopharmaceuticals Humulin® and Flucelvax® are produced using *E. coli* and MDCK cells, respectively. Humira® is produced by large-scale culture of genetically engineered Chinese hamster ovary (CHO) cells [14]. Again, the complexity of the molecules that are the active ingredient in biopharmaceuticals necessitate that they be produced by or from living organisms or cells of living organisms – because chemical synthesis of such complex structures is neither economically viable nor, in many cases, currently possible.

3. *Biopharmaceuticals are largely administered to patients parenterally,* [15] which literally means other than through the gastrointestinal tract. They are not taken orally. Common parenteral routes of administration for biopharmaceuticals are intravenous, subcutaneous, and intramuscular injection and intravenous infusion. Why parenteral delivery? Because most biopharmaceuticals are proteins or contain protein components as their active ingredients. If administered orally, these proteins would be digested in the stomach due to low pH and/or digestive proteases, thereby rendering the biopharmaceutical ineffective. In addition, their large size would make transport of the molecule across the intestinal wall difficult [16]. Small-molecule drugs, on the other hand, are primarily administered orally – as a tablet, capsule, or liquid – but also topically and parenterally. All parenteral drugs, including biopharmaceuticals, must be sterile because the parenteral route of drug delivery bypasses the body's natural defenses [17].

Other differences between biopharmaceutical and traditional small-molecule drugs are that biopharmaceuticals are typically more immunogenic – that is, likely to create an unwanted immune response – and less stable than small-molecule drugs.

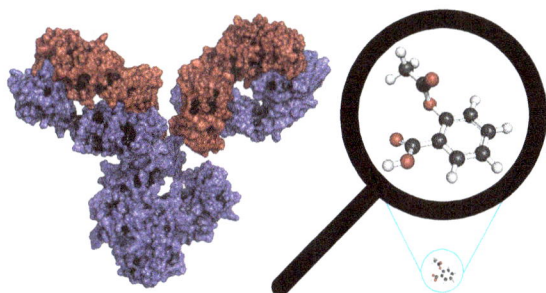

Figure 1.4: A comparison of an IgG1 monoclonal antibody, like the active ingredient in Humira®, to acetylsalicylic acid, the chemically synthesized active ingredient in aspirin. Note that the IgG1 model shows the antibody heavy chains in blue and light chains in red. In the aspirin representation, white represents hydrogen, black carbon, and red oxygen. IgG1 image by molekuul © 123RF.com. Aspirin image, © NC State University, includes ball-and-stick illustration by molekuul © 123RF.com.

Table 1.1: A size comparison of some well-known small-molecule drugs to biopharmaceuticals.

Product example	Active ingredient	Biopharmaceutical or small molecule	Molecular weight (Da)[a,b]	Method of production[b]
Aspirin [20]	Acetylsalicylic acid	Small molecule	180	Chemical synthesis
Tylenol® [20]	Acetaminophen	Small molecule	151	Chemical synthesis
Lipitor® [20]	Atorvastatin calcium trihydrate	Small molecule	1,209	Chemical synthesis
Humulin® [4]	Insulin	Biopharmaceutical	5,808	Produced by growing genetically engineered *E. coli*
Humira® [12]	Anti-TNF monoclonal antibody	Biopharmaceutical	148,000	Produced by growing genetically engineered CHO cells
Novoeight® [21]	Factor VIII	Biopharmaceutical	166,000 (does not include post translational modifications)	Produced by growing genetically engineered CHO cells
Flucelvax® Quadrivalent [5]	Subunits of influenza virus	Biopharmaceutical	Subunits of 100 nm virus particles	Produced by propagating influenza virus in MDCK cells

Table 1.1 (continued)

Product example	Active ingredient	Biopharmaceutical or small molecule	Molecular weight (Da)[a,b]	Method of production[b]
Luxturna® [22]	Adeno-associated virus serotype 2 (AAV2) genetically modified to express the human RPE65 gene	Biopharmaceutical	~3,750,000 [18]	Produced by transfection of HEK293 cells with plasmid DNA
Kymriah® [23]	T cells	Biopharmaceutical	7 µm T cells [19]	Produced by genetically modifying T cells

[a] For the small-molecule drugs, molecular weight values were obtained using the PubChem database [20].
[b] For the biopharmaceuticals, method of production was taken from their package inserts. Molecular weights (or size) were also taken from package inserts, with the exception of Luxturna® and Kymriah®, as indicated in the table.

1.3 Biopharmaceutical product classes and examples

There are a variety of different types of biopharmaceuticals, beyond the examples presented up to now. They are powerful medicines that treat, prevent, or cure numerous diseases. The majority of biopharmaceuticals can be classified broadly as protein therapeutics, vaccines, gene therapies, and cell therapies. Each is considered in greater detail below.

1.3.1 Protein therapeutics

As the name implies, protein therapeutics are proteins that treat an existing disease. Insulin and Humira®, both discussed previously, are two examples. Further, even though the U.S. FDA makes a distinction between proteins and peptides based on the number of amino acids, as described previously, we will not make that distinction here. There are many peptide therapeutics, like Ozempic®, on the market that fit our definition of biopharmaceuticals. And given that both proteins and peptides are made up of amino acids, we will include peptide therapeutics within the broader category of protein therapeutics.

Why do proteins make good medicines? The thousands of proteins in your body [24] serve a variety of functions. The many biological activities carried out by proteins lead to a number of ways they can be used for therapeutic benefit [25], including:
- Replacing a protein that is lacking or abnormal. Insulin for treatment of diabetes is an example; a variety of insulin-based biopharmaceuticals, such as Humulin® and Lantus®, are commercially available. Factor VIII for treatment of hemophilia

A – a disease in which blood does not clot normally, caused by a deficiency of the protein clotting factor Factor VIII – is another example; a number of different Factor VIII-based biopharmaceuticals are on the market, such as Esperoct®.

– Augmenting an existing biological pathway. Neulasta® is an example. It is a granulocyte colony-stimulating factor (G-CSF) manufactured by recombinant DNA technology. In the human body, G-CSF causes neutrophils, a type of white blood cell that fights infection, to grow in the bone marrow. In patients being treated with chemotherapy or radiation, neutrophils may stop being produced, which increases a patient's risk of infection. Neulasta® stimulates production of neutrophils thereby helping patients undergoing chemotherapy or radiation to fend off infection [26].

– Interfering with a molecule or organism. Avastin® is an example. It is a recombinant humanized IgG1 antibody produced in CHO cells that is used to treat various cancers. The antibody binds vascular endothelial growth factor (VEGF). In the human body, VEGF is involved in the formation of new blood vessels in cancer cells, allowing them to grow and metastasize. To work, it must bind to receptors on cells; however, because the Avastin® antibody binds VEGF, VEGF cannot bind to the cell receptors, which inhibits formation of new blood vessels, thereby inhibiting cancer development and growth [27].

– Providing a function that does not normally exist. Botox® is an example. It is a botulinum toxin type A, produced from fermentation of Hall strain *Clostridium botulinum* type A. It is used to treat a number of conditions, including chronic migraines and upper limb spasticity. It is also used for cosmetic purposes. It "blocks neuromuscular transmission by binding to acceptor sites on motor or autonomic nerve terminals, entering the nerve terminals, and inhibiting the release of acetylcholine" [28].

– Delivering other compounds. Antibody-drug conjugates – a monoclonal antibody covalently linked to a small, cytotoxic compound – are an example. The monoclonal antibody binds to specific receptors found on certain types of cells, such as cancer cells. The linked drug enters these cells and kills them without harming other cells.

Protein therapeutics can be classified according to the type/function of the protein as shown in Table 1.2.

1.3.2 Vaccines

A vaccine is a biopharmaceutical used to prevent a disease by improving immunity to that disease. It typically contains an agent that either: resembles a disease-causing microorganism or virus, resembles a portion of the microorganism or virus, or has the ability to produce a part of the disease-causing microorganism or virus. The agent (or its action) stimulates the body's immune system to recognize and destroy the disease-causing microorganism or virus if encountered again. Vaccines take many forms, as

Table 1.2: Different classes of protein therapeutic biopharmaceuticals.

Class	Description	Examples of commercial products/company/primary indication
Blood clotting factors	Proteins in the blood, such as Factor VIII and Factor IX, that control bleeding. Most are for treatment of hemophilia A or B.	Esperoct®/Novo Nordisk/Hemophila A
Growth factors	Molecules that simulate cell growth. Common examples include erythropoietin (EPO), a protein that plays a role in production of red blood cells, and colony-stimulating factor (CSF), which promotes the growth and differentiation of stem cells. Numerous EPO-based biopharmaceuticals are available for the treatment of anemia. There are also numerous CSF-based biopharmaceuticals for the treatment of low white blood cell count.	Epogen®/Amgen/anemia Neulasta®/Amgen/Reduces risk of infection during chemotherapy
Hormones (or hormone-mimicking therapeutics)	Molecules that serve as messengers, controlling and coordinating activities throughout the body. Insulin, human growth hormone, and follicle-stimulating hormones are protein hormones produced as biopharmaceuticals.	– Humulin® R/Eli Lilly/Diabetes – Lantus®/Sanofi/Diabetes – Ozempic®/Novo Nordisk/Type 2 diabetes[a]
Interferons, interleukins, tumor necrosis factors	Relatively small proteins that impact immune response. Among this group, interferon-based biopharmaceuticals are the most common. They are used to treat hepatitis C and multiple sclerosis among other diseases.	Plegridy®/Biogen/Multiple sclerosis

Monoclonal antibodies and monoclonal antibody-based products	Proteins that have a high affinity for the same epitope on an antigen. They work through multiple mechanisms in providing treatment for diseases. This category also includes monoclonal antibody-based products such as antibody-drug conjugates and bispecific antibodies. Many of the world's largest-selling drugs are monoclonal antibodies. Antibody-drug conjugates (ADCs) are monoclonal antibodies covalently linked to a small, cytotoxic compounds. The monoclonal antibody binds to specific receptors found on certain types of cells, such as cancer cells. The linked drug enters these cells and kills them without harming other cells. All currently approved ADCs are cancer treatments. Bispecific antibodies are antibodies that can bind to two specific antigens at the same time.	– Keytruda®/Merck/Melanoma, lung cancer, various other cancers[a] – Humira®/AbbVie/Rheumatoid arthritis, numerous others[a] – Dupixent®/Regeneron/Atopic dermatitis, numerous others[a] – Stelara®/Johnson & Johnson/Plaque psoriasis, psoriatic arthritis, others[a] – Opdivo®/Bristol-Myers Squibb/Metastatic melanoma, other cancers[a] – Darzalex®/Johnson & Johnson/Multiple myeloma[a] – Skyrizi®/AbbVie/Plaque psoriasis, numerous others[a] – Avastin®/Genentech (Roche)/Metastatic colorectal cancers, other cancers[a]
Therapeutic enzymes	This category represents a variety of recombinant protein enzymes – molecules that catalyze reactions.	Kyrstexxa®/Savient Pharmaceuticals/Gout
Thrombolytic agents	Drugs that dissolve clots; many are based on the protein tissue plasminogen activator. Most are for treatment of heart attacks, strokes, and other conditions caused by blood clotting.	Activase®/Genentech (Roche)/Acute ischemic stroke, myocardial infarction
Fusion proteins	Combined proteins created by combining genes originally coding for separate proteins. Frequently, one partner has a molecular recognition function while the other may impart added stability or novel targeting.	Eylea®/Regeneron/Macular degeneration[a] Enbrel®/Amgen/Rheumatoid arthritis, numerous others

[a] Drugs among the top 15 biggest selling (based on revenues) in the world in 2023 [11].

shown in Table 1.3, including viral (based on attenuated – i.e., weakened – whole virus, inactivated whole virus, or portions of a virus), microbial (based on an attenuated whole microorganism, portion of the microorganism, or inactivated toxin from a microorganism), and nucleic acids such as RNA.

Table 1.3: Examples of vaccines by type.

Type	Commercial example	For the prevention of	Host (production system)
Viral			
Whole:			
– Live attenuated	FluMist® [29]	Flu	Chicken eggs
– Inactivated	Ipol® [30]	Polio	Vero cells
Subunit:			
– From actual virus	Flucelvax® Quadrivalent [5]	Flu	MCDK cells
– Recombinant protein	Recombivax HB® [34]	Hepatitis B	*S. cerevisiae*
– Virus-like particle	Gardasil® 9ª [33]	Diseases by human papillomavirus	*S. cerevisiae*
Microbial			
Whole:			
– Live attenuated	Vaxchora™ [31]	Cholera	*V. cholerae*
Subunit:			
– Recombinant protein	Bexsero® [37]	Meningococcal disease	*E. coli*
– Polysaccharide/ conjugated polysaccharide	Prevnar 20® [35]	Diseases caused by *S. pneumoniae*	*S. pneumoniae and C. diphtheriae*
Toxoid	Daptacel® [32]	Diphtheria, tetanus, and pertussis	*C. diphtheriae, C. tetani, and B. pertussis*
Nucleic acid vaccines	Comirnaty®ª [36]	COVID-19	NA

Note: Indication and host information taken from package inserts for each vaccine.
ªAmong the top 15 biggest-selling drugs in the world in 2023 [11].

Numerous vaccines are available in addition to the flu vaccine discussed previously. Examples are given in Table 1.3. You are likely familiar with at least some of these from vaccinations that you or loved ones have received. Vaccines have gotten a lot of attention over the past few years in light of the coronavirus disease 2019 (COVID-19) pandemic. One of the vaccines approved in the United States for use against COVID-19 is Comirnaty®. It is a relatively new type of vaccine based on messenger RNA rather than a weakened form of the virus [36].

Messenger RNA provides a blueprint to produce proteins in cells through the translation process mentioned in the previous discussion on recombinant medicines. Viruses contain a protein coat that surrounds their genetic material. If the virus's genome – that is, the hereditary information in the form of DNA or RNA that is responsible for producing the protein coat – is known, then genes can be determined that encode for the specific virus proteins. Once these genes are known, messenger RNA sections can be designed that, along with our own cells, produce a viral protein. The immune system recognizes the protein and mounts an immune response that builds the body's defenses against the pathogen, like the SARS-CoV-2 virus that causes COVID-19. One advantage to RNA vaccines is the relatively short time for development.

Note that while most vaccines are preventive, some vaccines are therapeutic. In therapeutic vaccines, antigens associated with an illness are introduced to a patient to stimulate the body's immune system to fight an illness that is present, hence use of the term *therapeutic*. Because they work to stimulate a patient's immune system, these therapies are referred to as vaccines.

1.3.3 Gene therapies

Gene therapy is a technique that uses genetic material to treat, prevent, or even cure a disease. This can be done by replacing a faulty or missing gene with a healthy copy, inactivating a gene that is functioning improperly, introducing a new gene into the body to help fight a disease, or by editing existing genes within the body. Genes are typically delivered to patients using viruses that act as vectors, such as adenovirus, lentivirus, or adeno-associated virus. Viruses make good vectors – a "vehicle" used to transport DNA into a host cell – because they have the ability to deposit DNA or RNA into cells by infection. Gene therapy offers the potential to treat a range of disorders that result from defects in a single gene (i.e., monogenic disorders). Importantly, the U.S. FDA has recently approved gene therapy treatments for certain monogenic disorders, such as sickle cell disease and hemophilia, demonstrating the significant advancements in this field [38]. It also offers the potential to treat polygenic disorders such as heart disease, cancer, diabetes, and infectious diseases such as HIV.

Broadly, gene therapies may be classified as in vivo or ex vivo. In vivo gene therapy delivers a gene (or some other form of genetic manipulation) directly to cells in a patient's body. Ex vivo gene therapy involves removing cells from a patient's body, manipulating the cells in a lab, and then transplanting them back to the patient. Figure 1.5 illustrates the differences in the two therapies.

A list of some U.S. FDA-approved gene therapy products is provided in Table 1.4.

In Vivo

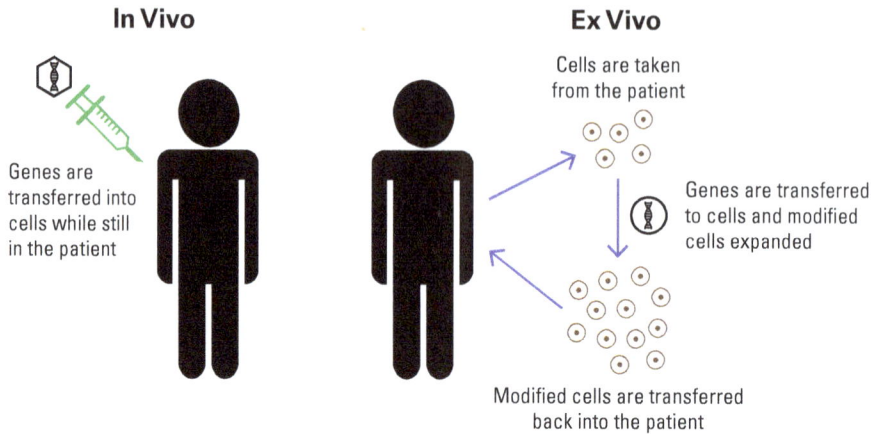

Genes are transferred into cells while still in the patient

Ex Vivo

Cells are taken from the patient

Genes are transferred to cells and modified cells expanded

Modified cells are transferred back into the patient

Figure 1.5: In vivo versus ex vivo gene therapy. Image © NC State University; reprinted with permission.

Table 1.4: Some in vivo and ex vivo gene therapies approved by the U.S. FDA [38].

Product	Company	Indication	Type	Vector
Zolgensma® [39]	Novartis	Spinal muscular atrophy (SMA) with bi-allelic mutations in the survival motor neuron 1 (SMN1) gene, patients less than 2 years old	In vivo	AAV9
Luxturna® [22]	Spark Therapeutics	Retinal dystrophy	In vivo	AAV2
Kymriah® [23]	Novartis	B-cell precursor acute lymphoblastic leukemia and large B-cell lymphoma	Ex vivo	Lentivirus used to genetically modify a patient's T cells
Yescarta® [40]	Kite Pharma	Relapsed or refractory large B-cell lymphoma	Ex vivo	Retrovirus used to genetically modify a patient's T cells

Information for each taken from package inserts.

1.3.4 Cell therapies

Cell therapy refers to the use of living, whole cells for treatment of a disease. A variety of cells are used including T cells and various stem cells. Cell therapies have the potential to treat numerous disorders, including various cancers, autoimmune diseases, urinary problems, and infectious diseases. Cell therapies can be categorized according to whether cells are taken from and administered to the same individual – referred to as autologous cell therapy (this includes ex vivo gene therapies described previously) – or derived from a donor or multiple donors other than the patient who receives the cells – referred to as allogeneic cell therapy.

Kymriah® and Yescarta®, described in Table 1.4, are classified as autologous cell therapies as well as ex vivo gene therapies. They are both prepared from a patient's own T cells, which are key to immune response in humans. A gene is introduced ex vivo into the T cells using a viral vector, such as lentivirus, that expresses a chimeric antigen receptor (CAR). The CAR on the surface of the T cells helps them find and kill cancer cells more effectively than if the CAR is not present. This type of therapy is referred to specifically as CAR-T cell therapy.

If you are interested in reading more about the products listed in this section or other biopharmaceuticals, perform an Internet search for "full prescribing information" (used interchangeably with the term "package insert" in this book) for the product of interest. The resulting document contains a plethora of information on the product, including indications, contraindications, dosage and dosage forms, a general product description, basic information on the process used for production, the drug's mechanism of action, and results from clinical studies.

1.3.5 Biopharmaceutical impact

As you can see from Tables 1.2–1.4, the health benefit of biopharmaceuticals is significant, with biopharmaceuticals treating or preventing a variety of diseases, including a number of cancers, stroke, heart attack, multiple sclerosis, diabetes, flu, hepatitis, rheumatoid arthritis and a host of other autoimmune diseases, and various eye diseases.

The biopharmaceutical examples listed in the previous tables represent only a fraction of all biopharmaceuticals approved for marketing throughout the world. In the United States and European Union alone, 443 biopharmaceuticals have active licenses as of June 2022. (Note that the reference from which this information is taken defines biopharmaceuticals as recombinant proteins, including recombinant antibody-based products, nucleic acid-based and genetically engineered cell–based products. It does not include biomanufactured medicines produced by nonrecombinant methods.) Among all new drugs approved in the United States between January 2018 and June 2022, 29% were biopharmaceuticals, with monoclonal antibodies the dominant class. A look at biopharmaceutical approvals by the U.S. FDA and European Medicines Agency (EMA) from January 2020 to June 2022 shows that 54% of first-time approvals in the United States and European Union were for monoclonal antibody products. Hormones, nucleic acid/gene-therapy-based products, vaccines, cell-based products, and enzymes also had a significant number of approvals during the January 2018 to June 2022 survey time frame. Based on products in clinical trials, it appears that monoclonal antibody-based products will continue to dominate new biopharmaceutical approvals. It is also worth noting that approval of biosimilars – a biological product highly similar to and interchangeable with an approved biological product – is rapidly increasing, with 94 approved in the United States and/or European Union since 2006 [41].

In addition to the health benefits, economic impact of biopharmaceuticals is significant. In 2023, 15 of the top 20 selling medicines worldwide were biopharmaceuticals [11], each with annual sales in excess of US$5 billion and combined sales of more than US$150 billion. Among the 15, eight are monoclonal antibodies, two are fusion products each of which includes a portion of an antibody, four are vaccines, and one is a hormone-like product. A recent study estimates the biopharmaceutical industry in the United States exceeded US$800 billion in "direct output" in 2022 [42].

1.4 Biopharmaceutical drug product: dosage forms and containers

To understand the processes involved in manufacturing biopharmaceuticals, it is worth starting at the end of the process by considering the finished biopharmaceutical product. The term *drug product* is used to describe this finished dosage form that contains the active ingredient (i.e., the drug that is administered to or by the patient) [43]. Generally speaking, drug products can take a variety of forms, including tablets, capsules, solutions, and lyophilized (also referred to as freeze-dried) cake or powder. For many biopharmaceuticals, drug product is a liquid solution or lyophilized powder, both for parenteral administration. The liquid solution is made up of the active ingredient dissolved in water and formulated with a number of other ingredients. Lyophilization is a process by which the biopharmaceutical is dried – that is, the water is removed – by freezing it, then reducing the pressure to allow the frozen water to sublimate directly into vapor, bypassing the liquid state. If heat were used to evaporate the water, the product might be denatured or degraded. Lyophilized drug product is a powder or cake. To be administered parenterally, as most biopharmaceuticals are, the drug product must be in a liquid state, which means lyophilized products must be reconstituted with water prior to injection.

So why are some biopharmaceuticals lyophilized while others are liquid solutions? The answer relates to product stability. Biopharmaceuticals are prone to undergo chemical and/or physical changes over time; for example, a biopharmaceutical may denature, aggregate, or undergo chemical degradation while the drug product is in storage. Generally speaking, biopharmaceuticals are more stable as lyophilized products than as liquid solution [44]. Stability is critical at the drug product stage, as most biopharmaceutical drug products are stored prior to use and commonly have a shelf life for at least two years at 4 °C [45]. When liquid solution cannot provide the drug product stability needed for adequate shelf life, then lyophilization is a likely option. Images of liquid and lyophilized products in vials are shown in Figure 1.6.

(a)   (b)

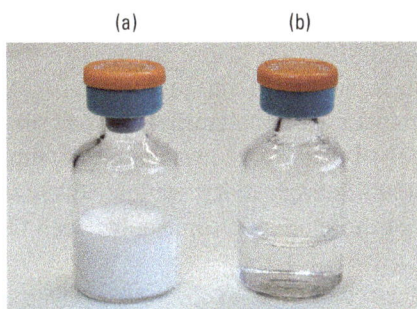

Figure 1.6: (a) Lyophilized (freeze-dried) and (b) liquid solution forms of biopharmaceutical drug product. Photo © NC State University; reprinted with permission.

While details of biopharmaceutical stability are beyond the scope of this book, it should be noted that a main objective of drug product formulation is maintaining chemical and physical stability of the product. When a drug product liquid formulation can be developed that provides the necessary product stability, then liquid form is usually desirable because it can be administered directly to a patient, without reconstitution, and enables a wider range of delivery devices to be used. Drug delivery devices are designed to create a patient-friendly delivery method that supports patient compliance with the prescribed regimen and to accommodate the broad dosage range typical of biopharmaceutical products. Historically, biopharmaceutical drug products have been delivered intravenously, with the product contained in vials as shown in Figure 1.6, and administered with a sterile needle/syringe or by intravenous infusion. While intravenous administration offers the advantage of 100% bioavailability (bioavailability refers to the percentage of drug that enters systemic circulation and is therefore able to have the intended effect), it is inconvenient as it typically requires administration by medical personnel and is painful.

A growing number of liquid formulations are now available, especially for use in prefilled syringes for subcutaneous administration and in autoinjectors. Autoinjectors are prefilled devices with a spring-loaded syringe that delivers a dose of medication when pressed against the body [46]. These prefilled syringes and autoinjectors are convenient and effective for self-administration of the drug, which better enables patients to comply with their drug prescription. For example, Enbrel® is available in a prefilled syringe or autoinjector and continues to be available as a lyophilized powder in a vial for reconstitution [47]. Much work is happening in the field of drug delivery to engineer more patient-centric delivery methods, thereby improving patient compliance with the prescription regimen. For a good review, see Anselmo et al. [48].

A variety of components is used to formulate biopharmaceutical drug product. Generally speaking, beyond water and the active ingredient, components are chosen to provide product stability and clinical efficacy. Components that are not the active ingredient are referred to as excipients. Typical components of biopharmaceutical drug product in their liquid form include the following [49, 50]:

- WFI (water for injection). A type of purified water used to formulate parenteral products. It is discussed in greater detail in the next chapter.
- The active biomolecule.
- Buffers. These are agents used for pH control such as phosphate and acetate.
- Product-stabilizing agents. Nonionic surfactants such as polysorbate 20, polysorbate 80, and poloxamer 188 are commonly used. They serve to stabilize proteins by reducing interfacial stresses that may result in product aggregation [50]. Sugars such as sucrose, trehalose, maltose, and lactose are also commonly added to enhance product stability.
- Tonicity agents. They are added to increase the concentration of solutes to ensure that the formulation has the same osmotic pressure as cells. Sugars such as sucrose and polyols such as sorbitol are used.
- Antioxidants. These are added to inhibit product oxidation. The amino acid methionine is an example.
- Preservatives. Benzyl alcohol is an example.
- Product- and process-related impurities (at low levels). These are soluble components not intended to be part of drug product. Impurities are discussed in greater detail in Chapter 8.

As an example, a list of the components in Avastin® drug product is shown in Figure 1.7. The active ingredient in Avastin®, as previously described, is a humanized IgG1 antibody, presented in a liquid formulation. Avastin® is manufactured by Genentech, a member of the Roche group.

Components of Avastin drug product

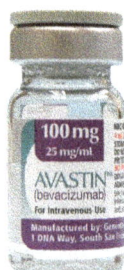

- Water for injection (WFI)
- mAb at a concentration = 25 mg/mL

  100 mg/4 mL single vial or
  400 mg/16 mL single vial

  For the 100 mg vial:

- Sodium phosphate (monobasic, monohydrate) (23.2 mg)
- Sodium phosphate (dibasic, anhydrous) (4.8 mg)
- α, α-trehalose dehydrate (240 mg)
- Polysorbate 20 (1.6 mg)

Figure 1.7: Components of Avastin® drug product [27]. Image © NC State University; reprinted with permission.

Note that while most biopharmaceutical drug products are liquid solution (like Avastin) or freeze dried, some are not. For example, Kymriah®, the ex vivo gene therapy described previously, looks very different from the final product images in Figures 1.6 or 1.7. It comes as a frozen cell suspension in an infusion bag and is intravenously infused into patients [23].

1.5 Biopharmaceutical drug product: quality

An essential characteristic of drug products is quality . But what is meant by quality? The guidance document from the International Council for Harmonization of Technical Requirements for Pharmaceuticals for Human Use (ICH), ICH Q6A, Specifications: Test Procedures and Acceptance Criteria for New Drug Substances and New Drug Products: Chemical Substances (October 1999), defines it as follows: "The suitability of either a drug substance or drug product for its intended use. This term includes attributes such as identity, strength, and purity" [51].

Drug substance and drug product are terms frequently used throughout this book. So let's differentiate between them. We've already defined drug product as the finished dosage form – tablet, capsule, or in the case of biopharmaceuticals, a liquid solution or lyophilized product – that contains the active ingredient. Drug substance is an active ingredient that is intended to furnish pharmacological activity or other direct effect in the diagnosis, cure, mitigation, treatment, or prevention of disease, but does not include intermediates used in the synthesis of such ingredient [43]. Drug product is made from the drug substance, with additional excipients added as needed to form the final product. We elaborate on the distinction between drug substance and drug product from a processing perspective in the next chapter.

As part of the process of developing drug products, a quality target product profile (QTPP) is needed. From the document ICH Q8(R2): Pharmaceutical Development (November 2009), the QTPP is: "[a] prospective summary of the quality characteristics of a drug product that ideally will be achieved to ensure the desired quality, taking into account safety and efficacy of the drug product" [52]. Note that the use of the term prospective in that definition sets the expectation that the QTPP is to be defined at the outset of the development process. Components of the QTPP include [52]:

- Intended use in clinical setting
- Route of administration, dosage form, and delivery systems
- Dosage strength(s)
- Container closure system
- Therapeutic moiety release or delivery and attributes affecting pharmacokinetic characteristics appropriate to the drug product dosage form being developed
- Drug product quality criteria (e.g., sterility, purity, stability and drug release) appropriate for the intended marketed product

As part of the QTPP, a list of critical quality attributes (CQAs) is developed. Again from ICH Q8(R2), the definition of a CQA is as follows: "a physical, chemical, biological or microbiological property or characteristic that should be within an appropriate limit, range, or distribution to ensure the desired product quality." CQAs are generally associated with the drug substance, excipients, intermediates (in-process materials), and drug product [52].

When discussing CQAs in this book, we are most often referring to attributes of the drug substance and drug product. CQAs are developed from a variety of sources – the QTPP, prior product knowledge, scientific literature, etc. – and vary from product to product. We've already discussed at least one CQA common to biopharmaceutical drug products, sterility, which is critical to ensuring safety of a parenteral drug. There are numerous others to consider. Generally speaking, they can be categorized as follows:

- General quality
- Identity
- Quantity
- Potency
- Purity
- Safety

Biopharmaceutical processes must be designed and executed to ensure the CQAs are consistently met. And how the manufacturing steps are conducted is as important as what the steps are. Biopharmaceutical products are manufactured using current good manufacturing practice (CGMP) [53, 54]. CGMP are requirements enforced by regulatory authorities throughout the world that assure proper design, monitoring, and control of manufacturing processes and facilities, thereby ensuring that products are consistently produced with quality [55]. These requirements are important for any type of drug, whether small molecule or biopharmaceutical. There are many important aspects to CGMP, and these are explored more fully in Chapter 3.

CGMP requires that drug products be tested to ensure batch-to-batch consistency and product quality. Specifications, including both testing methods and acceptance criteria, are established to routinely confirm the quality of product; conformance to specifications is required for the product to be deemed acceptable for release and for use. The quality control (QC) lab is typically responsible for this testing. These specifications are tied to the CQAs that have been determined for a product, although several CQAs may be measured by a single test method. An example of typical release testing for a monoclonal antibody product is shown in Table 1.5.

Table 1.5: An example of specifications for a monoclonal antibody drug product.

Test category	Examples of specific tests for each category	Range
General quality	Appearance	Colorless. No visible particles
	pH	6.0–6.5
Identity	Capillary zone electrophoresis	Conforms to standard
	Peptide mapping	Conforms to standard
Quantity	Protein concentration by UV absorbance at 280 nm	50–60 mg/mL
Potency	Antigen binding assay	80–120% compared to reference standard
Purity (and impurities)	SE-HPLC for aggregates	Aggregates < 2.0%
	Percent deamidation by IEC	≤5.0%
	Immunoassay for host cell proteins	≤100 ng/mg
	PicoGreen for residual DNA	≤20 pg/mg
Safety	LAL assay for endotoxin	≤0.1 EU/mg
	Sterility	Sterile (USP)

Adapted from Anurag Rathore, Setting Specifications for a Biotech Therapeutic Product in the Quality by Design Paradigm, BioPharm International, 23(1) [56].

Note that details of QC/analytical testing are beyond the scope of this book and are not covered.

Now that we have covered biopharmaceutical products, including their quality attributes, we shift focus in the subsequent chapters to the methods, equipment, facilities, design principles, and regulations used in production of these products from biological sources.

1.6 Summary

In this book, we define biopharmaceuticals as medicines inherently biological in nature and manufactured by or from living organisms. Biopharmaceuticals are among the biggest-selling medicines in the world. In fact, Humira®, a monoclonal antibody produced in CHO cells, was the biggest-selling drug (based on revenues) worldwide from 2012 to 2020, with sales of US$20 billion in 2020 [11]. Biopharmaceuticals are different from more traditional chemically synthesized drugs like aspirin. In particular, they are:

1. large, complex molecules, viruses, or cells, while more traditional pharmaceuticals are small molecules with well-defined structures,
2. produced using relatively complex processes that are biological in nature, and
3. largely administered to patients parenterally.

Major classes of biopharmaceutical include protein therapeutics, vaccines, gene therapies, and cell therapies. Many products within these classes are produced by recombinant DNA technology, but many, including a number of vaccines, are not. Table 1.6 provides examples of biopharmaceuticals in each major category.

Table 1.6: Summary of biopharmaceutical categories, with example products.

Biopharmaceutical category	Example	Description	Use
Protein therapeutic	Humira®	Human immunoglobulin (IgG1) monoclonal antibody	Rheumatoid arthritis and a number of other autoimmune diseases
Vaccine	Flucelvax®	Fragments of inactivated influenza virus	Prevention of influenza disease (i.e., the flu) caused by influenza virus type A and type B
Gene therapy	Luxturna®	Live, non-replicating adeno-associated virus serotype 2 that has been genetically modified to express the human RPE65 gene	Retinal dystrophy
Cell therapy	Kymriah®	T cells genetically modified to produce anti-CD19 chimeric antigen receptor	B-cell precursor acute lymphoblastic leukemia and large B-cell lymphoma

Most biopharmaceutical drug products, with the notable exception of cell therapies, are either a lyophilized powder or liquid solution. The latter is preferred due to ease and flexibility in administration; however, the former is sometimes required to provide necessary shelf life, typically on the order of two years. Drug products must be designed and manufactured for their intended use, or, alternatively, we would say they must have suitable quality. Quality attributes for a biopharmaceutical product must be developed and translated to specifications, which include both test methods and acceptance criteria. Drug product is tested to these specifications to ensure batch-to-batch consistency and product quality. An example of specifications for a monoclonal antibody product is given in Table 1.5.

Subsequent chapters cover the regulations that must be followed to ensure biopharmaceutical product quality, along with an in-depth look at the principles, design, and operational considerations of key unit operations required to produce a quality product.

Chapter 2
An overview of processes and facilities for biopharmaceutical production

Having described biopharmaceutical products, our attention turns to their manufacture, with an emphasis on product quality and producing the required quantity. Producing quality product in the required quantity depends significantly on the process – including the equipment and methods – used in the manufacture of the product, the facility in which manufacturing takes place, and (in the case of quality) the practices implemented to comply with current good manufacturing practice (CGMP) requirements. This chapter provides an overview of the processes for manufacture of biopharmaceuticals, including their design and the facilities that house them. More in-depth coverage of individual unit operations used to carry out these processes is provided in Chapter 4 and beyond. The topic of CGMP is covered in detail in the next chapter.

Specific questions that this chapter addresses are:
– Many biopharmaceutical processes can be viewed as consisting of four stages. What are these stages and the objectives of each?
– What unit operations are involved at each stage?
– What is the difference between the upstream and downstream biopharmaceutical manufacturing process?
– How is drug product produced from drug substance?
– What advantages does continuous processing offer over the more common batch processing in biopharmaceutical production? Disadvantages?
– What advantages does single-use equipment offer over reusable equipment? What are the disadvantages?
– What are the measures of product output for a process?
– What are the basic steps involved in designing a biopharmaceutical process, including establishing the design space for a process?
– What important concepts are used to guide the design of a biopharmaceutical facility?

Before going further, let's take a minute to explain a couple of terms that are used throughout this chapter and the remainder of the book: *unit operation* and *product*. Unit operation refers to any distinct step in the process that serves a specific function. A unit operation is often associated with a single piece of major equipment. For example, centrifugation is a unit operation commonly used in biopharmaceutical manufacturing. Centrifuges are used to separate solids from liquids in order to isolate the product. In some cases, product may be in the liquid phase; in other cases, product may be in the solid. Either (or both) can be collected from a centrifuge. Individual unit operations, such as centrifugation, are connected to create the overall process.

https://doi.org/10.1515/9783111112459-002

The second term, *product*, seems clear enough but deserves some explanation. In the previous chapter, the term "drug product" was defined as the final dosage form that contains the active ingredient plus a number of other ingredients, referred to as excipients, which have no pharmacological effect. It is the form of the drug administered by or to the patient. In this and subsequent chapters, we frequently use the term *product* (without "drug" as a descriptor) by itself. When used in the context of the biopharmaceutical process, product has several meanings. It may refer to the specific active ingredient (e.g., protein or virus that causes the desired effect of the medicine) at any step in the process. It may also refer to the intermediate form of the product that includes the active ingredient dissolved in the process liquid at any process step. Throughout the book, we typically refer to this form of the product as the intermediate, product intermediate, or process intermediate.

2.1 Biopharmaceutical manufacturing processes – an overview

This section of the chapter provides an overview of topics related to processes for manufacturing biopharmaceuticals, offering context for chapters that follow. We begin with a description of the main process steps required to make drug product from cells or other starting materials, followed by activities that support the main process – such as preparing solutions for use in different unit operations. We then discuss features of the equipment used to carry out all process activities and define terminology related to the amount of product produced.

2.1.1 The main process

Two main goals are to be achieved by the processes used to manufacture biopharmaceutical products: to *consistently* produce drug substance and drug product with the intended quality [52, 57] and to produce the necessary quantity of drug product for clinical or commercial use. Producing a drug product from starting materials requires many steps. Rather than immediately considering the specific unit operations, which can vary depending on the product, let's first consider a generalized process divided into four stages. Each stage consists of multiple steps (or unit operations) with a common overall purpose. These four process stages are depicted in Figure 2.1 and are common to the manufacture of many products, with the exception of cell therapies which have distinct bioprocesses due to the unique nature of the product.

Stage 1 is the product synthesis or production stage. *The objective of this stage is to produce the product in quantities suitable for clinical trials and/or commercial distribution.* For the many biopharmaceuticals produced using genetically engineered cells, this stage involves cell growth in bioreactors. (Note that in this book, the term "cell growth" is used to refer to an increase in mass of an individual cell and/or an increase

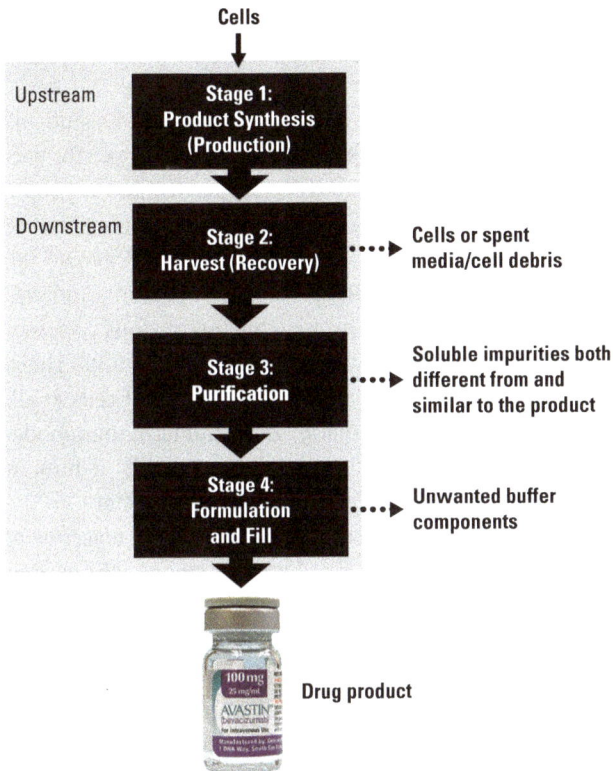

Figure 2.1: The four stages of biopharmaceutical production, using Avastin® as an example. As described in Chapter 1, the active ingredient in Avastin® is a humanized immunoglobulin G subclass 1 (IgG1) mAb produced by Chinese hamster ovary (CHO) cells through recombinant DNA technology. It is indicated for the treatment of metastatic colorectal cancer, among other cancers. Image © NC State University; reprinted with permission.

in the number of cells.) It is important to note that two different terms are commonly used to describe cell growth, based on the type of cells used:

- Cell culture: the growth of animal cells, including mammalian cells like CHO and insect cells like Sf9 (cells from ovaries of a Fall armyworm).
- Microbial fermentation (or just fermentation): the growth of microorganisms, like *E. coli* or the yeast *Saccharomyces cerevisiae*.

Both processes begin with a working cell bank (WCB) that has been prepared from a master cell bank (MCB). A WCB is typically made up of 1 mL cryogenically preserved vials of cells. This small number of cells must be expanded to a much greater number to produce the required amount of product. To produce an adequate number of cells for inoculation into a production bioreactor, the WCB is used to initiate a seed – also referred to as inoculum – train. That train may include multiple cell growth steps of increasing volume to generate the required number of cells. In both fermentation and cell culture steps, cells are typically grown in an aqueous medium that contains nutrients and other chemicals for growth.

It is worth noting that cell growth may also be required for biopharmaceuticals that are produced through nonrecombinant means. An example is Flucelvax® flu vaccine, described in Chapter 1. The influenza virus used to make the vaccine is propagated in Madin-Darby canine kidney (MDCK) cells, so virus stock must be added to the bioreactor containing MDCK cells for production [5]. Another example is Prevnar® 20, a conjugate vaccine produced using saccharides of Streptococcus pneumonia individually linked to a nontoxic diphtheria protein [58]. It protects against various pneumococcal bacteria that cause disease (such as pneumonia) in children and adults. There are even biopharmaceuticals whose production does not involve the use of cells at all. Gamunex®-C, an immune globulin injection used to treat primary humoral immunodeficiency, is an example [59]. It is produced by purifying IgG from pools of human plasma; thus, this product is synthesized by humans rather than in a bioreactor.

Stage 2 is the product harvest stage, also referred to as recovery. *The objective of this stage is to separate product from the production (i.e., host) system, as cells used in production cannot be part of the final drug product.* Additionally, stage 2 prepares the product intermediate for the subsequent purification stage by producing a clarified (i.e., particle-free) intermediate that is suitable for loading onto a chromatography column. The process intermediate from the harvest stage is commonly product dissolved in liquid – typically media from stage 1 – that has been cleared of cells and/or cell debris if cells are involved in the process. However, the clarified liquid from the harvest stage contains numerous soluble impurities that need to be removed.

There is a key distinction to make at this point regarding where product resides once produced: some remain within the cell and are referred to as intracellular products, while others are found dissolved in the aqueous bioreactor medium and referred to as extracellular products. This difference impacts the design of the harvest process and is discussed in greater detail in Chapter 5.

Stage 3 is the purification stage. *The objective is to purify the biopharmaceutical product; that is, separate the product from soluble impurities that are present.* This stage starts with a clarified aqueous solution (from the previous stage) in which the product is dissolved. As mentioned previously, intermediate from the harvest stage contains numerous soluble impurities that are not removed from the product as part of the harvest operations. Impurities common in many biopharmaceutical processes include host cell proteins, endotoxin, host cell DNA, and product aggregates. These and other impurities are discussed in much greater detail in Chapter 8. This stage is essential to ensuring that critical quality attributes (CQAs) related to purity, potency, and safety are met.

In biopharmaceutical manufacturing, a variety of unit operations is available for purification, with column chromatography being the most widely used. Because column chromatography typically involves a bed packed with resin beads that is easily clogged if particles are present in the feed, the product intermediate from stage 2 must be clarified. Usually, three or more distinct chromatography steps, each with a different interaction mode, are required to meet purity requirements. Other types of purification operations include precipitation, used to a much lesser extent than chromatography, and various viral clearance steps. Viral clearance steps are used to inactivate or remove adventitious viruses (i.e., viruses unintentionally introduced to the process) that could cause illness in patients if present in drug product. Viral clearance can involve non-chromatographic as well as chromatographic methods. Viral safety is an important consideration for any process that uses mammalian-derived components. Processes that use mammalian cell lines, such as CHO cells, typically require viral clearance steps. Examples of these steps include low pH incubation of the product to inactivate enveloped virus, virus filtration, which holds back adventitious virus while allowing product to flow through the filter, and certain types of chromatography. In most cases, processes that use bacterial cell lines like *E. coli* do not require viral clearance steps because *E. coli* presents no risk for adventitious mammalian viruses.

Stage 4 is the formulation/filling stage. *The objective of this stage is to produce the final dosage form, that is drug product*, and it involves a number of steps. These steps are central to ensuring acceptable quality attributes related to appearance, quantity, potency, and safety.

The term *formulation* refers to the steps needed to ensure that the purified product from stage 3 has the correct concentration, buffer system, and excipients (beyond just buffering agents) necessary for drug product production. Formulation is necessary because it is likely that the product concentration and buffer in the effluent from stage 3 are not what is required for final drug product but rather are dictated by the final chromatography step in the process. Figure 2.2 shows a common sequence of operations to meet these objectives: concentration/buffer exchange by ultrafiltration followed by addition of remaining excipients. After these steps, bulk filling takes place to produce drug substance, sometimes referred to as bulk drug substance.

Bulk filling prepares drug substance for storage prior to executing final fill-finish steps required to produce drug product. It is carried out by dispensing product intermediate from the UF step through a 0.2 µm filter (that is, a filter with 0.2 µm pores) into an appropriate storage container, which can have a range of sizes depending on the product. Examples of drug substance containers include disposable polymeric bags, rigid polymeric containers, or even stainless steel portable vessels. The bulk drug substance consists of the active ingredient in an aqueous buffered solution that is free of insoluble components and mostly free of soluble impurities. It is typically stored frozen or refrigerated until it is ready to be filled into the drug product container (e.g., vial, prefilled syringe, and autoinjector).

At this point, it is worth noting that drug substance is usually produced as a low-bioburden intermediate, while biopharmaceutical drug product is sterile, as discussed in Chapter 1, given that the common route of delivery is parenteral. Since we are using terms related to microbial contamination, let's take a minute to define these and a couple of other relevant ones.

(a) Production of bulk drug substance
(BDS) starting with purified product

(b) Production of drug product from drug substance

**Purified product
(from Stage 3)**

↓

| Concentration/buffer exchange (e.g., ultrafiltration) |

↓

| Excipient addition (if applicable) |

↓

| Bulk filling (through a 0.2 µm filter) |

↓

**BDS storage
(typically frozen)**

↓

To final fill-finish

BDS

↓

| BDS thaw |

↓

| Pooling and additional formulation |

↓

| Sterile filtration (through a 0.2 µm filter) |

↓

| Aseptic fill (including lyophilization if needed) |

↓

| Labeling and packaging |

↓

Drug product

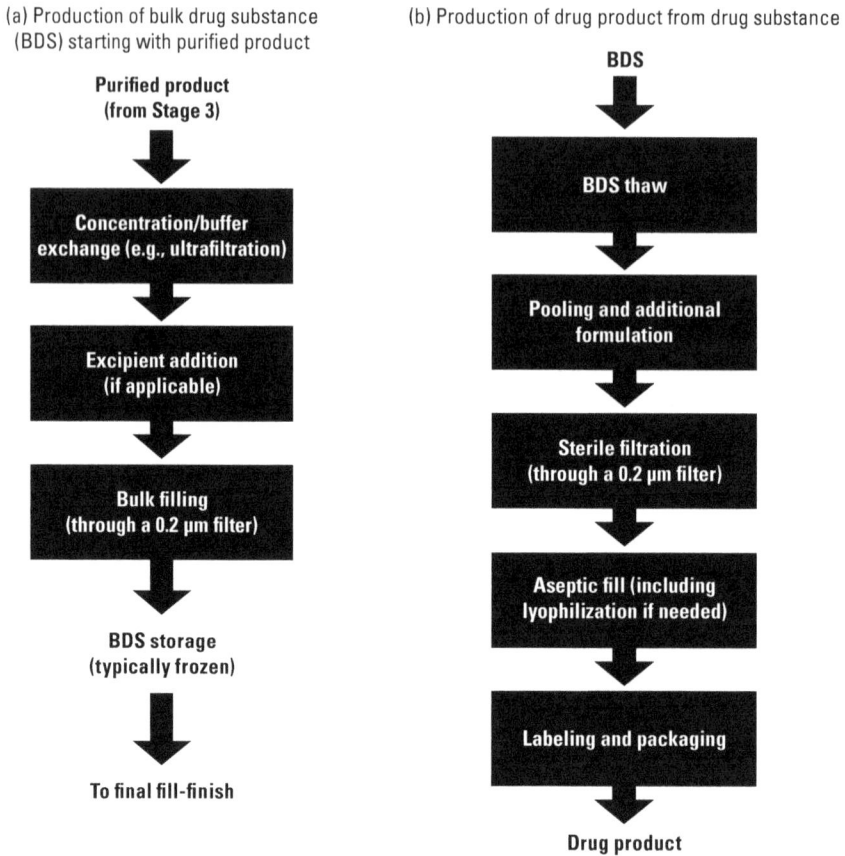

Figure 2.2: Process steps associated with stage 4: (a) steps starting from intermediate from the purification stage leading to bulk drug substance (BDS) and (b) steps used to produce drug product from drug substance. Note that these diagrams follow the path of the product and do not include other necessary steps, such as washing and sterilization of vials to be filled or preparation of vial stoppers and caps. Image © NC State University; reprinted with permission.

– Bioburden: the level and type of microorganisms that can be present in raw materials, process intermediates or drug substance. The term is usually applied to microbial contaminants. Bioburden should not be considered contamination unless the levels have been exceeded or defined objectionable organisms have been detected [61].
– Sterile: free from living organisms (i.e., free from bioburden). A sterile product, such as parenterally administered biopharmaceuticals, contains no bioburden. Further, bioreactors used to grow cells during the product synthesis stage are sterilized prior to inoculation with the cells.

- Axenic: free from living organisms other than the cells required for the process. When growth of cells is carried out in a bioreactor, the fermentation or cell culture is axenic, not sterile.
- Sanitized: a reduction in viable microorganisms to a defined acceptable level. Downstream process equipment is typically cleaned and sanitized, not sterilized.
- Aseptic: a process and/or facility designed to keep a product sterile.

The steps used to produce drug product from drug substance are also shown in Figure 2.2. Drug substance containers are thawed (if frozen) then pooled. Excipients are added to the drug substance as needed after pooling (referred to as additional formulation in Figure 2.2), followed by sterile filtration with a 0.2 µm filter and filling of product into its final container. Filling of biopharmaceutical drug product is usually done under aseptic conditions to ensure drug product sterility. To provide these conditions requires a space designed to maintain sterility of the product, delivery device, and equipment.

Note that the alternative method for ensuring sterility of a filled product is terminal sterilization. This term refers to filling a low-bioburden product into the final container followed by sterilization *after* filling. Sterilization can be accomplished in a number of ways, such as applying heat or irradiation, which would denature most biopharmaceuticals. Thus, aseptic filling rather than terminal sterilization is used for producing biopharmaceutical drug product.

It is important to note that in a manufacturing environment, unit operations for drug substance manufacturing are commonly categorized as being part of the upstream or downstream process, as shown in Figure 2.1. The upstream process includes product synthesis (stage 1) operations and may include harvest (stage 2), although some facilities may include harvest operations with the downstream process. The downstream process includes purification and formulation/filling operations, up to and including the bulk filling step.

Figure 2.3 brings together the individual unit operations in each stage in a process flow diagram for production of a monoclonal antibody (mAb). The process begins by thawing a vial of cells from the working cell bank. Those cells are expanded in number until there are enough to be inoculated into the production bioreactor. Once in the production bioreactor, cells are grown until they are no longer producing. In the case of CHO cells grown in fed-batch mode for production of a mAb, the production reactor step can last as long as two weeks. The product in the bioreactor is harvested by sending the broth, often referred to as harvest cell culture fluid (HCCF), to a series of clarification steps. Figure 2.3 shows clarification as being accomplished by a combination of centrifugation, for removal of most cells, followed by depth filtration for removal of residual cells and cell debris. The resulting process intermediate is often referred to as cell culture supernatant. That intermediate is fed to a series of chromatography and viral clearance steps as part of the purification stage. These steps remove the soluble impurities and contaminants (or inactivate virus in the case of low pH incubation), including adventitious virus that may be present. Note that ef-

Stage 1: Product synthesis	Stage 2: Harvest	Stage 3: Purification	Stage 4: Formulation/fill
WCB thaw *(Chapter 4)*	Centrifugation *(Chapter 6)*	Protein A affinity chromatography *(Chapter 8)*	Concentration/buffer exchange by ultrafiltration *(Chapter 9)*
Seed culture expansion *(Chapter 4)*	Depth filtration *(Chapter 7)*	Low pH incubation *(Chapter 8)*	Bulk filling *(Chapter 2)*
Production in bioreactor *(Chapter 4)*	To purification	Cation exchange chromatography *(Chapter 8)*	mAb drug substance
To harvest		Anion exchange chromatography *(Chapter 8)*	
		Viral filtration *(Chapter 8)*	
		To formulation/fill	

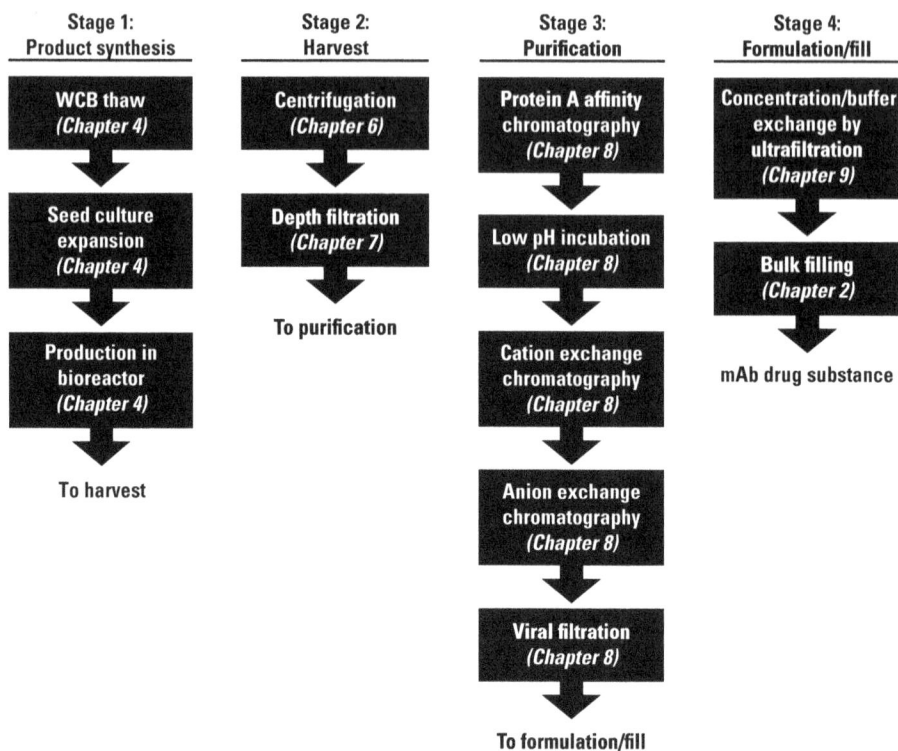

Figure 2.3: A process flow diagram showing individual unit operations in a process for drug substance production of a mAb from genetically engineered CHO [62]. Other process designs are also possible. The chapter in which each of these operations is covered is also shown. Note that cell lysis and tangential-flow microfiltration, which are not part of this mAb process, are covered in Chapters 5 (cell lysis) and Chapters 7 and 9 (microfiltration). Image © NC State University; reprinted with permission.

fluent from the chromatography steps is often referred to as column eluate. Filtrate from the viral filtration goes into the formulation/fill stage, where product concentration and buffer exchange take place by ultrafiltration. Once the proper concentration and buffer for drug product are achieved, the resulting drug substance is filled. The harvest, purification, and formulation/fill steps, stopping at filled drug substance, would typically require 5–7 days to complete.

This discussion highlights the multiple different, somewhat complex unit operations used in the manufacture of biopharmaceutical drug substance. In fact, biopharmaceutical drug substance manufacture involves significantly more steps than production of drug product from drug substance. This book provides greater detail on the operations required for the product synthesis, harvest, purification, and formulation/filling stages of drug substance manufacture. However, the process of converting drug substance into drug product is not covered in more detail than has already been discussed.

It is important to note that sampling and a variety of testing is performed throughout the process as part of the overall control strategy to ensure quality product is consistently manufactured. We discussed release testing of drug product against an item specification in Chapter 1. Likewise, drug substance is tested against an item specification as part of its release process. In addition, testing of raw materials and in-process material (i.e., product intermediate) is performed. Testing requirements vary from product to product. Most of this testing is performed by the quality control (QC) lab; so samples must be pulled, sent to the QC lab, then tested. In the future, it is possible that replacement of slower offline testing by rapid inline sensor measurement, particularly for in-process samples, will occur.

The unit operations used in drug substance production are most often run in batch mode, or in the case of a bioreactor step, fed-batch mode, rather than continuous mode. In batch mode, a finite amount of feed is "charged" to a unit operation, processed, and all product intermediate removed at some time later. The product intermediate is then charged to the next operation, where it is processed and removed at some time later. This continues at each processing step until drug substance is produced. In fed-batch operation of a bioreactor, cells are inoculated and grown in batch mode for a certain amount of time, while nutrients are fed (hence the term *fed-batch*) throughout the duration of the step to feed them. In continuous unit operations, input and output streams flow continuously. If you were to enclose a drug substance process comprised of only continuous unit operations within a box, you would observe a steady flow of materials into and out of that box.

Despite batch being the common mode of operation, some continuous processing is used in biopharmaceutical manufacturing, most notably perfusion systems for continuous cell growth. Perfusion systems are covered in greater detail in Chapter 4. Implementation of continuous processing in downstream operations has lagged in development due to the complexity of converting steps like chromatography and diafiltration to continuous operation. Further, to the best of our knowledge, no fully continuous process for biopharmaceutical drug substance manufacturing exists for an FDA-approved drug product. Regardless, there are a number of potential advantages that continuous processing offers over batch processing [63, 64]:

- Higher equipment utilization. Equipment is utilized constantly rather than only for a single step of every batch as in batch operation.
- Smaller equipment. For a fixed rate of production, relatively low flow rates are used in continuous operations, resulting in smaller equipment.
- Higher volumetric productivity.
- Lower capital investment in equipment. Smaller equipment means less up-front investment.
- Smaller, less costly facilities. Smaller equipment translates to a smaller, less expensive facility.
- Reduced risk of product quality issues as product residence time (within a bioreactor) and hold times are reduced.

The number of sessions being devoted to continuous processing at bioprocessing conferences suggests that much effort by industry, academia, and the vendor community is being placed on development of continuous processes for biopharmaceutical drug substance manufacture. It will become clearer in the coming years as to whether the promise held by continuous processing comes to fruition. And there are a few potential disadvantages that will have to be overcome, including an increase in the time required to design processes.

2.1.2 Supporting activities

A number of activities are required for production that are not explicitly shown in Figure 2.3 and yet are part of most biopharmaceutical processes. We provide an overview here so that the scope of these activities is clear.

The preparation of various solutions and media, commonly referred to as solution prep and media prep, respectively, are among the activities needed to support a biopharmaceutical process. We will focus on solution prep here. Biopharmaceutical processes require numerous solutions. For example, as will become clear in Chapter 8, chromatography steps are likely to require solutions of NaCl, NaOH, and various buffers, to name just a few. A single chromatography step may require eight different solutions. Looking at Figure 2.3, there are three separate chromatography steps in the process shown, which could require a total of as many as 20 different solutions. Other process steps require solutions as well. For a multi-product facility running at capacity, it is easy to see how the number of different solutions required may exceed 100. Consequently, solution prep tends to be a high-throughput operation in many facilities; further, failure to correctly prepare solutions can have negative consequences to process performance and product quality, as will become clear in later chapters. Solution prep is commonly performed by adding dry components (such as NaOH pellets) or solution concentrates (such as 5 M NaOH) to purified water at the required quantity and mixing in a vessel to produce a solution at the correct concentration and with the necessary volume. After mixing, solutions are transferred to storage vessels, often through a filter to reduce the likelihood of microbial contamination. Modifications to this conventional process to prepare solutions have been made. An example is the use of inline buffer dilution, using equipment like Asahi Kasei's MOTIV® Inline Buffer Formulation systems. These skid-based systems use stock concentrates as starting material and automatically dilutes them with water and adjusts pH (if required) to produce solutions of a specified volume, concentration, and pH [65].

Cleaning, sanitization, and sterilization operations are also critical to processes like the one shown in Figure 2.3. Reusable equipment must be cleaned after each batch to ensure that residues of product, buffers, cleaning agents, etc. from the previous use are removed. This cleaning is typically accomplished using clean-in-place (CIP) techniques. CIP is a cleaning mode in which removal of soil from product contact surfaces is carried out by spraying and/or flowing cleaning agents and water rinses over the surfaces to be cleaned without disassembling equipment. This method is in contrast to the less common method of disassembling equipment and manually cleaning in a location away from the processing area, referred to as clean out of place (COP). For vessels, such as bioreactors or portable process vessels, CIP is usually accomplished with a separate CIP system that delivers purified water and cleaning solutions to the vessel through a spray ball. The spray ball creates high velocity jets of water or cleaning solution that soak process fluid-contact surfaces in the interior of the vessel. Some process equipment, however, does not require a separate CIP system to carry out clean-in-place operations because pumps capable of delivering cleaning solution are built into the equipment. Examples include chromatography systems and ultrafiltration systems.

In addition to cleaning, some equipment is sanitized as well, particularly downstream equipment that does not undergo a separate sterilization step. Sanitization is frequently done by chemical methods and often in combination with cleaning. As an example, downstream equipment such as ultrafiltration systems are often simultaneously cleaned and sanitized with a 0.5 M NaOH solution. Sterilization of equipment, particularly reusable bioreactors that must be free from microbial contamination, is typically performed by heating up the equipment surfaces using clean steam, in a process referred to as steam in place (SIP). Cleaning, sanitization, and sterilization as they apply to the different unit operations are covered in more detail in subsequent chapters.

We should also add that for cleaning small parts used in a biopharmaceutical process, parts washers are used that work like large kitchen dishwashers. To sterilize small parts, an autoclave is used. The autoclave is basically a chamber into which small equipment is loaded and exposed to high-temperature steam for sterilization. Both parts washers and autoclaves are designed to run automated cycles. There are many good references that provide details on the topics of cleaning, sanitization, and sterilization, including the book *Clean-In-Place for Biopharmaceutical Processes* [66] and *Sterility, Sterilisation, and Sterility Assurance for Pharmaceuticals* [67].

To support operations related both to the main process presented in 2.1.1 and the supporting activities described so far, there are a number of clean utilities, also referred to as critical utilities, that must be generated. They are clean and critical because they have an impact on product quality in a significant way. Clean/critical utilities include the following:

–   Various purified waters. Biopharmaceutical processes require large volumes of water for cleaning and preparing media and solutions. The two types of purified water most commonly used begin as drinking water that has been purified to produce one type commonly referred to as purified water, which we will refer to as high purity water (HPW), and the second type known as water for injection (WFI), often referred to as "wiffey." In the United States, requirements for these purified waters are given in the U.S. Pharmacopeia (USP), a compendium that contains specifications for various ingredients used in preparation of drugs. (Other regions of the world have similar compendia; for example, there is a European Pharmacopeia and Japanese Pharmacopeia). For water, the relevant chapter is USP 1231, which describes processes and testing performed for various water types [68]. HPW and WFI must meet the same requirements for conductivity and total organic carbon. An important difference between the two is that WFI has an endotoxin specification that must be met, while HPW does not. In biopharmaceutical processes, HPW is often used for equipment cleaning. WFI is used as an excipient in production. It is also used for equipment cleaning and to prepare most solutions. The processes for producing either HPW or WFI are somewhat complex. For example, an HPW-generation process may consist of the following steps, starting with potable water feed: filtration for particulate removal; water softening for cation removal; carbon bed filtration for removal of chlorine and organic compounds; filtration for removal of carbon fines; treatment with ultraviolet (UV) light to kill microorganisms; reverse osmosis for removal of small molecules, including dissolved salts; a deionization step to further remove ions; and a 0.2 μm filtration step.

–   Clean steam, also referred to as pure steam. It is used for sterilization of equipment and/or process fluids. It is often generated from treated water, such as HPW.

–   Clean compressed gases. These include air, O_2, CO_2, and N_2. Air, O_2, and CO_2 are typically used for fermentation and cell culture. N_2 is used to blanket product when needed. These gases often come from a supplier. Many facilities generate their own compressed air with a system consisting of an air compressor, filters for removing particulate, and solid media for reducing humidity.

A number of other support utilities are required for operation of a biomanufacturing facility, including building heating, ventilation, and air conditioning (HVAC) systems, waste inactivation systems, chilled water generation systems, electricity and water. Details about these utilities are beyond the scope of this book.

2.1.3 Equipment for the main process and support activities

Specialized equipment is required to execute upstream processing, downstream processing, and the support activities discussed so far. This section provides general information about that equipment, while detailed equipment descriptions are included in the subsequent chapters on the various unit operations required for biopharmaceutical manufacturing.

Process equipment comprises more than just the main equipment component in which product is being processed. For example, cell culture requires more than just the bioreactor in which cells are grown. Likewise, performing a chromatography step requires more than just the packed column that performs the necessary separation. In addition to these main components, bioprocessing equipment includes lots of "plumbing," such as inlets that allow process fluids to be connected to the functional component and outlets that allow process fluids to be collected or discarded. Equipment also includes a variety of valves that direct flow and pumps that move fluids through the system. It also includes many sensors used to monitor parameters important to the operations and send input to the control software. For example, bioreactors are likely to be equipped with sensors to measure temperature, pH, dissolved oxygen, and dissolved carbon dioxide. Chromatography systems are equipped with sensors to measure conductivity, temperature, the presence of air bubbles in the liquid feeds, pressure, flow rate, UV absorbance, and pH.

Other important characteristics of process equipment include scalability, meaning the equipment must enable operations at all scales, from bench-scale development to full-scale manufacturing. Reusable process equipment must be cleanable, sanitizable, and sterilizable (if sterilization is necessary). For example, cleanable equipment is designed so that all fluid-contact surfaces are accessible during cleaning and that system pumps are able to deliver the necessary flow rate for cleaning solutions when an automated CIP system is not used. For more information on design of equipment for biopharmaceutical processing see the American Society of Mechanical Engineering Standard (ASME) on Bioprocessing Equipment [69].

Another common feature of equipment in biopharmaceutical processes, particularly downstream equipment, is that it is skid mounted. The term skid refers to a collection of components mounted on a frame to support a given activity. For downstream equipment, skids are often on wheels so that they are easily moved. Examples of a centrifugation skid and chromatography skid are shown in Figure 2.4. Each skid contains all of the inlets, outlets, valves, pumps, sensors and (in some cases) control hardware and software required to execute a centrifugation or chromatography step for a biopharmaceutical process.

Both single-use and reusable equipment are used in biopharmaceutical processes. With single-use systems, components that come in contact with process fluids are generally made from polymeric materials intended for one-time use and discarded afterwards. In contrast, reusable equipment is typically made from stainless steel. One key difference between single-use equipment and reusable equipment is that the latter needs to be cleaned, sanitized and/or sterilized between uses, while single-use equipment does not because it is disposed. Further, the cleaning, sanitization, and sterilization processes for reusable equipment must be validated for commercial manufacturing, a time-consuming and resource-intensive effort that is discussed in greater detail in the next chapter. A summary of the advantages of single-use equipment is given below:

(a) (b)

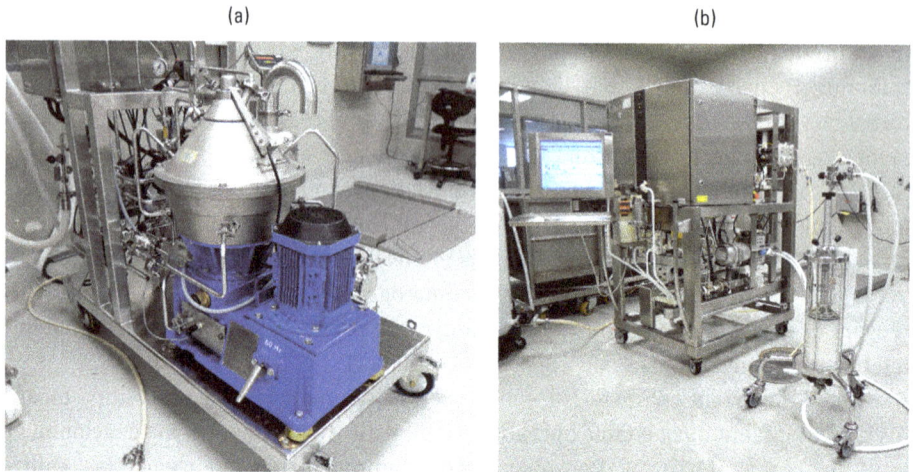

Figure 2.4: An example of a centrifugation skid (a) and chromatography skid (b). The centrifuge is an Alfa Laval LAPX 404 disc-stack centrifuge, and the chromatography system photo shows the side and back view of a Cytiva (formerly GE Healthcare) AKTAprocess system. Photos © NC State University; reprinted with permission.

– Speed to market
– Lower capital costs, particularly for a new facility
– Reduction or elimination of cleaning and sanitization, which reduces the cost of water, cleaning/sanitization agents, and cleaning validation
– Elimination of sterilization of bioreactors, which reduces the cost of water, power for steam generation, and sterilization validation
– More rapid, less expensive changeover between campaigns of different products, which can improve facility productivity
– Reduced probability of cross-contamination and microbial contamination

With all of those advantages, why isn't single-use equipment universally adopted? A few reasons are listed below:
– Scale limitations. For example, the largest single-use bioreactor currently available is 6,000 L [70].
– Lack of standardization of single-use components.
– Increased reliance on outside suppliers to provide the components required for each run.
– Cost of single-use components. A key question that must be addressed in deciding between single-use and reusable equipment is does the savings in capital costs, water, and labor associated with single-use equipment offset the cost associated with purchasing the single-use components required for each batch produced. For example, high costs of chromatography resins and tangential flow filtration

membrane modules often limit use of these unit operations in a truly single-use way.

- Leachables/extractables. Extractables are defined as compounds that migrate from any product-contact material when exposed to an appropriate solvent under exaggerated conditions of time and temperature. Leachables are compounds that are typically a subset of extractables that migrate from any product-contact material under normal process conditions [71]. Their presence potentially impacts the safety and quality of drug product.
- Concerns related to breakage of single-use components resulting in loss of production material.
- Increase in the amount of solid waste to be disposed.

Use of single-use equipment and components began in the early 1980s with the introduction of membrane filters within plastic capsules. The mid-1990s saw the introduction of "buffer" bags for storage of various solutions (and other process liquids), along with totes to contain and move the bags and solutions they held [72]. An example of buffer bags and totes is shown in Figure 2.5. Since the mid-1990s, single-use technology options have continued to evolve to include mixing systems, bioreactors, depth filters, filtration (including UF) units, and chromatography systems. Single-use versions of the latter two operations, in particular, have been enabled by the development of disposable flow paths. Single-use options for specific process steps are discussed in subsequent chapters that cover individual unit operations in more detail.

(a)   (b)

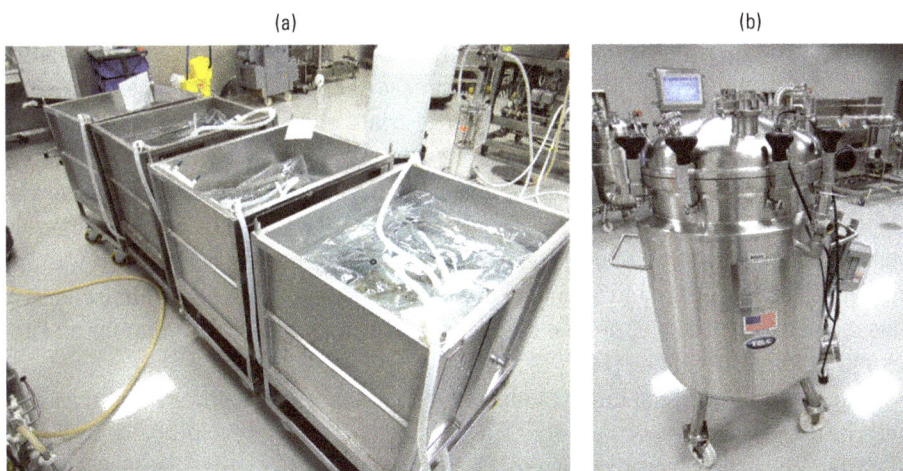

Figure 2.5: Options for storage of solutions in biopharmaceutical processes. (a) Buffer bags and totes used for storage of solutions for chromatography processing. A single-use bag holds the solution and is supported by a stainless-steel tote. (b) A stainless steel portable vessel. Photos © NC State University; reprinted with permission.

While the economic benefit of single-use equipment must be established on a case-by-case basis, implementation of single-use bags to supply solutions is generally considered to result in a significant savings compared to stainless steel storage tanks. A net present value analysis of a process requiring 100 or more fermentation batches annually demonstrated that bags for storage of media, solution, and intermediate product clearly win from a process economic perspective over stainless steel storage vessels [73].

Reusable equipment may be shared between different products or dedicated to one product. Either is acceptable. However, for multi-product operations in which reusable equipment is shared, steps must be taken to ensure that product-contamination risks, such as contaminating product A with product B, are minimized. These steps include development of effective cleaning procedures that can be validated to ensure that one product is completely removed from a piece of equipment prior to introducing a new product into that equipment. It is important to note that there are a number of process components that are not usually shared between products, such as chromatography resin, ultrafiltration membrane devices, filling equipment, and gaskets. While these components may be reusable, they are typically dedicated to a product due to concerns that if used across multiple products, the polymers from which they are made may desorb product A into a batch of product B.

2.1.4 Measures of product output

Now that we have addressed the process to make biopharmaceutical drug substance and drug product with the desired quality, let's shift focus to the quantity of product produced. For any given product, the amount needed depends on a number of factors. If producing for preclinical studies or clinical trials, the amount of product needed will likely be less than what is required for commercial distribution. And different types of products will require different amounts just to meet market demand. For example, market demand for all recombinant Factor VIII-based products, classified as blood clotting factors in Chapter 1 and used in the treatment of hemophilia A, is approximately 300 g of protein annually. By comparison, the single drug Humira® has a demand of about 700 kg of protein annually, significantly higher than that for all Factor VIII products combined [74].

There are a number of different measures used to quantify a process's product output. These are given below:

– Titer. The term generally refers to the concentration of a component in a solution. In biopharmaceutical manufacturing, it often specifically refers to the amount of product produced per unit volume of bioreactor broth, with units of g/L. As an example, the average titer for mAb products is estimated to be approximately 3.25 g/L in 2020 [75]. Titers for mAb products have grown through the years due

to improvements in engineering of cell lines, media optimization, and improved process control.
- Percent Yield. This term is defined by the following equation:

$$\% \text{ yield} = \frac{\text{amount of product at the end of one step or series of steps}}{\text{amount of product at the beginning of the step or series of steps}} \times 100$$

(2.1)

The "amount" is often in grams, but can be of any unit of measure relevant to a particular product. For example, in vivo gene therapies are often quantified in terms of vector genomes, which refers to the concentration of viral capsids carrying the gene of interest. Note that the term *percentage step yield* is used when equation (2.1) is applied to one step in the process and *percentage process yield* when applied to the entire process. *Percent recovery* is a term used synonymously with either percent step or process yield.

Even though CGMP regulations are not discussed until the next chapter, it is worth noting that several yield terms are defined in those regulations in the United States: actual yield, theoretical yield, and percentage of theoretical yield [53]. Actual yield refers to the quantity (e.g., grams) that is actually produced at any phase of manufacture, processing, or packing of a particular drug product. Theoretical yield is the quantity that would be produced at any phase of manufacture, processing, or packing of a particular drug product in the absence of any loss or error in actual production. And percentage of theoretical yield is the ratio of the actual yield to the theoretical yield applied to the same phase of manufacturing, stated as a percentage. Therefore, percentage of theoretical yield defined in 21 CFR Part 210 is analogous to the more commonly used terms *percentage step yield* and *process yield* as defined here.
- Productivity. The term refers to the amount of product produced per relevant volume per time (e.g., units of g product/L/h). For example, the productivity of a bioreactor could be given as the grams of product produced per liter of culture broth per hour.

It is also worth noting that when referring to the scale of production, we typically use the volume of a bioreactor. For example, a process that utilizes a 10,000 L bioreactor is referred to as a 10,000 L process.

Example: Estimating the number of bioreactor runs required to meet demand for a mAb product
A company wants to produce 1,000 kg annually of a mAb drug product using a process that delivers a titer of 3.25 g mAb/L bioreactor broth. The process uses bioreactors with a 10,000 L working volume, and the % process yield from the end of the bioreactor step to final fill-finish is 70%. How many bioreactor runs are required to meet the annual demand?

Solution

First, using the definition of titer and process yield, calculate the amount of mAb (in grams) produced as drug product from each bioreactor run given a titer of 3.25 g/L and process yield of 70%:

$$\text{Amount of mAb drug product} = \text{bioreactor broth volume} \times \text{titer} \times \% \text{ process yield}/100$$

$$= 10,000 \text{ L broth} \times 3.25 \frac{\text{g mAb}}{\text{L broth}} \times 0.7 \frac{\text{g mAb drug product}}{\text{g mAb in broth}}$$

$$= 22,750 \text{ g mAb}$$

Next, divide the total amount of mAb to be produced annually as drug product – 1,000 kg – by the amount of mAb produced (as drug product) from each bioreactor run:

$$= \frac{1,000,000 \text{ g mAb/year}}{22,750 \text{ g mAb/run}}$$

$$= 44 \text{ bioreactor runs/year}$$

Note that if a production run in the bioreactor takes 10 days, then one bioreactor is not enough to perform the required number of runs in a year. At least two bioreactors running simultaneously are necessary.

2.2 Process design

This section briefly discusses the methodology used to design processes for commercial production of biopharmaceuticals. We use the term *process design* to encompass both process development – those activities required to define the design space (a term discussed in more detail shortly) – and scale-up, among other activities described shortly.

The major objectives of process design efforts are to ensure that a sufficient amount of drug product is produced for commercial distribution; to establish a commercial manufacturing process capable of *consistently* supplying product of the intended quality [57]; and to reflect the design in planned master production and control records [76]. That latter goal makes clear that a key deliverable from design activities is a set of master batch records that provides detailed instructions for execution of a process. Batch records are considered in greater detail in the next chapter. There are other important deliverables from process design as well, including the actual physical process, a finalized list of CQAs, a complete process description showing each unit operation and associated material attributes and operating ranges, a list of raw materials and testing specifications, analytical methods and specifications for product release testing, in-process test methods and acceptance criteria, a complete process control strategy, and a host of reports to be written that document the results from and rationale for the various design activities that we will describe shortly.

Before discussing how process design is done, let's put it into the context of the life cycle for a biopharmaceutical product. Figure 2.6 shows some of the activities leading to commercialization of a biopharmaceutical product, including process design. Upon completion of drug discovery – the process by which new drugs are identified – the biopharmaceutical candidate goes through preclinical studies, which determine initial dosing and define pharmacological and toxicological effects. These studies usually involve animals. Following preclinical activities, the drug can move into clinical trials – studies that involve humans – that are conducted in phases. Success in one phase is required before progressing to the next. The number of subjects increases with each phase, and the final phase – phase 3 – looks at how the new treatment compares to the current treatment(s). Success in clinical trials may lead to commercial production and distribution. There are numerous sources available that describe clinical trials in greater detail [77].

As shown in Figure 2.6, production of material for human use must be done under CGMP. But even as CGMP manufacturing for clinical trial material takes place, process design activities are ongoing. Thus, the process used to produce phase 1 clinical trial material should be better defined and more robust by the time commercial manufacture begins. After the process design stage, process qualification takes place. Its main purpose is to bring the facility, utilities, equipment, and trained personnel together along with all applicable procedures to produce commercial batches to demonstrate that the process consistently performs as expected. Multiple production-scale runs (3–6 are common) are performed as part of this exercise. Once process qualification is successfully completed, and assuming success in phase 3 clinical trials and FDA approval, commercial production ensues, and the process is monitored throughout its lifetime. This monitoring is referred to as continued process verification, and its purpose is to ensure that the process remains in a state of control. The process design, process qualification, and process verification stages of the product life cycle are all part of a larger process validation program that is discussed in greater detail in the next chapter [76].

There is no template for designing a manufacturing process that works for every product. However, a general set of steps involved is shown in Figure 2.7. Discussion that follows makes reference to this figure.

To begin process design, it is necessary to know the quality target product profile (QTPP) and CQAs for the drug product because it is important to know the details of the product for which the process is being designed. For example, CQAs related to purity guide design of a chromatography step (e.g., choosing the type of resin to use or the composition of an elution buffer) so that the impurities to be removed and the required extent of removal are known in advance. Design of final fill-finish operations requires knowledge of the dosage form and delivery device, information that is typically part of the QTPP. In addition, demand for the commercial product should be understood so that production rates can be estimated. This information may influence choice of the expression system used as well as the scale of operations required. It is

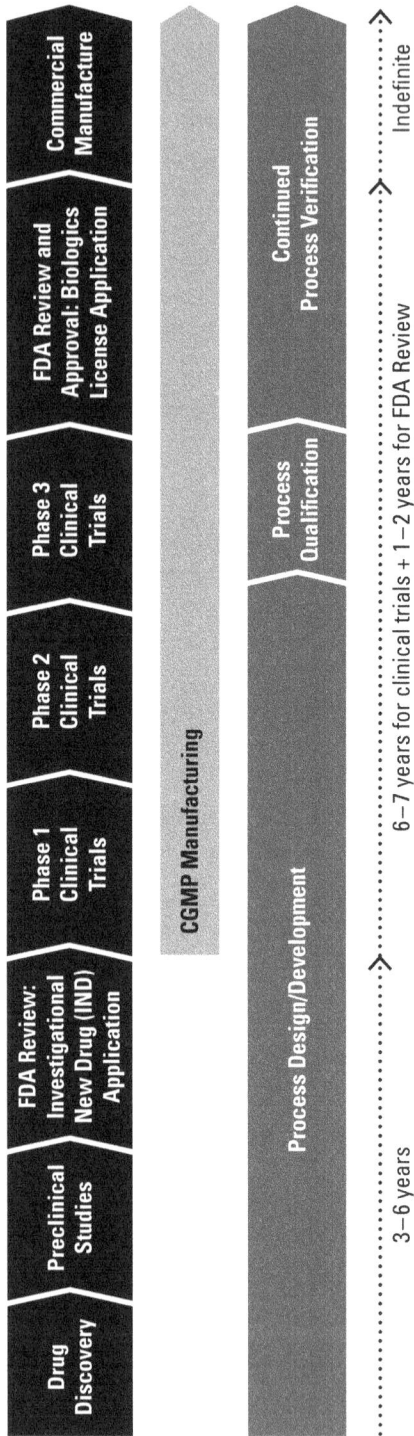

Figure 2.6: Activities leading to the commercialization of a biopharmaceutical [78, 79]. Image © NC State University; reprinted with permission.

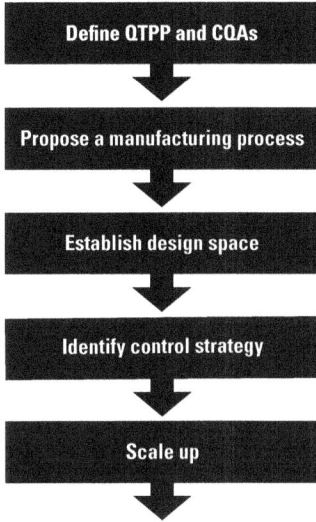

Figure 2.7: Steps involved in the design of a biopharmaceutical process. Even though each step is shown sequentially, there may be some iteration involved. For example, design space studies may result in reconsideration of the unit operations being used in the process. Image © NC State University; reprinted with permission.

also worth noting that defining the QTPP and CQAs likely started prior to process design, in the early stages of product development.

Armed with an understanding of the product through the QTPP and CQAs, a manufacturing process is proposed. A host cell system (when cells are used for production) is selected along with other raw materials and consumable items, and the unit operations to be used are identified. A preliminary list of process parameters, a term we define shortly, is put together for each unit operation along with an estimate of their allowable ranges. Proposing a process requires knowledge of the process stages from Figure 2.1, the unit operations available to meet these objectives, processes for similar products, and any existing process and product information, including physical and chemical properties of the active product molecule (or virus or cell). For example, selection of harvest operations is impacted by whether the product is intracellular or extracellular. The selection of purification steps is impacted by whether or not animal cells are used that require viral clearance steps. The unit operations selected may change as process design proceeds and process understanding deepens. Further, scalable operations, discussed previously, are chosen when large quantities of product are being produced for commercial distribution.

Once unit operations are proposed, including a list of process parameters for each and their ranges, effort focuses on defining the design space, as shown in Figure 2.7. Let's start the discussion by covering a few definitions that help in understanding the concept:

– Process parameter, also referred to as an operational parameter: an input parameter to a unit operation (e.g., the flow rate to a chromatography column) that is directly controlled. Process parameters can be further categorized as critical, key, and non-key. A critical process parameter is one whose variability has an impact on a CQA. A key process parameter is one whose variability is important to process performance but does not affect a CQA. Non-key process parameters are easily controlled or have wide limits and therefore pose little risk to impacting a CQA or performance parameter (defined below) [80].

– Performance parameter: an output parameter that cannot be directly controlled, but is indicative of process performance [80]. Note that CQAs are not usually categorized as performance parameters. An example of a performance parameter is the percentage step yield for a specific process step.

Every unit operation that makes up a biopharmaceutical process has inputs and outputs. The most obvious are the product into and out of the manufacturing step. In addition, as depicted in Figure 2.8, there are non-product materials, each with specific attributes, and process parameters that impact the operation. On the output side, the unit operation produces the product, characterized by its CQAs, and there are performance parameters that characterize the performance of a processing step. With this understanding, we can define design space. It is the multidimensional combination and interaction of the inputs that have been demonstrated to provide assurance of quality [52]. Defining the design space for a process involves identifying process parameters and material attributes that have an effect on CQAs and performance parameters as well as determining the relationships that link these inputs to the outputs.

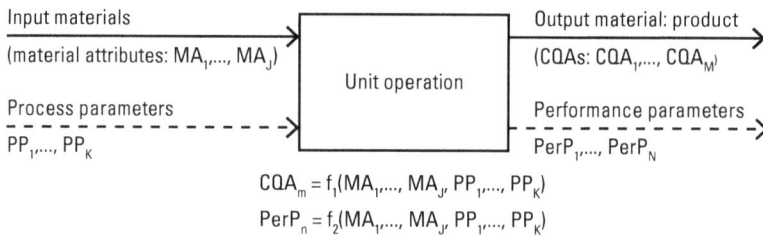

Input materials

(material attributes: $MA_1, ..., MA_J$)

Unit operation

Process parameters
$PP_1, ..., PP_K$

Output material: product

(CQAs: $CQA_1, ..., CQA_M$)

Performance parameters
$PerP_1, ..., PerP_N$

$$CQA_m = f_1(MA_1, ..., MA_J, PP_1, ..., PP_K)$$
$$PerP_n = f_2(MA_1, ..., MA_J, PP_1, ..., PP_K)$$

Figure 2.8: A diagram illustrating inputs to (including product) and outputs from a single unit operation and their relationship. Image © NC State University; reprinted with permission.

Some form of risk analysis is often used to determine which of the process parameters and material attributes identified as part of the proposed process have a high probability of affecting CQAs. Details of risk analysis methodology are not covered here, but the outcome would be an initial classification of process parameters and material attributes as critical, key, or non-key. Note that classification may change as process knowledge is gained through process development activities. The relationships between the inputs (material attributes, process parameters) and outputs (CQAs and performance parameters) are then established using a variety of methods, including studies conducted using a one-factor-at-a-time approach, studies utilizing a design of experiment (DoE) approach, and/or through the use of more fundamental mechanistic models. DoE methods typically adjust multiple parameters at once, in contrast with the one-factor-at-a-time method, resulting in fewer experiments and an understanding of parameters that may interact. As you can ascertain, the development activities to define the design space rely heavily on empirical studies. In addition, development activities usually focus on one unit operation at a time. To keep costs and time down, these studies are typically executed at small scale or even miniaturized scale operated in high-throughput mode. The resulting design space is used to define appropriate ranges for material attributes and process parameters to be used in process batch records.

Defining the design space can be challenging due to the numerous process parameters and material attributes involved in certain unit operations – particularly in cell culture, fermentation, and chromatography steps – that impact step performance. Let's consider the specific example of a cell culture step in the production bioreactor for a mAb in CHO cells, as might be used in the process shown in Figure 2.3. The A-Mab case study [62] provides a comprehensive example of how to design a process to produce A-Mab, a fictitious humanized IgG1 mAb intended as treatment for non-Hodgkin's lymphoma. Figure 2.9 shows a list of process parameters and material attributes considered in developing the design space to ensure that applicable CQAs and performance parameters are met for the production bioreactor step. This list was determined from knowledge of the production of other related mAbs, production of A-Mab for preclinical studies and phase 1 and 2 clinical trials, and previous process optimization studies for A-Mab. Risk analysis was then used to assess the potential of each process parameter to affect a CQA or performance parameter. The results of this exercise are also shown in Figure 2.9. Reducing the number of parameters in this way makes design space studies more manageable. Following risk analysis, DoE studies were performed using an appropriate scale-down bioreactor model to determine process parameters that deserved further consideration and to define the relationship between these process parameters and CQAs. Plots showing the resulting design space are also shown in Figure 2.9. Specifically, those plots show values for each process parameter ultimately determined to potentially impact CQAs – osmolality, dissolved CO_2, temperature, pH, culture duration – and the probability that all the quality attributes included in the model will be within the acceptable limits. This example shows that defining the design space – particularly for unit operations like a production bioreactor step that have a large number of potential process parameters – can be a complex undertaking.

With an understanding of the design space, defining the control strategy is the next design step as shown in Figure 2.7. The goal of the control strategy is to minimize variability, which can be significant, and ensure that CQAs for drug substance and drug product are met. Elements of a control strategy typically include the following: monitoring process parameters and controlling them within acceptable ranges, monitoring performance parameters and ensuring acceptable ranges are met, testing and accepting raw materials, sampling and testing process intermediates to ensure acceptable ranges

(a) Process parameters, material attributes

- Inoculum viable cell concentration
- Inoculum viability
- Inoculum in vitro cell age
- Osmolality
- Antifoam concentration
- Nutrient concentration in medium
- Medium storage temperature
- Medium hold time before filtration
- Medium filtration
- Medium age
- Timing of feed addition
- Volume of feed addition
- Component concentration in feed
- Timing of glucose feed addition
- Amount of glucose fed
- Dissolved oxygen
- Dissolved carbon dioxide
- Temperature
- pH
- Culture duration (days)
- Remnant glucose concentration

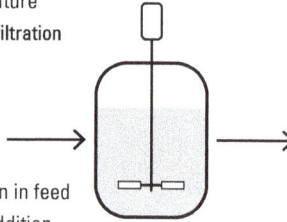

Critical quality attributes

- Product aggregate concentration
- Fucosylation
- Galactosylation
- Deamidation
- Host cell protein concentration
- Host cell DNA concentration

Performance parameters

- Product yield
- Cell viability at harvest
- Broth turbidity at harvest

(b) Perform risk analysis to determine those process parameters likely to have an impact on CQAs

↓

Process parameters considered in the design space study

- Inoculum viable cell concentration
- Osmolality
- Nutrient concentration in medium
- Volume of feed addition
- Component concentration in feed
- Dissolved oxygen
- Dissolved carbon dioxide
- Temperature
- pH
- Culture duration (days)
- Remnant glucose concentration

Perform DoE studies to determine design space

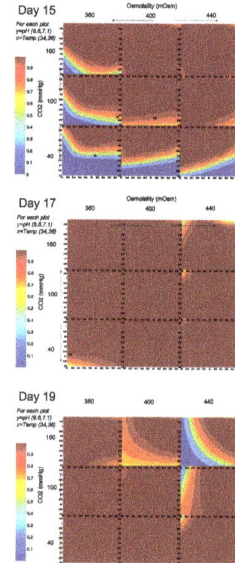

Figure 2.9: Defining the production bioreactor design space for A-Mab production in CHO. (a) A list of process parameters, material attributes, CQAs, and performance parameters initially considered.

are met, maintaining and calibrating equipment used in manufacturing, and release testing on drug substance and drug product to ensure CQA specifications are met. A key aspect of the design of the control strategy is setting appropriate ranges on the critical process parameters and controlling within the set ranges. For the A-Mab study, the authors [62] ultimately developed a set of inequalities – one for each CQA limit – that describe the combination of process parameter values that ensure that each CQA limit is met. From the complex design space, simpler operational ranges for process parameters can be established for routine manufacturing. These operational ranges must remain within the boundaries of the design space. Below is one possible set of ranges, derived from the curves in Figure 2.9, that ensure with greater than 99% probability that the bioreactor product will meet CQA limits:

- pH: 6.85–7.10
- Temperature: 34.5–35.5 °C
- Dissolved CO_2: 100–160 mm Hg
- Osmolality: 360–400 mOsm
- Culture duration: 15–17 days

These parameters could be controlled by a combination of process instructions, such as a batch production record, and equipment control features. It is also important to note that ranges are typically specified for key and even some non-key process parameters to ensure process consistency from batch to batch of product.

The last step in Figure 2.7 is scale-up, required because the studies used to define the design space are typically conducted at a scale significantly smaller than production. There are different approaches to scale-up for the different unit operations; these approaches are discussed in subsequent chapters. Generally speaking, scale-up is the process of transferring (1) material attributes and process parameters and (2) equipment dimensions from a small scale to larger scale to increase throughput for commercial production while preserving the quality of performance achieved at small scale.

Also worth noting is that the development of platform processes can greatly reduce the cost and increase the speed of process design and potentially improve quality of the biopharmaceutical drug products manufactured. The term *platform* refers to the use of common methods, procedures, and equipment in the development of processes for or manufacture of different products [81]. Products with active ingredients that are similar lend themselves to platform processes. Examples include production of mAbs or mRNA

──────

Figure 2.9 (Continued)

(b) Traces the process of reducing the number of process parameters considered to arrive at the design space through risk analysis and characterization studies [62]. Note that regions in dark red have >99% probability of meeting all CQA limits. Images (a) and (b) © NC State University; reprinted with permission. Graphs in image (b) reprinted from "A-Mab: a Case Study in Bioprocess Development," by CMC Biotech Working Group, p. 84. 2009.

vaccines. In both cases, development of a platform is enabled by the physical and biochemical similarities in the product being produced; for example, mAb products based on IgG1 antibodies all have similar chemical and physical properties. Therefore, upstream and downstream processes can be similar from one mAb-based product to another, which enables more rapid and less costly process design and implementation once the platform is in place.

2.3 Manufacturing facilities

In the beginning of this chapter, we emphasized the importance of facility design in meeting product quality and quantity requirements. In this section, we review some of the basic facility design considerations that ensure quality and quantity requirements are met.

At the most basic level, facilities must be designed to house all of the equipment required for both the main process and for supporting activities. Therefore, the processes (including supporting activities) housed within the facility and the amount of product to be produced, which dictates the scale of operations, will impact the final design. In addition to the supporting activities previously mentioned, space will be needed for other support activities, including a warehouse, weigh and dispense areas, quality control labs, and office space for groups such as process engineering, quality, and environmental health and safety. Because product types, production capacity, and processes differ from one facility to another, facilities come in a variety of shapes and sizes. Further, the facility for manufacturing drug product (from drug substance) is often separate from drug substance manufacturing facilities.

A clear understanding of the design of the process and demand for the product is required before embarking on designing the facility. Some specific examples of how the design of the process impacts the design of the facility include the following. A process designed with completely single-use systems will not require space for CIP skids for cleaning because equipment cleaning between batches or between product campaigns is not required. Space for clean steam generators for SIP steps would also not be required because single-use components for operations that require sterilization are generally received sealed and sterile. Likewise, a process designed for continuous operation would lead to a relatively small facility due to the reduced equipment footprint relative to a process for the same product designed for batch operation. Facility design also requires a clear understanding of the demand for the product to allow for proper equipment sizing, which will impact the space required to house the equipment.

There are also a number of considerations in the design of the facility that focus on "protecting the product," a paradigm that is often referred to in facility design literature. Protecting the product is most often used to refer to steps taken to keep microbial contamination out of the process and to avoid cross-contamination. Let's consider each separately.

Microbial contamination refers to the presence of unwanted microorganisms, such as bacteria and fungi, in a product. As discussed in Chapter 1, biopharmaceutical drug products must be sterile, given that they are typically administered parenterally. Therefore, controlling bioburden during the production of sterile drug products from drug substances (see Figure 2.2) is essential. The need for microbial contamination control also applies to fermentation and cell culture steps, in which contaminating microorganisms compete for nutrients and adversely impact the growth of product-generating cells.

Additionally, microbial contamination control is also required for steps downstream of fermentation or cell culture in the production of drug substance. Despite the fact that product is likely to see multiple 0.2 µm filtration steps aimed at removing unwanted microorganisms in the drug substance process, conditions during drug substance manufacturing favor microbial growth. Aqueous environments along with suitable temperature and pH levels can facilitate the proliferation of microorganisms. Additionally, there are plenty of sources of contamination within manufacturing facilities, including personnel, equipment, raw materials, the facility environment (such as airborne particulate matter), and clean utilities. This contamination can lead to the degradation of the product through microbial enzyme release or the production of endotoxins, exotoxins, and other by-products that may not be entirely eliminated in later processing steps, ultimately posing safety risks for patients [82, 83]. The take-home message is that end-to-end control of microbial contamination is needed in biopharmaceutical processes.

The potential for cross-contamination also exists within a process and can take many forms. For example, product A contaminating product B in multi-product facilities is a form of cross-contamination, as is batch 001 of product A contaminating batch 002 of the same product. Product intermediate from the bioreactor step making its way into intermediate from a chromatography step would likewise be considered cross-contamination. In addition to microbial and cross-contamination, there are other types of contamination as well, such as residues like lubricants and oil from equipment making their way into product.

Before getting into details of facility design features that protect the product, it's worth considering closed versus open processing, as the implementation of one versus the other can have a significant impact on facility design considerations related to contamination. Definitions for each are given below:

–   Closed system. A system designed and operated so that product is not exposed to the surrounding environment during processing [84]. There are numerous examples in the industry of steps that have been closed, including bioreactor steps, fluid transfer systems, and bulk filling steps. Expectations for facility design features that minimize contamination are lessened in areas that house closed systems.

– Open system. A process that is not closed and therefore must be performed in an environment where the probability of contamination is low. For example, chromatography and ultrafiltration steps in downstream processes are often open. When connections are made to the chromatography system in Figure 2.4, the lines connected are typically filled with room air; thus, a product-contact surface has been exposed to the environment. If after making all connections, the chromatography skid were sterilized, or all flow paths were initially sterile and connected to feeds (i.e., solutions and product load required for the step) in a way to maintain sterility, then the system could be considered closed.

There are a number of ways to close a system. Those details are beyond the scope of this book, but it is important to recognize that a claim of system closure must be adequately validated. For more details on closure methods and demonstration of closure, the ISPE Baseline Guide *Biopharmaceutical Manufacturing Facilities* is a good resource [84].

Numerous design practices are used in biopharmaceutical manufacturing facilities to mitigate the risk of both microbial and cross-contamination of product. Table 2.1 describes some of these. For processes that use reusable equipment, space for CIP systems and clean steam generation must be included. Because personnel are a main source of microbial contamination, gowning areas are needed within the facility to implement suitable gowning procedures. Gowning refers to the apparel that staff entering a facility must wear to protect the product. Note that the more closed a process is, the less stringent the gowning requirements are. There are also a number of design features that fall under the category of segregation, which can be spatial, temporal, procedural, or carried out by environmental control. In the context of this discussion on facility design, spatial and environmental are most common. Open processes are executed in cleanrooms designed to keep particulate – both viable and nonviable – concentration in the surrounding environment low. This type of segregation would be categorized as environmental control. There are also a variety of spatial segregation concepts listed in Table 2.1 that can be implemented to reduce the likelihood of contamination. Air locks are designed as a buffer that keeps air from flowing from less clean to cleaner areas. Separating each unit operation into its own dedicated space is a common approach for reducing risk of cross-contamination. In a multi-product facility, clear physical separation of the process for product A from product B may be used.

Because cleanrooms are a particularly important method of environmental control, we provide a few more details here. A cleanroom is a room in which the concentration of airborne particles is controlled and that is constructed and used in a manner to minimize the introduction, generation, and retention of particles inside the room [85]. Cleanrooms are achieved by:

– Filtering air entering from outside with high-efficiency particulate air (HEPA) filters.
– Maintaining a positive air pressure relative to less clean surrounding space so that air flows from cleaner to less clean rooms. A pressure differential of 0.02–0.08 inches of water is recommended [86].

Table 2.1: Facility design features used to protect product.

Design feature	Microbial or cross-contamination	Required for open systems, closed systems, or both	Segregation category	Comments
Space for CIP skid and parts washing systems	Cross	Both	Enables procedural	Needed to house equipment necessary for cleaning process equipment and small parts. Unnecessary for single-use equipment.
Space for clean steam generation	Microbial	Both	Enables procedural	Needed to house equipment required to generate steam for sterilization. Unnecessary for single-use equipment that comes sterilized.
Gowning areas	Both	Both	Enables procedural	Allows for gowning procedures to be implemented, which are critical to microbial contamination control. Less gowning is likely required when using closed systems.
Cleanrooms	Microbial	Open	Environmental control, procedural	Rooms in which the concentration of airborne particulate is controlled, primarily through filtration of outside air.
Appropriate facility finishes	Microbial	Open	Environmental control, procedural	Enables effective cleaning and disinfection of manufacturing spaces and reduces risk of microbial growth. Examples include floors, walls, and ceilings with smooth, hard surfaces.
Air locks	Microbial	Open	Environmental, spatial	Separates areas of different cleanliness levels by creating a buffer between clean and less clean areas.

(continued)

Table 2.1 (continued)

Design feature	Microbial or cross-contamination	Required for open systems, closed systems, or both	Segregation category	Comments
Separate space for individual unit operations for one product	Cross	Open	Spatial	Reduces risk of contaminating one step of a process with product intermediate from another step. The more of the process that is closed, the less separation into different rooms required. For example, if all upstream and downstream processing steps were closed, they could all take place in the same room.
Separate space for processes for different products running concurrently	Cross	Both	Spatial	Reduces risk of contaminating product A with product B. Multi-product facilities can also be operated using temporal segregation that relies on equipment cleaning and changeover procedures to minimize risk of cross-contamination between products.
Separating pre-viral clearance and post-viral clearance operations for one product	Microbial	Open	Spatial	Reduces risk of contaminating product intermediate that has undergone viral clearance with product intermediate that has not yet undergone viral clearance.
Separating supply and return corridors	Cross	Open	Spatial	Combined with appropriate flow of personnel, equipment, and materials, reduces risk of cross-contamination.

- Including air locks within the facility. An air lock is a room that typically has two doors in series to separate a cleanroom from a less clean environment, such as a corridor. The two doors are typically interlocked to avoid being opened at the same time [87].
- Using appropriate facility finishes. For example, floors, walls, and ceilings should have smooth hard surfaces that are easily cleanable and do not harbor microbial contamination.

There are several cleanroom classification systems in use. The two most common are the International Organization for Standardization (ISO) system, often used in the United States, and the European Union (EU) system. The specific classification under each system is based on the number of particles of a particular size permitted per volume of air. For example, a cleanroom meeting ISO 5 classification would have no more than 3,520 particles/m^3 of size greater than or equal to 0.5 μm. The corresponding classification in the EU system is Class A. In the ISO system, higher numerical designations mean a higher particle concentration in the environment; for example, ISO 8 allows a higher maximum concentration for a given size particle than ISO 5, 6, or 7 classifications. In the EU system, higher letters correspond to a higher particle concentration; for example, Grade D allows for a higher particulate concentration than Grades A, B, and C. Specific limits for particulate concentration in the EU classification system can be found in Annex 1 of the EudraLex Volume 4 [289]. For the ISO system, the limits are contained within the ISO 14644 standard [85]. Generally speaking, rooms get cleaner farther downstream in the process, as the product gets closer to drug product.

To illustrate how the facility design concepts in this section are put into practice, consider Figure 2.10, which shows a facility layout for production of a protein therapeutic drug substance. Take note of the following features of the design, which follows the list provided in Table 2.1:

- The layout features space for both the main process and supporting activities. There are rooms for media prep/storage and buffer prep/storage. There is also space included for a parts washer, autoclave, and/or CIP skid to clean equipment. And there is space for generating clean utilities such as HPW, WFI, and clean steam.
- Spaces for support activities are located close to the process areas they support. This is referred to as defining adjacency. For example, buffer prep and storage are located close to the DS processing areas in which the buffers are used.
- Gowning rooms for both the upstream and downstream processing areas are included so that gowning procedures can be implemented.
- Support activities and the main process are conducted in cleanrooms. Cleanrooms from ISO 8 down to ISO 5 are included. Note that generally, as the process moves toward drug substance, cleanroom classification becomes more stringent.
- Rooms with a cleaner classification are separated from areas with a less clean classification by an air lock.

- Separate rooms have been included to separate processing steps. The two purification suites can be used to house operations that are pre- and post-viral clearance, respectively. Note that two purification suites may not be enough to separate every viral clearance and chromatography step, so each purification suite may contain multiple unit operations. For example, Purification 2 may house the final two chromatography steps (post viral clearance) in the process. In this case, two batches of the same product – one being processed with the first (of the final two) chromatography steps in the space and one process with the second – would not be allowed in the room at the same time. This type of segregation is temporal.
- Separate supply and return corridors exist.

Note also in Figure 2.10 that product flow is unidirectional, always moving forward, which minimizes the likelihood of a "purer" form of the product intermediate being in the same space as a less "pure" form. Like product, flow of people, materials, and equipment is typically unidirectional. "Clean" people, materials, and equipment enter the facility and move through it via the supply corridor, also referred to as the clean corridor. Personnel would enter into the room in which they work, execute their tasks, and then leave the room and enter the return corridor, sometimes referred to as the dirty corridor.

Note that if multiple products are to be produced within a facility that uses open processing, either spatial or temporal segregation may be used to keep products separated. To segregate spatially, many of the process areas in Figure 2.10 would have to be replicated within the facility. Temporal segregation would involve running processes for products A and B, for example, in campaigns (i.e., a series of runs to produce a specified amount of one product) separated by time to avoid cross-contamination. Between campaigns, equipment would go through changeover. Changeover refers to the replacement of wetted soft parts on equipment – such as gaskets, O-rings, and valve diaphragms – to avoid the possibility of release of one product that may have been absorbed into the soft parts into the next product.

Much of what has been discussed previously in this section has focused on protecting the product. One other consideration with the potential to impact facility design is biohazard containment. When working with some biological agents, steps must be taken to protect the worker and the environment. These steps can involve process design, facility design, and procedures. Biosafety refers to the degree of precautions required to isolate dangerous biological agents. The Centers for Disease Control and National Institutes of Health have developed four biosafety levels (BSL): BSL 1, 2, 3, and 4 [89, 90] and the levels are designated before work with the infectious agent actually begins. BSL 1 applies to agents posing the least risk to human health and have the least stringent safety requirements. BSL 4 applies to biological agents that pose the highest risk to human health and requires the most stringent containment measures. Similar concepts are in place in most countries. The final assignment of biosafety level for a particular agent should be done through risk analysis as described in the U.S. Department of Health and Human

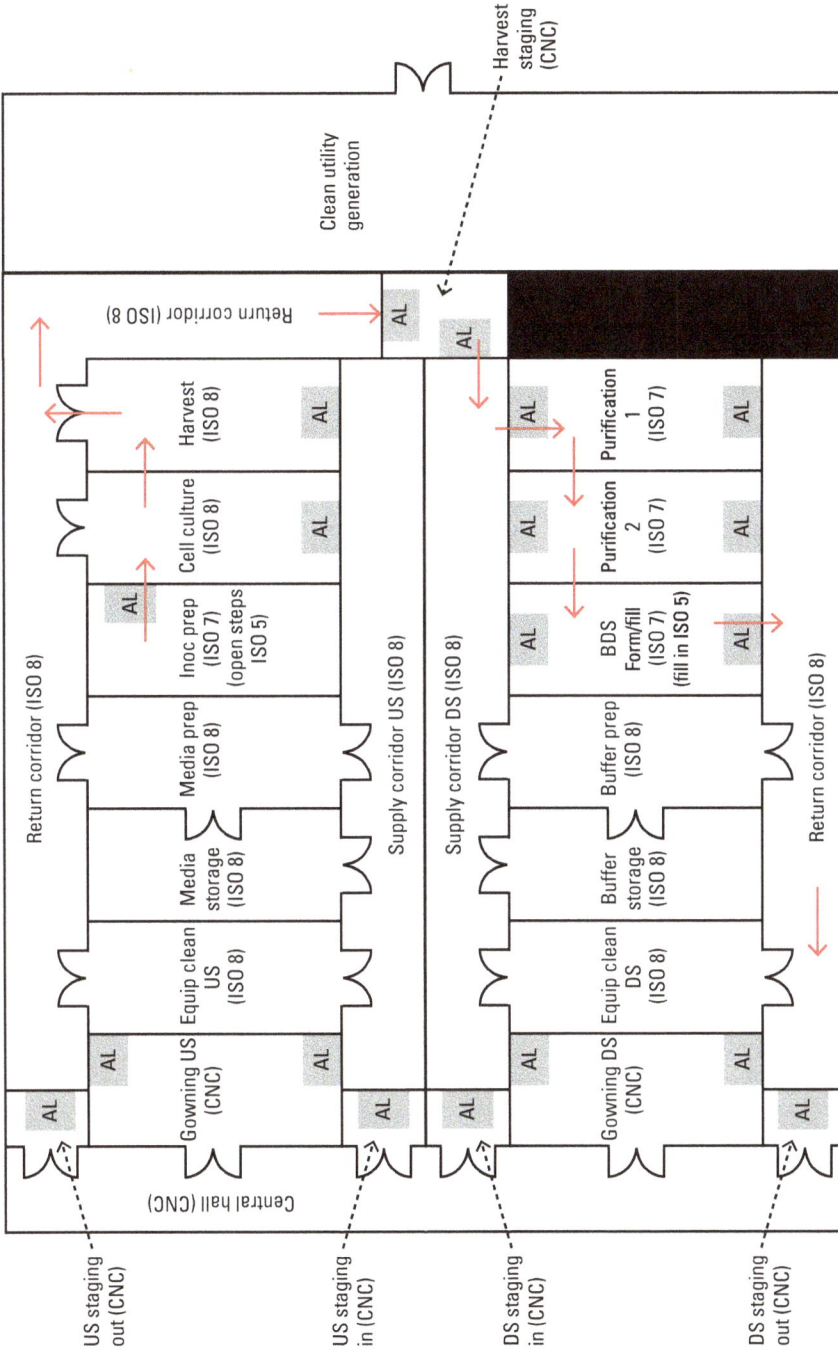

Figure 2.10: A facility layout for production of protein therapeutic drug substance, such as a mAb, with open processes and reusable equipment. Red arrows represent the product path through the

Services' publication *Biosafety in Microbiological and Biomedical Laboratories* [89]. From that document, the primary risk criteria used to define which of the four ascending levels of containment to assign are infectivity, severity of disease, transmissibility, and the nature of the work being conducted. Generally speaking, facility design and procedures put in place in a BSL-1 facility will not differ significantly from what is designed for standard CGMP operations. Examples of agents in which BSL-1 is likely to apply include CHO cells and *E. coli* strains typically used for biopharmaceutical manufacturing. Agents typically requiring BSL-2 facilities include influenza virus, commonly used in production of flu vaccines, and lentivirus, commonly used in production of gene and cell therapies. A BSL-2 facility requires limited access to the space and physical containment devices, such as a biosafety cabinet, for all manipulations of agents that cause splashes or aerosols of infectious materials.

There are numerous resources that provide more depth on the topic of facility design than we have been able to cover in this chapter [84, 91]. Please refer to them for additional details.

Finally, rather than building a facility, a biopharmaceutical company may choose to outsource its manufacturing to a contract manufacturing organization. Outsourcing is common and growing [92]. Companies may choose outsourcing because they lack necessary existing facilities or lack funding to build a new facility, which can cost up to US$2 billion and take 5-10 years to become operational [93]. Other reasons to outsource include insufficient qualified personnel in-house, pressure to reduce time to market, and alignment with a company's business model (e.g., a virtual company).

2.4 Summary

Well-designed biopharmaceutical processes and facilities are essential to ensuring that drug product is manufactured with the intended quality and in the required quantity. This chapter provided an overview of both. Most biopharmaceutical processes require four stages: production, harvest, purification, and formulation/fill. Each stage has a common objective carried out by multiple unit operations.

– In the *production stage*, the product is synthesized. For most biopharmaceutical products, this involves cell growth in bioreactors. At the end of this stage, the cells are part of a broth that includes spent aqueous media.
– In the *harvest stage*, the product is separated from the production system as cells used in production cannot be part of the drug product (cell therapies being an

---

Figure 2.10 (Continued)
facility. CNC stands for controlled not classified. AL stands for air lock. BDS is bulk drug substance, and US and DS stand for upstream and downstream. Image © NC State University; reprinted with permission, and adapted from Biopharmaceutical Processing Development, Design, and Implementation of Manufacturing Processes, 2018 [96].

exception). For intracellular products, a cell lysis step such as homogenization must be used to release the product into the surrounding liquid. Regardless of whether product is intracellular or extracellular, removal of solids from the aqueous process stream – either cells or cell debris – is required. This can be accomplished by a variety of unit operations, including centrifugation, depth filtration, and tangential-flow microfiltration. By the end of this stage, the clarified process stream with the product in aqueous solution may be essentially free of solid particles, but it still contains numerous soluble impurities that must be removed for the sake of patient safety.

– The *purification stage* removes these impurities most often by chromatography. Processes that rely heavily on mammalian-derived components, such as CHO-based processes for mAb production, also require viral clearance steps aimed at inactivating or removing viruses from the process stream. Upon completion of the purification stage, the product is still in a clarified liquid solution, but that solution is now free from most of the soluble impurities that were present at the harvest stage. However, the buffer system in which the product is dissolved is dictated by the last chromatography step and is likely not what is needed for drug product.

– During the *formulation/fill stage*, the product matrix is adjusted through unit operations such as ultrafiltration, other excipients may be added, and the product aseptically filled into its final container to produce the drug product described in detail in Chapter 1.

Generally speaking, steps involved in production of drug substance and drug product are operated in batch, rather than continuous, mode. And drug substance is a low-bioburden intermediate, while drug product is sterile. It is important to note that not all drug substance is produced in exactly the same way. In this chapter, we provided details in Figure 2.3 on a process to produce a mAb product. Processes for other products may look different. For example, Gamunex®-C is an immune globulin injection produced by purifying IgG from large pools of human plasma [59]; thus, product is synthesized in humans rather than in cells grown in a bioreactor. Kymriah®, is an ex vivo gene therapy described in Chapter 1. T cells from a patient are transduced with a lentiviral vector, expanded in culture, washed to remove impurities, formulated for cryopreservation, and then reinfused into the patient [23]. Thus, the final product is a suspension of modified cells rather than an active ingredient dissolved in an aqueous solution. Additionally, the original cells required for production are taken from the patient. The processes required for drugs like Gamunex®-C and Kymriah® would obviously be different than those for a mAb.

Equipment for biopharmaceutical processes is often skid mounted; that is all parts needed to support a given activity (e.g., chromatography) are mounted to a frame that is often on wheels. Further, there are processes that use reusable, typically stainless steel equipment, others that rely on single-use equipment (scale permitting), and others that use both. Additional equipment details are presented in subsequent chapters. The numerous activities required to support the main process were also discussed in this chapter. These include solution and media preparation, clean-in-place and steam-in-place operations, and operations to generate clean utilities like high purity water, water for injection, and clean steam.

There are many steps involved in the design of a process for biopharmaceutical production. Understanding what the product is – through the QTPP and CQAs – is an important first step in the design process. From there, significant effort is devoted to bench-scale studies to develop the design space. The design space is the combination of process parameters and material attributes that ensure quality of the biopharmaceutical product. Once the design space is established, a control strategy is developed. Central to that control strategy are batch production records that provide processing instructions and tracking of data for the process.

The facility in which the main process and supporting activities are conducted is also key to ensuring product quality and quantity needs are met. A paradigm central to facility design is to protect the product. To meet this objective, a good design:

– Allocates space for CIP skids and parts washing systems
– Allocates space for clean steam generation
– Includes air locks and gowning areas
– Includes cleanrooms for processing (particularly important for open processes)
– Uses appropriate facility finishes
– Provides separate spaces for individual unit operations for one product
– Provides separate space for processes for different products running concurrently
– Separates pre-viral clearance and post-viral clearance operations for one product
– Separates supply and return corridors

The extent to which these design features are implemented depends on whether a process or processing step is open or closed. A validated closed process – one that is designed and operated so that product is not exposed to the surrounding environment – allows for much more flexibility in design. For example, a validated closed system does not necessarily require operation in a clean room or segregation of unit operations for a given product.

2.5 Review questions

1. The stage 3 and 4 steps used to produce Xembify®, a 20% IgG solution for treatment of primary humoral immunodeficiency disease, are shown below as reported by Alonso et al. [94]. The IgG for Xembify® is from human plasma, the starting material for the manufacturing process. List *all* steps used in the production of Xembify® by stage, keeping in mind that the steps in the diagram below apply only to stages 3 and 4.

Plasma collected from healthy donors

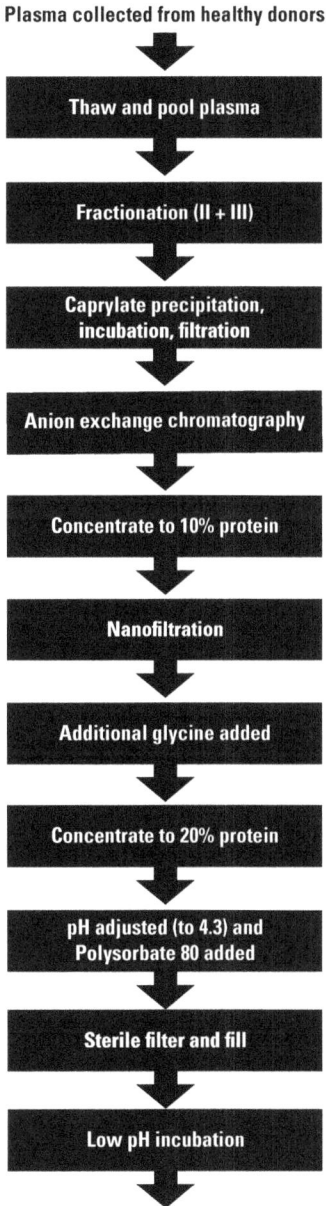

Figure 2.11 flow chart:

- Thaw and pool plasma
- Fractionation (II + III)
- Caprylate precipitation, incubation, filtration
- Anion exchange chromatography
- Concentrate to 10% protein
- Nanofiltration
- Additional glycine added
- Concentrate to 20% protein
- pH adjusted (to 4.3) and Polysorbate 80 added
- Sterile filter and fill
- Low pH incubation

Figure 2.11: Stage 3 and 4 steps used to produce Xembify®. Image © NC State University; reprinted with permission.

2. Human embryonic kidney 293 (HEK293) cells are used for the production of adeno-associated virus serotype 2 (AAV2), a viral vector commonly used for gene therapy. Because a significant amount of the AAV2 produced is intracellular, cells must be lysed within the bioreactor, which is commonly done by treatment with nonionic surfactant. The concentration of capsids containing the gene of interest is

1.11×10^{11} vector genomes (vg)/mL, while the concentration of vector genomes in the filtrate from the depth filter is 1.00×10^{11} vg/mL. The vector genomes are measured using quantitative polymerase chain reaction (qPCR).

a) If the volume fed to the depth filter is 50 L and the volume of filtrate collected is 48 L, calculate the step yield.

b) List three issues that could contribute to the step yield being <100%.

3. Table 2.2 shows the step yields for the mAb process shown in Figure 2.3.

Table 2.2: Step yields for the mAb process shown in Figure 2.3.

Step	% Step yield of target protein
Centrifugation/depth filtration	90
Protein A chromatography	95
Low pH viral inactivation	98
Cation exchange chromatography	80
Anion exchange flowthrough chromatography	95
Virus filtration	95
Concentration/buffer exchange by ultrafiltration	95
Bulk fill	98

Protein production is extracellular. The 5,000 L bioreactor step used to produce the protein has a titer of 4 g mAb/L bioreactor volume.

a) What is the % process yield, through bulk drug substance?

b) How many grams of product are bulk filled?

c) The final concentration of product is 110 mg mAb/mL. You fill 1 L bags to a volume of 0.9 L during the fill. How many bags are required to perform the fill?

4. You are working on the design of a process and facility to produce 2 metric tons of a mAb product (as drug substance) annually. Information about the process is given below:

– Titer in the bioreactor is 5 g/L

– Production bioreactors are on a 12-day cycle (10 days for production, 2 days for turnaround)

– DS process is on a 3-day cycle

– % yield for the DS process is 70%

a) How many kg of mAb will have to be produced in the bioreactors annually for a 2 metric ton production rate (drug substance)?

b) What total volume of bioreactor fluid is required?

c) How many bioreactors are required to keep the DS process running nonstop (i.e., to feed cell culture broth to DS every 3 days)? Assume you will operate 365 days/year, but then explain why this is likely an overestimate.

d) How large will each bioreactor be? Will it be stainless steel or single use?

5. The image below shows different rooms planned for a multi-product facility (Products A and B) for production of drug substance by open processing.

a) Arrange the rooms in an order that aligns with the design concepts presented in this chapter.

b) Draw arrows representing the product flow through the facility.

c) Draw arrows representing non-product flows, such as equipment, media, and solutions.

d) To reduce size and cost of a new facility, you choose to have only one process train that can be used for both Products A and B. Discuss differences and similarities between the two facilities (i.e., one with equipment dedicated to each product, and another with equipment shared between products) related to design features, operation, and cost.

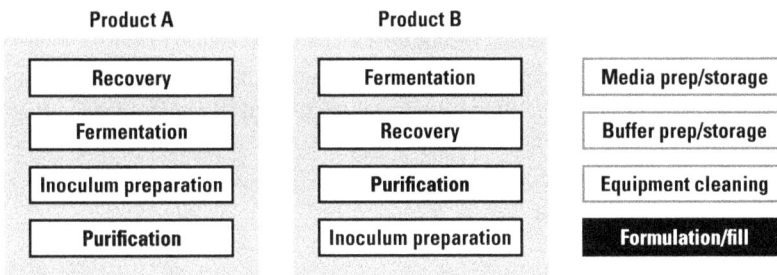

Product A	Product B	
Recovery	Fermentation	Media prep/storage
Fermentation	Recovery	Buffer prep/storage
Inoculum preparation	Purification	Equipment cleaning
Purification	Inoculum preparation	Formulation/fill

Figure 2.12: Rooms in a multi-product facility. Image © NC State University; reprinted with permission.

6. Describe five changes that could be made and their advantages to the facility design in Figure 2.10 if the process for manufacturing drug substance was closed rather than open and all equipment was single use.

Chapter 3
Current good manufacturing practice (CGMP) for biopharmaceutical production

As discussed in Chapter 1, biopharmaceutical drug products must have quality; that is, they must be suited to their intended use. In Chapter 2, we learned about meeting quality requirements through the biopharmaceutical manufacturing process and design of the manufacturing facility. This chapter focuses on another element for ensuring product quality, current good manufacturing practice (CGMP).

Questions explored in this chapter include:

– What is CGMP and how is it defined?
– What role do people play in meeting CGMP requirements?
– What are the CGMP expectations for people and for equipment used for biopharmaceutical processing?
– Why are documents so important to biopharmaceutical manufacturing, what types of documents are used, and what are the expectations regarding good documentation practice?
– What role does analytical testing by the quality control lab play in CGMP production of biopharmaceuticals?
– What does it mean to qualify equipment, systems, and utilities? And what are cleaning validation and process validation?
– What is involved with release of a biopharmaceutical batch after manufacture?
– How does CGMP apply even after biopharmaceutical products are on the market?

3.1 CGMP – an overview

To understand the importance of quality as it relates biopharmaceuticals, it is first important to acknowledge the human element that is involved. We described in Chapter 1 that many of the biggest-selling drugs worldwide are biopharmaceuticals; therefore, their use directly helps many patients. However, to be of benefit to patients, biopharmaceuticals products must be safe and efficacious. While it is indisputable that biopharmaceuticals made with quality benefit people, it is also important to recognize that it is people who design and execute the manufacturing processes and design and build facilities for production. The way in which these activities are conducted impacts product quality, and this chapter focuses on the standards that ensure appropriate systems are in place to manufacture products that truly benefit patients.

CGMP defines the minimum requirements that must be met to ensure that drug products are consistently produced with the required safety, identity, strength, purity, and quality (SISPQ). CGMP requirements achieve this by laying out expectations for

https://doi.org/10.1515/9783111112459-003

proper design, monitoring, and control of manufacturing processes and facilities. CGMP is designed to minimize the risks involved in biopharmaceutical production, risks that cannot be eliminated through testing of a small sample of the final product alone. Further, their focus is on what to do, but not specifically how to do it.

Current good manufacturing practice addresses general practices to be followed across the biopharmaceutical industry. These requirements are intentionally flexible to allow for the adoption of new innovations, which arise frequently, as they emerge. Additionally, the requirements themselves may evolve over time. The U.S. FDA includes the word 'current' to emphasize that manufacturers are expected to use the latest technologies and systems to comply with these standards and that manufacturers must keep up with the latest changes in requirements to ensure compliance. Throughout this book, we use "current good manufacturing practice" to emphasize the dynamic and evolving nature of these quality expectations. However, "good manufacturing practice" (GMP) and the abbreviation "cGMP" are also commonly used.

Just as CGMP requirements change over time, they may also vary from country to country, although at a high level, basic CGMP concepts are similar throughout the world. Examples of regulatory agencies outside of the United States that enforce CGMP include the European Medicines Agency (EMA), the Pharmaceuticals and Medical Devices Agency in Japan, or Health Canada.

With so many different biopharmaceutical products being produced, variation in CGMP expectations from country to country, and a variety of agencies enforcing CGMP, a company has many considerations in how best to implement CGMP. It is the responsibility of the drug manufacturer to not only define the methods for meeting CGMP requirements but to justify their approach to the appropriate regulatory agencies. Failure to meet CGMP requirements can result in regulatory observations that a company must remediate prior to receiving approval for marketing in a particular country. Other enforcement actions that may be taken by regulatory agencies when CGMP is not followed include seizure of product (by court order), an injunction preventing individuals or companies from specific actions that violate CGMP, fines, and for the most serious offences, criminal prosecution. Understanding the various CGMP requirements throughout the world is therefore especially important so that companies can comply.

3.2 CGMP regulatory framework

CGMP requirements for drugs take many forms, including regulations, guidelines, guidances, and standards, reflecting the intricate nature of the CGMP regulatory framework. Examples from various regions throughout the world are presented in Table 3.1.

Note that there are numerous guidance documents available to help companies determine the best practices to use for achieving CGMP requirements. While there may be legal ramifications for not following the regulations and some guidelines, guidance documents are not usually enforceable by law. They do, however, provide

Table 3.1: Examples of CGMP guidelines and guidance documents.

Selected CGMP regulations and guidelines throughout the world	Relevant guidance documents and standards
21 CFR Part 210 – Current Good Manufacturing Practice in Manufacturing, Processing, Packing or Holding of Drugs; General (United States)	ICH Q7 – Good Manufacturing Practice for Active Pharmaceutical Ingredients
21 CFR Part 211 – Current Good Manufacturing Practice for Finished Pharmaceuticals (United States)	ICH Q8(R2) – Pharmaceutical Development
EudraLex Volume 1 – Pharmaceutical legislation for medicinal products for human use (European Union)	ICH Q11 – Development and Manufacture of Drug Substances
EudraLex Volume 4 – Guidelines for good manufacturing practices for medicinal products for human and veterinary use (European Union)	United States Pharmacopeia – National Formulary (USP-NF)
Good Manufacturing Practices Guidelines for Active Pharmaceutical Ingredients (GUI-0104) (Canada)	European Pharmacopoeia (Ph. Eur.)
Ministerial Ordinance on Standards for Quality Assurance for Drugs, Quasi-Drugs, Cosmetics and Medical Devices (Japan)	FDA Guidance for Industry – Process Validation: General Principles and Practices
Ministerial Ordinance on Standards for Manufacturing Control and Quality Control for Drugs and Quasi-Drugs (Japan)	FDA Guidance for Industry – Sterile Drug Products Produced by Aseptic Processing – Current Good Manufacturing Practice

a level of detail not typical in regulations that is helpful for determining how to put CGMP into actual practice. Regulatory agencies such as those mentioned above publish guidance documents in addition to regulations. The International Council for Harmonization of Technical Requirements for Pharmaceuticals for Human Use (ICH) is an organization that provides guidance by bringing together requirements from multiple regulatory agencies and establishing harmonized guidance that can be applied worldwide. A few relevant ICH documents are listed in Table 3.1 along with other CGMP guidance documents well known in the biopharmaceutical industry. Additional information about best practices for implementing CGMP can be found through publications from trade organizations and industry associations such as the International Society for Pharmaceutical Engineering (ISPE) and the Parenteral Drug Association (PDA). Both organizations are good sources for guidance documents related to biopharmaceutical manufacturing in general, including CGMP.

CGMP encompasses all aspects of biopharmaceutical manufacturing, including personnel, facilities and equipment, documentation, production, quality control, and quality systems. Chapter 2 provided an overview of facility design considerations to ensure product quality requirements and conformance to CGMP. We examine some other CGMP expectations in the following sections.

3.3 Notable CGMP requirements for biopharmaceutical manufacture

In this section, we look at some of the notable CGMP requirements and how they apply to biopharmaceutical manufacturing. The basis for this discussion is primarily the CGMP regulations from the U.S. Code of Federal Regulations and guidelines from the European Union's EudraLex, although as mentioned previously, CGMP requirements throughout the world are similar at a high level. In addition, information from relevant ICH guidance documents is included.

3.3.1 Personnel

People play an important role and are necessary in some way for essentially every step in biopharmaceutical manufacturing processing. People are needed for all types of tasks, such as execution of manufacturing processes, testing of product samples, review of manufacturing documentation, cleaning of manufacturing facilities, maintenance and calibration of manufacturing equipment, and management of other personnel, just to name a few. Because personnel are so essential to successful execution of a biopharmaceutical manufacturing process, it is not surprising that the various documents that describe CGMP requirements include sections specific to personnel.

For example, it is a requirement that biopharmaceutical companies employ an appropriate number of qualified personnel to manufacture a quality product [61, 95].You might think that the focus of this requirement would be the personnel who are directly responsible for manufacturing and testing of product. However, it is equally important to ensure sufficient ancillary personnel are in place to provide oversight and support of manufacturing operations. The EMA requirements go so far as to require organizational charts showing the relationships between different departments and specifying general responsibilities for production and quality [97]. It is also a requirement that personnel are qualified for their particular roles, having the appropriate education, training, and experience to perform their assigned functions [95]. From the previous chapter, it is clear that biopharmaceutical manufacturing is complex and therefore requires input from many different people with many different backgrounds. Having an adequate number of qualified personnel in place is key to ensuring SISPQ.

Every biopharmaceutical company must also establish a training program that provides initial and ongoing training for their CGMP personnel. Even with appropriate education and potentially prior experience in the biopharmaceutical industry, employees still need training on company-specific procedures and practices. In addition, it is an expectation that employees are provided training on topics related to CGMP and quality systems at some frequency to reinforce basic principles and to provide awareness of new regulatory requirements, company procedures, or industry best practices.

Expectations regarding personnel hygiene and cleanliness are also part of CGMP requirements. From Chapter 2, one of the key inputs to product quality is design of the manufacturing facility with a focus on protecting the product from contamination. Protecting the product from contamination from personnel is important as well, given the risk to contamination that personnel pose. A large number of personnel enter a facility every day. They have close proximity to the product being manufactured, and people can shed 1,000,000,000 skin cells per day, of which approximately 10% may contain microorganisms [98]. No matter how well a facility is designed to prevent microbial and cross contamination, if the cleanliness of the people inside the facility, especially those with close proximity to manufacturing and support operations, is not controlled, the product is not adequately protected. Specific ways in which product is protected from employees include:

- Restricting access to CGMP areas for any employees who are sick or have open wounds/lesions
- Ensuring personnel wash their hands and follow good sanitation practices
- Ensuring that employees who access CGMP areas are gowned appropriately, which is discussed in greater detail in the next section
- Controlling cleanroom behaviors

The recent coronavirus pandemic has brought an increased awareness of the importance of practicing good hygiene to prevent spread of infectious diseases. These practices, although important in day-to-day life when faced with a global crisis such as a pandemic, have always been important in the manufacture of biopharmaceuticals, where it is critical to protect the product from contamination. To this end, CGMP requires companies to prohibit any person with an apparent illness or open lesions from direct contact with product or product materials and components and that personnel are to report any health conditions that may adversely affect drug products to supervision [61, 99]. A biopharmaceutical manufacturing employee is also required to wash his or her hands frequently, especially prior to entering the manufacturing areas and before donning an initial set of gloves. Rules around good hygiene also typically prohibit wearing makeup, jewelry, and perfumes and storing or consuming food and drink in the CGMP working areas.

3.3.2 Personnel gowning

Personnel gowning, which refers to the special protective apparel that staff entering the manufacturing facility must wear primarily to protect the product, is a basic

CGMP requirement for biopharmaceutical manufacturing; however, the specific gowning requirements will differ based on the operations within a manufacturing area. In general, as a process moves closer to drug substance or drug product, more protective gowning is required. Likewise, open operations are conducted with more gowning than closed. Some examples of protective gowning typical in this industry are hairnets, gloves, shoe coverings, frocks/lab coats, coveralls/jumpsuits, masks, sterile sleeves, and safety glasses/goggles. In addition, some companies restrict employees from wearing street clothes – that is, their personal clothing – in the manufacturing areas and instead require employees to put on scrubs or cleanroom clothing and even dedicated facility shoes. Whatever the requirements, it is important to gown according to approved facility procedures, which are discussed shortly, using the correct gowning supplies and donning gowning in the correct order.

As a first gowning step when entering manufacturing areas, personnel will likely don laundered scrubs or a plant uniform followed by facility-dedicated shoes or shoe coverings. Donning shoes/shoe coverings is generally done such that the gowning only touches the "clean" side of the gowning area. Initial gowning also likely includes donning a hairnet and pair of gloves. As personnel make their way into specific manufacturing suites where more critical operations taken place, they will be required to add additional layers of protective gowning. For example, when entering an ISO 8 area, it is likely that personnel would add a coverall over their scrubs, a pair of shoe covers over existing shoe covers, and a pair of gloves over existing gloves. If transitioning to an ISO 5 cleanroom from an ISO 8 area, more stringent gowning would be required as ISO 5 areas are required to meet a tighter criterion for particulates than an ISO 8 cleanroom (see Chapter 2). Personnel would likely add a frock over their coveralls, a hood over their hairnet, boot covers over their shoe covers, etc. No skin should be exposed when working in an ISO 5 area. Typical gowning items required for ISO 8 and ISO 5 areas are described below.

Example ISO 8 Gowning Procedure
1. Apparel from initial gowning
2. Safety glasses
3. Hairnet (if not already present) – ensure all hair is contained within the hairnet
4. Beard cover (if applicable)
5. Coverall – do not let coverall touch the floor or any unclean surfaces, ensure any personal clothing is covered
6. Shoe covers – don one at a time while crossing from dirty side of room to clean side
7. Gloves – spray hands with 70% isopropyl alcohol or use hand sanitizer after donning

Example ISO 5 Gowning Procedure
1. Sterile mask – ensure nose and mouth are covered
2. Hood – tuck bottom of hood into top of coverall
3. Frock – added over existing gowning (i.e., coverall)
4. Boot covers – don one at a time while crossing from dirty side of room to clean side
5. Sterile sleeves – avoid touching outside of sleeves when donning
6. Sterile gloves – ensure sterile sleeves are captured under each glove
7. Sterile goggles

A comparison of typical gowning for ISO 8 operations versus ISO 5 operations is shown in Figure 3.1.

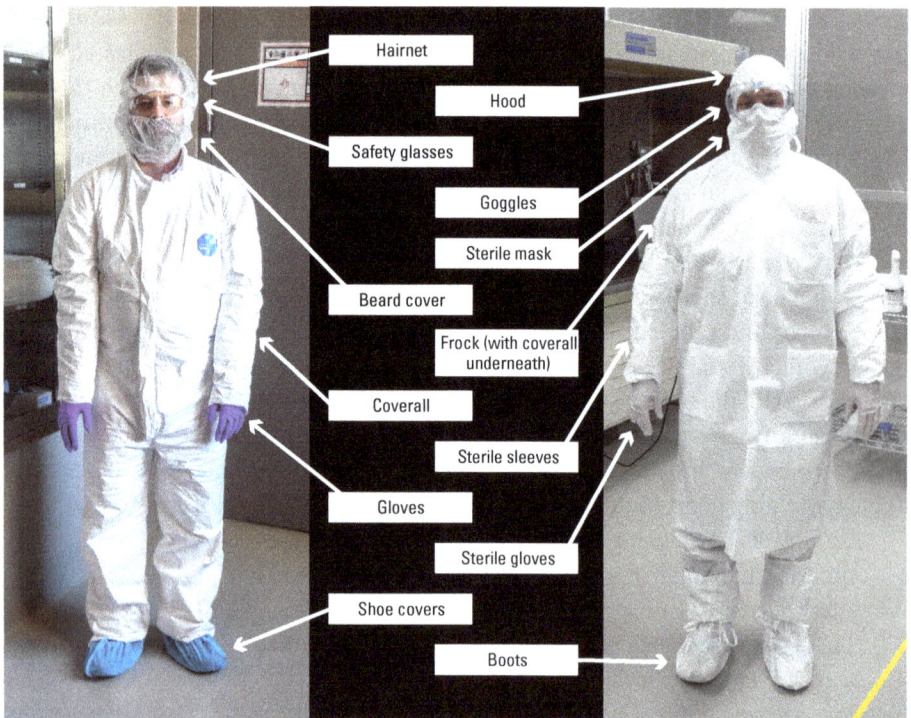

Figure 3.1: Gowning for ISO 8 (on left) operations versus ISO 5 (on right) operations. Photos © NC State University; reprinted with permission. Note that ISO 5 gowning is typically donned on top of existing gowning (e.g., gowning from ISO 8). Note also that gowning components are not listed in the order they are donned.

3.3.3 Cleanroom behaviors

Once gowned and in the cleanroom areas, personnel must take additional precautions to remain as "clean" as possible and to reduce the risk of contaminating the environment and the product. Often referred to as aseptic technique, these precautions include the following:

- Refrain from touching clean items and surfaces unless necessary. Minimizing contact between the hands and other items/surfaces leads to less opportunity for contamination.
- Frequently sanitize gloved hands, typically with an alcohol-based solution. The frequency may vary based on the area in which you are working, but a good rule of thumb is to sanitize gloves every 10–15 min or after touching any other surface.
- Replace gloves if heavily soiled or damaged. Once gloves are damaged, they are no longer effective for containing any contamination or shedding skin cells on the hands.
- Avoid leaning or sitting on clean surfaces. Avoiding contact with these areas with any part of your gowning helps to minimize contamination of critical areas.
- Use slow, deliberate movements. Rapid movements can disrupt the laminar airflow in a cleanroom, causing the air, and any particulates, to swirl and move in a more disorganized manner throughout the room.
- Avoid movements directly above open containers or near open processing. Since the airflow in the room is likely flowing from top to bottom, placing your hand directly above an open processing container could result in any particulates or organisms on your hand falling into the vessel.

Learning appropriate aseptic techniques is an important part of good cleanroom behavior. It is likely your company will require you to complete training in these techniques specifically for their particular process and the tasks you are responsible for executing.

3.3.4 Equipment

CGMP applies not only to the personnel involved with biopharmaceutical manufacturing but to the processing equipment as well. The focus of CGMP related to equipment is to protect the product and to ensure that equipment functions properly for its intended use, both of which ensure product quality. Protecting the product means minimizing potential for contamination, which can be accomplished in a number of ways. From 21 CFR 211 subpart D [100], key equipment considerations related to CGMP are as follows:

- Equipment should be designed and located appropriately for its intended use and allow for cleaning and maintenance.
- It is important to ensure the equipment on its own does not introduce any contaminants to the product stream. Equipment surfaces that come in direct contact

with the product should be non-reactive, and not be a source of foreign materials, residues, microbial contamination, or particulates.
–   The product contacting surfaces should also be easy to clean and compatible with various cleaning and sanitization solutions. Common materials of construction for biopharmaceutical equipment include stainless steel, glass, and thermoplastics.

Unless disposable and intended for one-time use, equipment must be cleaned between uses and therefore methods for cleaning must be developed. Cleaning of equipment can be performed manually or by an automated means. Automated cleaning may be performed either while the equipment stays in place – referred to as clean in place or CIP, as described in Chapter 2 – or by transferring the equipment to a separate location specifically for cleaning, such as a wash room containing a parts washer. In addition, depending on the purpose of the equipment in the process, it may be necessary to establish a sanitization or sterilization procedure as well. We won't go into details here given that these topics were covered in the previous chapter.

Regardless of the exact process for cleaning and sanitization, CGMP requires that written procedures be established so that the same process can be consistently executed. Procedures should also include details related to required frequency of cleaning and a requirement for inspection of equipment for cleanliness prior to use. It is expected that a cleaning history or record is maintained for process equipment. We'll discuss CGMP requirements regarding documentation later in this chapter, but a general rule of thumb is that if it isn't documented, it didn't happen. For this reason, it is important to record each time cleaning and sanitization/sterilization is performed. In general, the frequency of cleaning and sanitization/sterilization should be established to ensure there is no build-up of residual product/process components and no carryover of potential contaminants between equipment uses. Cleaning procedures are often validated – meaning that testing is performed under a protocol to demonstrate that the cleaning process consistently meets expected results. More specific information related to execution of cleaning validation is covered in Section 3.3.7.

Maintaining equipment in an appropriate state of repair is also part of CGMP. Doing so ensures that the equipment operates consistently during multiple production batches and is accomplished through routine inspection, calibration, and preventive maintenance. Written procedures as well as records of execution are also required for these activities. The intended use of the equipment influences the schedule for performing calibration and maintenance activities, with more critical equipment needing the most frequent oversight. In addition, it is expected that calibrations are performed against established acceptance criteria and that equipment is prevented from use when these criteria are not met. To that end, most companies employ a means of designating the current calibration status, either by physically labeling the equipment or through an electronic asset management system where personnel can easily find calibration status and calibration due dates.

Example: Calibrating a flow meter on a chromatography system
A calibration check on a chromatography system flow meter is scheduled at six-month intervals. The procedure involves pumping water through the chromatography system at flow rates that cover the range of operation for the skid. The flow rate reading on the control screen (transmitted by the inline flow meter) is recorded during the procedure. The actual flow rate (i.e., the standard against which the control screen reading is compared) is measured by weighing the amount of water collected over a one-minute period and converting to a flow rate in units of L/h, assuming 1 kg = 1 L. The percent deviation between the reading from the control screen and the standard is determined. This percent deviation must fall within +/- 2.5% to be deemed acceptable. Data from the most recent calibration check is shown in Table 3.2.

Does the flow meter require any adjustment to meet the allowable range for percent deviation?

Table 3.2: Data collected for a calibration check on a chromatography flow meter.

Flow set point (L/h)	Control screen flow Rate reading (L/h)	Volume water collected (L)	Time for collection (s)	Calculated flow rate (L/h)	% Deviation
36	35.8	0.61	60		
72	71.3	1.19	60		
108	108.6	1.82	60		
144	145.0	2.41	60		
180	179.5	3.01	60		

Solution
Calculation of the actual flow rates based on the volume of water collected over a 60 s period result in the following values, listed from the low to high set point: 36.6, 71.4, 109.2, 144.6, and 180.6 L/h. The calculation for the 36 L/h set point is shown below as an example:

Volumetric flow rate at 36 L/h set point = 0.61 L/60 s × 60 s/min × 60 min/h = 36.6 L/h

The corresponding percent deviation values are 2.19, 0.14, 0.55, −0.28, and 0.61%, respectively. Therefore, the flow meter is accurate based on the percent deviation criteria required, and no adjustment to flow rate readings is required. As an example, the calculation for the percent deviation at 20% set point is shown below:

% deviation = (36.6−35.8)/36.6 × 100 = 2.19%

There are additional CGMP considerations for equipment that involve automation or computerized systems. For example, computer systems associated with this equipment are validated to demonstrate that the computer system hardware and software performs tasks consistently and as expected [61, 101]. In addition, access to a computerized system should be controlled to prevent unauthorized access, and to restrict the ability to change data or records generated by the system [102]. In many cases, the data generated by the computerized system becomes part of the official manufacturing batch documentation, so it is important that data are verified for accuracy, complete, and protected from modification or deletion. CGMP also requires that companies have processes in place to routinely back up such data as a safeguard against data loss. Written procedures describing the specific requirements for each computerized system are expected.

3.3.5 Documentation system: procedures, records/reports, and specifications

As mentioned in the previous sections, a familiar phrase in biopharmaceutical manufacturing is "If it wasn't documented, it didn't happen." Further, use of procedures, records, and specifications is critical to controlling the process and product. EudraLex, Volume 4 indicates that the main objective of a documentation system is to "establish, control, monitor and record all activities which directly or indirectly impact on all aspects of the quality of medicinal products" [103]. As a result, multiple regulations, guidelines, and guidance documents focus on CGMP requirements related to documentation.

Documents are generated to execute the manufacture of a biopharmaceutical product and may be captured in paper or electronic format. In general, documents fall within one of three categories – procedures, records and reports, and specifications.
- Procedures are documents that provide guidance or instructions to execute a specific task or process. In a sense, procedures are an input to a task. Procedures define what is supposed to happen, in what order, and what the expected outcomes are. Written procedures provide control by ensuring that an activity is performed the same way every single time. They must be written in clear and unambiguous language and be accurate. The use of pictures/diagrams/flow charts where appropriate can be helpful in making a procedure easier to follow. Examples of procedures are included in Table 3.3.
- Records and reports document the output of the executed tasks. Personnel record information directly within these documents. Specific examples are included in Table 3.3.
- Specifications are similar to a procedure in that they can be viewed as an input to a task. However, specifications are generally limited to stating requirements or acceptance criteria and serve as a basis for quality evaluation [103]. Specifications do not generally include individual instructions. For example, an analytical testing specification might include the acceptance criteria for several different tests but would not include instructions for how to complete each assay. The analyst would instead have to reference the specific test procedures to understand how to complete each test. Note that analytical specifications are written for a variety of samples, including raw materials, process intermediates, drug substance, and drug product.

Because procedures, records/reports, and specifications are an important means of controlling a biopharmaceutical manufacturing process, and thus important for ensuring product quality, CGMP requires these documents be approved by appropriate person-

Table 3.3: Example document types.

Document type	Document category
Standard Operating Procedure (SOP)	Procedure
Master Production Record (the version of a production record from which the batch production record is created)	Procedure
Batch Production Record	Procedure and Record
Validation/Qualification Protocol	Procedure and Record
Validation/Qualification Report	Report
Forms	Record
Logbook	Record
Analytical Testing Report	Record/Report
Status Tag	Record
Stability Protocol	Procedure and Record
Stability Report	Report
Material Release Specification	Specification
Product and Sample Labels	Record
Complaint Files	Record
Certificate of Analysis	Record/Report

nel and assigned an effective date. Any revisions to procedures are also controlled and must be justified and preapproved prior to use by personnel. Figure 3.2 illustrates a typical controlled document process flow. It is never acceptable to knowingly deviate from approved instructions and procedures. In fact, a common observation by regulatory agencies during inspections is companies failing to follow their written procedures.

Records and reports provide evidence that tasks were performed correctly and per approved procedures and specifications. This document category includes not only the manufacturing documents, such as batch records and test results, but also documents related to supporting activities, such as validation, training, investigations, storage, maintenance, and cleaning. How information is documented is an important concept in CGMP. In fact, this is such a crucial topic in the industry that there are standards related to how to document results, referred to as good documentation practice or GDP.

GDP applies to both written and electronic records/reports supporting the manufacture of biopharmaceuticals. When dealing with written documentation, some specific best practices are universal [103]. For example, no areas where data is expected should be left blank. Handwritten entries should be made at the time of performance;

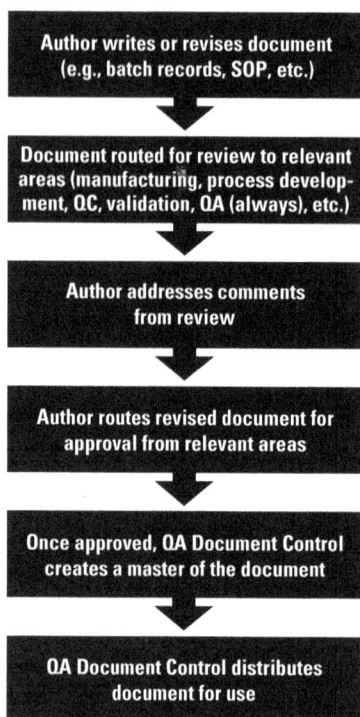

Figure 3.2: Controlled document process flow. Image © NC State University; reprinted with permission.

the individual making the entry and when it was made should be able to be traced. This is generally accomplished by requiring the person making the entry to provide their initials and the date, where the date includes the day, month, and year and does not use slash marks or dashes. All entries should be clear and legible and should be captured in indelible ink so that records are permanent and cannot be changed. Numerical entries should be recorded to the same number of significant figures as associated acceptance criteria or specifications. Military time is generally preferred for time entries as it does not require designation of am versus pm. Writing over previous entries to make corrections is not allowed. If corrections or changes are made to original documentation, GDP requires that these are made in a certain way as explained below:

- What changed?
  - Draw a single line through the error so that the original entry is not obscured and can still be easily read.
  - Do not erase, use white-out or otherwise obliterate the existing information.
- Who made the change?
  - Document with your initials.
  - Ensure that initials can be traced to a single individual.
- Why was the change made?
  - Document an explanation for the change.

- Legibly add the correct information as close to the original entry as space allows.
- If there is not enough space near the original entry to make the correction, include a footnote at the location requiring correction and enter a corresponding footnote and explanation in the margin or at the bottom of the page where space allows.
- When the change occurred
  - Document the date the correction is made.
  - Use the date format required by your SOPs.

Example: Finding documentation errors in an executed batch record

Figure 3.3 provides an excerpt from an executed batch record used in the production of tris solution, a buffer commonly used in chromatography steps for biopharmaceutical processes. Focusing on the written entries made by operators, identify specific examples of poor documentation practice.

ACME Biopharmaceuticals			Page number: 3 of 5
Title: Tris, pH 8.0 Solution	**Document type:** Batch Record	**Document number:** SR-083	
Authored by: Bill Smith **Title:** Manufacturing Associate	**Date:** June 23, 2020 **Department:** Solution Prep	**Revision:** 04, **Effective date:** July 2, 2020	
Approved by: Susan James **Title:** Manager, Solution Preparation	**Date:** July 2, 2020 **Department:** Solution Prep	**Lot number:**	

Step #	Task	Item	Data	Performed By Init/Date	Verified By Init/Date
2	Fill V-4100 with 300 +/- 5 L of high purity water (HPW) per SOP CS-007.	HPW volume added (L)	~~351~~ 302 (300 +/-5 L)	BRm 02NOV2020	CG 02NOV2020
3	Add the Tris base weighed out for this prep and recorded in Table 2.2 to V-4100 and close the addition port.	✓	X	BRm 02NOV2020	G6 02NOV2020
4	Turn on the V-4100 agitator to 1200 +/- 10 rpm. Mix for not less than 5 minutes.	Agitator speed (rpm) Start mix time End mix time	1210.5 (1200 +/-10 rpm) 1020 10:33	BRm 03NOV2020	G6 02NOV2020
5	Obtain 2 L of 25% HCl and record the requested information in Table 2.2.	✓	✓	Bem 11/02/2020	G6 02NOV2020
6	Adjust the pH of the Tris solution to 8.00 +/- 0.20 per SOP GL-028. Record the ID of the meter used, the date of standardization of the meter, and the final pH value.	pH meter ID Standardization date pH	1037 02NOV2020 8.2 (8.00 +/- 0.20)	Bem 02NOV2020	G6 02NOV2020
7	Add HPW to V-4100 per SOP CS-007 to achieve the final solution volume of 400 L +/- 5.0 L.	Final HPW volume (L)	~~404~~ 405 (400 +/-5 L) ①	BRm 02NOV2020	G6 02NOV2020

① WRONG NUMBER

Comments: N/A BRm 02NOV2020

  N/A BRm 02NOV2020

Figure 3.3: Excerpt of an executed batch record for preparing tris solution. Image © NC State University; reprinted with permission.

Solution

Documentation errors in the above record include:

Step 2 – The original data entry is obscured, and the corrected value does not include a reason for why the data was changed, initials of who made the correction, and a date to indicate when the correction was made.

Step 3 – The data field was completed incorrectly. An "X" was used in data field rather than "✓" as indicated in the "Item" field.

Step 4 – The incorrect number of significant figures was used for agitator speed, and, when rounded, the result does not meet the acceptance criteria for this parameter. Instead of correcting the "Start Mix Time" data entry, the operator attempted to write over the original data (1026 changed to 1028). The "Performed By" entry appears to be a day later than when all other steps were performed. The same person initially signed for "Performed By" and "Verified By." Correction of the "Verified By" entry was not done correctly; the original entry is obscured and the correction does not include why, who, and when.

Step 5 – The date entry in the "Performed By" field is not in the correct format. The slashes (/) may be mistaken for the numeral 1. A more common format would be 02Nov2020.

Step 6 – Data for the pH meter ID is not legible (could be 1637 or 1037). Entry for pH is not recorded to the correct number of significant figures, which prevents determination of whether the value is within the acceptable range.

Step 7 – The correction to the data entry was made correctly; however, the footnote explaining the correction at the bottom of the page should also be initialed/dated. In addition, because the original value recorded for "Final HPW Volume" was outside the acceptable range, a more descriptive note would be necessary to explain why the value was changed.

The acronym ALCOA or ALCOA+ is commonly used to summarize principles associated with GDP, where ALCOA stands for: Attributable, Legible, Contemporaneous, Original, and Accurate. The + has been added more recently to indicate that documentation must also be Complete, Consistent, Enduring, and Available. Table 3.4 provides additional information about each of these document characteristics.

Table 3.4: ALCOA+ principles for good documentation practices.

Letter	Principles	Description
A	Attributable	Entry is traceable to the person who performed the action or generated the data.
L	Legible	Entry is easy to read and understand.
C	Contemporaneous	Entry was made or data generated at the time of the activity.
O	Original	Data is from original source.
A	Accurate	Entry is valid, correct, and contains no errors.

Table 3.4 (continued)

Letter	Principles	Description
+	Complete	All data is included, including repeat analysis.
	Consistent	Date/times associated with data follow expected sequence.
	Enduring	Entry/data is permanent and sustainable.
	Available	Data can be accessed for review or inspection over the lifetime of the record.

A notable trend in the biopharmaceutical industry is the increasing use of electronic records and reports, including electronic batch records. Electronic systems for documentation may help to minimize the occurrence of documentation errors and has the added benefit of allowing data to be readily accessible for review and analysis. Electronic documentation, however, is subject to the same requirements as handwritten documentation, and ALCOA+ principles still apply. Electronic documentation has the added component of electronic signatures in lieu of a handwritten signature/date or initials/date. The requirements for the use of electronic signatures are defined in several regulations, guidelines, and guidance documents, including 21 CFR Part 11. Electronic signatures must capture the identity of the signer, the date and time when the signature was made, and the meaning or purpose of the signature. In addition, electronic signatures must be made using at least two distinct identification components (e.g., username and password) [104].

3.3.6 Testing: raw materials, in-process, drug substance, and drug product

There are many points throughout a biopharmaceutical manufacturing process where testing is performed. On the front end of the process, CGMP requires that each lot of raw materials and components used in the manufacture of the product are sampled, tested, or examined and released by the Quality unit prior to use [61, 105]. Minimally, this testing usually includes a confirmation of identity, which may only require inspection of the certification of analysis for the raw material/component. However, depending on the criticality of the material and its use in the manufacturing process, more comprehensive testing may be required. Approved material specifications indicating the testing requirements and associated acceptance criteria are generated for each material.

Once the manufacturing process is initiated, various tests are performed along the way on process intermediates to ensure the process is performing as expected or to prepare product streams for further processing [61, 106]. A variety of analytical methods may be employed for measuring characteristics of intermediates such as protein concentration, product purity, impurities, pH, conductivity, endotoxin, or bioburden. The specific testing performed depends on the type of biopharmaceutical being produced. The requirements and acceptance criteria for such in-process testing are generally specified in the manufacturing batch records or associated specifications. Other types of testing that may occur during processing of a biopharmaceutical manufacturing batch include environmental monitoring (EM), which is performed to verify the manufacturing environment in which the production process takes place is under a state of microbial control.

At the end of the manufacturing process, testing of the bulk drug substance and drug product is performed [61, 107]. As we learned in Chapter 1, CGMP requires that drug products are tested to ensure batch-to-batch consistency and to verify product quality. These tests are commonly referred to as release testing because the drug substance or drug product is required to conform to the approved acceptance criteria for the material to be released by the Quality unit. Approved product specifications are typically generated to identify the testing requirements and acceptance criteria for this type of testing. Table 1.5 provides an example specification for a monoclonal antibody drug product.

3.3.7 Qualification and validation

Another important CGMP concept is qualification/validation. It is a CGMP requirement that manufacturers control the critical aspects of their particular operations through qualification and validation activities over the lifecycle of the product and process [88]. Qualification and validation are performed to provide evidence that a specific system or process (including analytical tests) is capable of meeting predetermined requirements or acceptance criteria that demonstrate that the system or process works as expected. Qualification/validation can apply to many aspects of biopharmaceutical manufacturing including facilities, utilities, equipment, computerized systems, manufacturing processes, cleaning processes, and analytical test methods. These two terms are often used interchangeably, however, there are typically differences between qualification and validation activities. Although both qualification and validation are executed to ensure a system or process is fit for intended use, validation usually has an added factor of demonstrating consistency – that is, the particular requirements for a specific intended use can be consistently fulfilled [108]. In many cases, qualification activities are a prerequisite to validation activities. Another distinction between the two terms is that qualification is often associated with equipment, facilities, and utilities whereas validation is associated with processes (cleaning, testing, and manufacturing).

Qualification and validation activities are executed under protocols that are preapproved by the appropriate functional areas, including quality assurance. After testing is executed, it is typical to summarize the results in a report that includes a discussion of any testing failures or departures from the approved protocol. Specific activities related to qualification of equipment, facilities, and utilities include the following [88]:

- **Writing a Validation Master Plan (VMP)** – a document that summarizes how qualification/validation activities will be conducted for the entire process for a particular product, or for a particular project.
- **Writing a User Requirements Specification (URS)** – a document to summarize the minimum requirements that the system must meet and which becomes the basis for subsequent qualification testing.
- **Design Qualification (DQ)** – demonstration that the design of equipment, facilities, and utilities complies with CGMP and meets user requirements.
- **Factory Acceptance Test (FAT)** – testing performed at the system manufacturer's site to verify user expectations are met prior to delivery.
- **Site Acceptance Test (SAT)** – testing performed at the user's site to verify no negative impact during transportation and to confirm user acceptance.
- **Installation Qualification (IQ)** – verification that the system has been installed correctly, consists of all expected components and instrumentation, consists of the expected materials of construction, and that all associated documents, drawings, specifications are available.
- **Operational Qualification (OQ)** – verification that the system operates as designed, including verification of acceptable operation at the upper and lower operating limits. The OQ is generally performed after IQ completion but may also be performed as part of a combined Installation/Operational Qualification (IOQ).
- **Performance Qualification (PQ)** – verification that the system, in conjunction with ancillary systems and associated personnel and procedures, performs appropriately during normal operation with normal operating limits.

Cleaning validation is performed for multi-use equipment (dedicated to one product or used for multiple products) to demonstrate that a cleaning process is consistently able to remove process soils and cleaning agents to acceptable levels between uses. Carryover of processing and product materials from one batch to another has impact to product quality and therefore must be minimized. Cleaning validation is also performed under a preapproved protocol with established acceptance criteria. It is typically initiated during manufacturing of product for Phase 3 clinical trials and may continue into commercial manufacturing. Until a cleaning procedure is validated, cleaning verification must be conducted that demonstrates equipment is clean prior to moving forward with processing. Once cleaning is validated, cleaning verification is not required.

Execution of cleaning validation involves taking samples during the final water rinses of the cleaning cycle (referred to as rinse water samples) and from the product contact surfaces of the equipment after the cleaning cycle is complete. Rinse water samples may be analyzed for pH, conductivity, total organic carbon (TOC), endotoxin, and bioburden. TOC, as the name suggests, measures the amount of organic carbon present, which serves as a nonspecific indicator of product left behind. The surface samples are typically collected using swabs to physically remove any soil that may be remaining on the surface. Swab samples are usually analyzed for TOC or for the specific product that is being manufactured. Regardless of the analytical method used to test cleaning validation samples, it is important to set acceptance levels that provide assurance that the equipment has been adequately cleaned prior to its next use. To demonstrate consistency of a cleaning procedure, cleaning validation requires that a specific piece of equipment be successfully cleaned (i.e., meets established acceptance criteria) multiple consecutive times using the applicable procedure. Three consecutive cleanings is the minimum number commonly used.

Process validation is performed to ensure that a biopharmaceutical manufacturing process is capable of consistently producing a product of appropriate quality. Process validation is not a single activity or protocol but instead consists of multiple activities that are performed as three stages [76]:

- **Stage 1 – Process Design**: Based on knowledge gained throughout development and scale-up activities, the commercial process is defined during this stage. Critical aspects of the process are defined and operating ranges established to ensure manufacture of a quality product. The steps illustrated in Figure 2.9 for designing an antibody process serve as an example.
- **Stage 2 – Process Qualification**: During this stage, the process is confirmed as being capable of reproducible commercial manufacturing. This stage includes execution of multiple batches, usually a minimum of three, at full manufacturing scale, which are referred to as consistency batches or process performance qualification (PPQ) batches. PPQ runs combine the actual facility, utilities, qualified equipment, and trained personnel to produce commercial batches. There is more sampling and testing conducted than during routine commercial production in order to sufficiently establish consistency and product quality throughout the process. Qualification of equipment, utilities, and systems must be performed prior to PPQ runs. In addition, PPQ runs must be completed successfully before product can be commercially distributed.
- **Stage 3 – Continued Process Verification**: Ongoing assurance is gained during routine production that the process remains in control. Continued process verification (CPV) is often accomplished through routine trending of process and test data to detect shifts in process performance.

In Figure 2.6, we showed that the three process validation stages span the entire biopharmaceutical product lifecycle, from drug discovery to commercial manufacture and beyond. Many different specific studies may be required to support validation of a biopharmaceutical manufacturing process. Table 3.5 includes some of the most common that are performed for biopharmaceutical processes along with a basic description of the study purpose. These studies are typically performed once during the product lifecycle. Timing for completion of these studies may occur during Stage 1 (Process

Design) as a pre-requisite for Stage 2 (Process Qualification) or may occur concurrently with Stage 2. Generally, a process validation master plan (PVMP) is written to summarize all activities to be completed as part of the process validation effort and to identify the timing for study execution.

Table 3.5: Process validation studies.

Study type	Description
End of production cell bank testing	Demonstrates that the cells used to produce the product are stable and free of contamination.
Media hold	Shows that media maintains sterility for a specific duration and promotes cell growth in the bioreactors used for production.
Resin/membrane reuse (lifetime)	Determines number of times chromatography resin and microfiltration or ultrafiltration membranes may be reused for processing of the same product while maintaining expected performance with minimal degradation. Requires studies at both bench- and production-scale.
Resin/membrane cleaning validation	Validates that the cleaning process defined for resins/membranes results in minimum product carryover. Requires studies at both bench- and production-scale.
Viral clearance studies	Validates that process steps inactivate or remove viruses to acceptable levels. Conducted by spiking model viruses into a bench-scale version of manufacturing steps to measure clearance.
Residual/impurity removal	Demonstrates consistent removal of process residuals or product impurities to acceptable levels. Typically executed at bench scale.
Process intermediate hold times	Determines the amount of time a product intermediate pool can be held without impacting product quality.
Sterile filter validation	Demonstrates a 0.2 µm filter is able to reduce microbial load in a given process stream to an acceptable level and that there are no adverse reactions to the product or filter.
Shipping validation	Demonstrates the procedure for shipping product to other locations has no impact on product quality.
Leachable validation	Characterizes the amount/types of impurities that may be leached from plastic/thermoplastic components used during processing.
Solution hold time	Determines the amount of time a solution used in the manufacturing process can be held without impacting quality. Typically requires studies at both bench and production scales.

Once qualification and validation are completed, CGMP requires that any planned changes to the facilities, equipment, utilities, and processes that may affect the quality of the product should be formally documented and the impact on the validated status or control strategy assessed [88]. The system put in place for this documentation and assessment is referred to as change control or change management quality system. The so-called change controls are created for each proposed change according to a company's individual procedure and assessed by subject matter experts to determine impacts of the change. Evaluation may result in a determination that requalification/ revalidation must be completed to return the system or process to an acceptable validated state. For example, consider a case where a new high-density shelving system is to be installed in a qualified walk-in cold room. When the cold room was originally qualified, temperature mapping was performed to verify that the cold room can maintain temperature between 2 and 8 °C throughout all locations for a 24-h period. This mapping was performed both in an empty chamber and while holding materials representing a typical load. How might this new shelving system affect the original qualification performed? Depending on the configuration, the racks may change the airflow dynamics within the chamber, which may impact the temperature distribution within the cold room. In addition, high-density shelving will likely allow for an increased amount of materials to be stored in the walk-in cold room; therefore, the temperature mapping performed previously for a loaded chamber may no longer be representative. In this case, the unit will likely need to undergo requalification to maintain a qualified state. The change control system would not only identify actions needed to ensure the unit remains appropriately qualified, but also any other actions necessary to implement the change, such as updates to SOPs, changes to cleaning procedures, and changes to preventive maintenance and/or calibration requirements. Implementation of planned changes through a formal change control system helps to ensure that changes are made in a controlled manner, that all potential impacts are appropriately planned for, and that a qualified/validated state is maintained.

3.3.8 Deviations and corrective and preventive actions (CAPA)

CGMP requires that a biopharmaceutical manufacturer has a system in place to investigate when events deviate from what is planned or expected. A variety of terms are used to refer to such events, including deviations, exceptions, nonconformances, failures, or atypical events. ICH Q7, CGMP Guidance for Active Pharmaceutical Ingredients, defines a deviation as a "departure from an approved instruction or established standard" [61]. FDA regulations require any "unexplained discrepancy or the failure of a batch or any of its components to meet any of its specifications shall be thoroughly investigated" and a "written record of the investigation shall be made and shall include the conclusions and follow-up" [109]. Some examples of deviations that may occur during biopharmaceutical processes are summarized in Table 3.6.

Table 3.6: Deviation examples.

Deviation category	Deviation examples
Production/process controls	Deviation from batch record instructions, such as executing 3 homogenizer passes when batch record requires 2. Failure to meet an in-process limit, such as pH, conductivity, or step yield.
Documentation	Failure to complete documentation at the time of execution. Use of a non-effective or superseded version of controlled document.
Materials management	Failure to release raw materials prior to use in batch. Use of expired material in a product batch.
Facilities/equipment	Power outage. Use of equipment that is past the calibration due date.
Quality control	Failure of bulk product sample to meet release criteria. Analytical test method system suitability failure.

Note that the categories indicate the type of deviation and not necessarily the root cause.

When documenting a deviation, the following information is typically included:
- A full description of the atypical event that occurred, including details around what happened, who was involved, when the event occurred, where the event occurred, and how it was different than what was expected. Information regarding root cause is typically left out of the description since an investigation has yet to be performed.
- Any immediate actions taken to correct the event. Examples include containment of materials/batches impacted, restricting access to impacted facilities/equipment, or performing specialized cleaning in response to the event.
- Results of the investigation to determine the root cause of the event, which is important for determining how to prevent similar deviations from occurring in the future. The root cause investigation should be systematic, where all possible "whys" are evaluated and either confirmed or ruled out as a root cause. Fishbone analysis, 5-Why Analysis, and Cause and Effect analysis are some common investigation techniques for determining root cause. Details on how these methods are used for root cause determination can be found elsewhere [110, 111].
- An assessment of impact to product quality.

Once a root cause is identified for a deviation, the logical next step is to determine if there is a way to eliminate that cause to prevent recurrence of the deviation, referred to as a corrective action [293]. In addition, it is important to consider if additional actions can be taken preemptively to prevent a similar deviation from occurring in the future, referred to as a preventive action [293]. These actions are generally captured through CAPAs – corrective and preventive actions. To fix/prevent a deviation, CAPAs must be directly related to the root cause(s) identified during the deviation investigation. CAPAs are typically managed by a company's quality organization to ensure the proposed actions are completed according to a defined timeline and that the actions are effective. If a corrective action is ineffective, meaning that it does not prevent the deviation from recurring, then it may be that the "true" root cause of the deviation determined during the original investigation is not correct. For this reason, if a corrective action is found to be ineffective, further root cause investigation is typically required.

Example: Identifying the root cause and CAPAs for a step-yield deviation

Consider a deviation in which the step yield for a chromatography run is 40%, while the required range in the batch record is 70%–90%. The root cause investigation has determined that the volume reading from the level indicator on the process vessel used to collect eluate (i.e., product) from the step is erroneously low, which has led to the out-of-range step yield. (Recall that step yield is defined by equation (2.1). To calculate the amount of product at the end of the chromatography step – the numerator in equation (2.1) – the eluate volume is multiplied by the product concentration in the eluate.) The erroneously low volume reading was confirmed by putting the process vessel on a floor scale, adding a known amount of water by weight (and assuming 1 kg = 1 L), and observing that the level indicator on the vessel read less than the amount of water added.

What are possible root causes and associated CAPAs for this deviation?

Solution

There are a number of possible root causes, and more investigation is required to determine the true cause. Some possibilities include:

- Possible root cause #1: The investigation determines that the level sensor that provides the volume reading from the process vessel was just calibrated. Interviews with the calibration technician uncovered that he failed to follow the applicable SOP during the procedure in a way that led to the low volume reading.

  Corresponding CAPAs: The level sensor should be recalibrated to read accurately and, given that this is a personnel error, the technician should be retrained on the procedure.

- Possible root cause #2: The investigation reveals that the level sensor was calibrated just prior to the deviation, and interviews with the calibration technician reveal that the applicable SOP is missing a key instruction necessary for proper sensor calibration.

  Corresponding CAPAs: The level sensor should be recalibrated to read accurately and, given that the root cause is related to a method, the applicable SOP should be revised to include the missing instruction. Further, all calibration technicians would be retrained on the revised SOP.

- Possible root cause #3: The investigation reveals that the level sensor malfunctioned during the chromatography run leading to an erroneously low eluate volume reading.

  Corresponding CAPAs: Given that the root cause is related to equipment, the level sensor should be repaired or replaced with a new sensor. Further, the level sensor should be placed on a preventive maintenance program.

3.3.9 Batch disposition

Every batch of a biopharmaceutical manufactured is ultimately dispositioned by the Quality unit, meaning it is assigned a specific status or product usage. This applies to batches of both drug substance and drug product but may also apply to product intermediates, solutions that are used in the drug substance or drug product process, raw materials, and consumable items. The Quality unit is responsible for determining the final disposition status, which is commonly described as one of the following:

- Approved or Released – Material is approved for further processing or distribution.
- Conditionally Approved or Conditionally Released – Material is approved for further processing, but additional action is required for final disposition.
- Rejected – Material does not meet specification requirements and is not intended for CGMP use in further processing or distribution.

We'll focus on disposition of drug substance and drug product in the remainder of this section. For each batch of drug substance or drug product, CGMP requires that all of the records associated with that batch are reviewed to ensure compliance with established specifications and procedures. The records covered in this review include any completed batch production records, solution preparation records, equipment preparation records, laboratory testing records, and packaging and labeling records. The batch review likely also includes a genealogy review to verify that all of the inputs into the process – for example, raw materials, consumables, product intermediates, and buffer solutions – have been properly released and are within expiration at time of use, and to verify that every input can be traced back to its original lot information. Most records undergo an initial review by the department responsible for the execution (e.g., manufacturing reviews batch production records and quality control reviews lab testing records) followed by a final review by quality assurance.

Additionally, the batch review process includes a review of all deviations associated with the batch, including those related to any materials, product intermediates, and buffer solutions used in that batch. It is important to ensure that all deviation investigations have been completed prior to release of a batch, with an assessment that the deviations are deemed to have no product quality impact. A review of any batch-related change controls is also performed to ensure no impact to batch release. Other items that may be reviewed as part of the batch release process are environmental monitoring results, equipment work orders, and facility alarm reports. A manufacturing facility will have an approved procedure to define what must be considered in the record review. It will also have a procedure describing the steps involved in batch disposition. An example of checklists used for record review and batch disposition in CGMP facilities is provided in Table 3.7.

Table 3.7: Example of a document review checklist and batch disposition checklist used to determine the final disposition of drug substance and drug product.

Document review checklist (complete checklist for all records associated with the batch)	
Are all pages present and header information correct?	☐ Yes ☐ No
Is all equipment used within the record within calibration?	☐ Yes ☐ No
Are all rooms used within the record released per appropriate procedures?	☐ Yes ☐ No
Have all steps been initialed/dated by both the performer and verifier?	☐ Yes ☐ No
Has the record been signed by Manufacturing for review?	☐ Yes ☐ No
Has the record been reviewed for GDP standards?	☐ Yes ☐ No
Have all calculations been verified to be correct and any values carried to subsequent step(s) correctly?	☐ Yes ☐ No
Has the record been signed by QA for review?	☐ Yes ☐ No
Batch disposition checklist	
Have all records associated with the batch been reviewed by Manufacturing and QA?	☐ Yes ☐ No
Are all deviations related to the batch closed?	☐ Yes ☐ No
Have all environmental monitoring results been reviewed by QA and any related deviation investigations closed?	☐ Yes ☐ No
Has the batch genealogy been created and approved by QA?	☐ Yes ☐ No
Have all change controls associated with the batch been reviewed by QA?	☐ Yes ☐ No
Has a Certificate of Analysis (COA) been generated for the batch and approved by QA?	☐ Yes ☐ No
Final batch disposition	
☐ Approved ☐ Conditionally Approved	☐ Rejected
QA Signature/Date:	

3.3.10 Post-marketing surveillance

You might think that CGMP would no longer be applicable following completion of biopharmaceutical manufacturing and final batch disposition. However, even after biopharmaceutical drug product has been released and distributed for patient use, European CGMP guidelines require that "a system and appropriate procedures should be in place to record, assess, investigate and review complaints including potential quality defects and if necessary, to effectively and promptly recall . . . medicinal products from the distribution network" [112]. Similar language can be found in the U.S. Code of Federal Regulations [113]. Investigations related to post-marketing complaints are handled much the same way as those described in Section 3.3.8 for deviations and include assessment of impact and determination of root cause(s) and corrective/preventive actions. However, because these investigations relate to product already on the market and thus may have impact to patient safety, there is an increased level of urgency to resolve these issues quickly and thoroughly. In addition, complaints may require notification to regulatory authorities, and when there is high risk to patient safety, may require recalling product lots from the market. A product recall is a profoundly serious situation that may have significant impact to patients as well as to the biopharmaceutical company. Use of CGMP throughout the biopharmaceutical manufacturing process ensures that products are consistently produced with SISPQ, minimizing the risk for product recalls.

An additional post-marketing responsibility of the biopharmaceutical manufacturer is to provide ongoing reporting of new information regarding the approved product and manufacturing process. The FDA requires these reports to be provided annually and include a summary of significant new information that might affect product safety, effectiveness, or labeling of the drug product [114]. In addition, the annual report may also include summaries of any stability data, investigations, and change controls generated over the past year as well as manufacturing process data trended to demonstrate the manufacturing process continues to perform consistently.

3.4 Summary

In this chapter, we learned how current good manufacturing practice (CGMP) is essential to ensuring production of biopharmaceuticals of appropriate quality. CGMP standards define the minimum requirements necessary to ensure production of products with the appropriate safety, identity, strength, purity, and quality (SISPQ). These standards encompass many aspects of biopharmaceutical manufacturing and are intended to reduce risks that can be encountered during biopharmaceutical production. Compliance with CGMP is enforced by various agencies throughout the world, such as the U.S. Food and Drug Administration (FDA), the European Medicines Agency (EMA), and Health Canada. Biopharmaceutical companies must therefore be aware of the CGMP requirements in the area where their products are marketed. As the industry evolves and new technologies are introduced, processing improvements are made, and regulations are updated, companies may have to adapt the ways they meet CGMP. This is the reason that the U.S. FDA includes the word current in current good manufacturing practice.

This chapter touched on some of the main aspects of CGMP and specific requirements biopharmaceutical companies must adhere to in areas such as personnel, equipment, documentation, testing, qualification and validation, deviations and CAPA, and batch disposition. Best practices for meeting CGMP requirements in these areas were also discussed.

- *Personnel* are a critical factor throughout all steps of a biomanufacturing process. CGMP requires that companies employ an appropriate number of people with the right qualifications to adequately support both the manufacturing process and supporting functions to produce quality product. Training programs must be established to ensure personnel follow company-specific policies and procedures and to reinforce basic CGMP principles. Because personnel can be a major source of product contamination, CGMP also includes requirements around personnel hygiene, gowning, and appropriate behavior while in the manufacturing cleanrooms.
- *Equipment* used for manufacture of biopharmaceuticals must function properly for its intended use and not introduce contamination to the product, such as foreign materials, residues, adventitious microorganisms, or particulates. Reusable equipment must be cleaned between runs, which is most often accomplished by automated means. Equipment must also be maintained appropriately through routine maintenance, calibration and inspection. CGMP requires that the history of equipment use, cleaning, and maintenance is tracked and documented over the equipment lifetime.
- *A document system* must be implemented that allows for control, monitoring and recording all activities that directly or indirectly impact the quality of biopharmaceutical products [103]. Various documents are generated and used as part of CGMP manufacturing. Procedures specify instructions for execution of specific tasks. Records and reports capture information or data as tasks are executed. A common phrase in biopharmaceutical manufacturing is "if it wasn't documented, it didn't happen" – records and reports provide proof of activities. Specifications provide acceptance criteria or serve as a basis for evaluating quality at certain steps within the manufacturing process [103]. Control of documentation requires that changes are not made without appropriate justification, review, and approval.

    Good documentation practice (GDP) is an important concept within CGMP that applies to both written and electronic records and reports. There are many best practices within the industry related to GDP, including specific ways to correctly record information and make corrections to original documentation. Documentation must be attributable, legible, contemporaneous, original, accurate, complete, consistent, enduring, and available (ALCOA+). When electronic records are used, there are additional CGMP requirements regarding electronic signatures and maintenance of data integrity.
- *Testing* is required throughout the biopharmaceutical process to verify compliance with established acceptance criteria and specifications. Such testing is required for process inputs such as raw materials, process consumables, and process solutions. Testing is also performed throughout the manufacturing process on product intermediates to monitor process performance. Testing of the product – both drug substance and drug product – at the end of the process is also necessary to ensure conformance to approved specifications for product SISPQ.
- *Qualification and validation* are performed to provide evidence that a specific process or system is capable of meeting predetermined requirements to demon-

strate that it functions as expected. These requirements apply to facilities, utilities, equipment, computerized systems, manufacturing processes, cleaning processes, and analytical test methods. Various documents are generated to plan for and to execute qualification and validation activities, including master plans, specification and design documents, and protocols for testing execution and documentation. Validation of cleaning processes for multi-use equipment is necessary to demonstrate process and product residues are removed to acceptable levels, to ensure carryover between batches is minimized. Manufacturing processes are validated to demonstrate consistent production of quality product. In general, process validation is separated into three stages: process design, process qualification and continued process verification. In addition to demonstrating consistency of the manufacturing process through execution of multiple successful at-scale runs, multiple specific studies, many executed at bench scale, are required to demonstrate an understanding and control of the process and to establish a complete process validation package.

– *Batch disposition* is the responsibility of the Quality unit, and refers to drug substance or drug product being assigned a specific status such as released, conditionally released, or rejected. The decision as to batch disposition depends on the outcome of review of all records associated with the batch manufacture for compliance with documentation procedures, a genealogy review to verify that all inputs to the process were within expiry and are traceable to their original source, and a review of any testing results and deviations associated with the batch. If the results of the batch review are acceptable, the batch is dispositioned as released and may be used for further processing or for distribution to patients. If there are issues identified during batch review, the batch might be conditionally released pending additional actions to disposition as approved or may be rejected barring any further CGMP use.

– *Post-marketing activities* take CGMP beyond just the manufacture and release of biopharmaceutical products for use by patients. Biopharmaceutical manufacturers must have systems and procedures in place to address complaints or potential quality issues after their product is marketed. In addition, companies have a responsibility to continue to provide updated information regarding their product to regulatory agencies on a routine basis. The updates may take the form of an annual report that summarizes additional process and stability data, changes, and deviations generated during the previous year.

3.5 Review questions

1. What does CGMP stand for?
   A. Clean good manufacturing practice
   B. Corrective good manufacturing practice
   C. Current good manufacturing practice
   D. Current good manufacturing procedures

2. Fill in the blank for the following statement. Adherence to CGMP should lead to products that meet the requirements of SISPQ. The acronym SISPQ stands for safety, identity, strength, _____, and quality.

3. What is the purpose of personnel gowning in CGMP manufacturing?
   A. To protect personnel from the environment
   B. To prevent contamination of the product by personnel
   C. To differentiate between manufacturing areas
   D. To minimize the cost of cleaning equipment

4. True or false? Personnel must avoid entering cleanrooms if they have an illness or open wounds to prevent contamination.

5. What are critical requirements for equipment used in biopharmaceutical production under CGMP? Select all that apply.
   A. It must be designed for easy cleaning and maintenance.
   B. It must be disposed of after every use.
   C. It should only require manual cleaning.
   D. It must be constructed so that contaminants are not introduced into the process stream.
   E. It must be operated without human intervention.

6. An operator uses the incorrect format when recording the date in a batch record. What should the operator do?
   A. Nothing. Original documentation cannot be changed.
   B. Erase the entry and write in the date using the correct format.
   C. Use white-out to cover up the original entry and write in the date using the correct format.
   D. Draw a single line through the date, make the correction, provide an explanation, and initial and date.

7. What does the acronym ALCOA, which is used to summarize the principles associated with good documentation practice, stand for?
    A. Attributable, Legible, Complete, Organized, Accurate
    B. Accessible, Legible, Complete, Original, Accurate
    C. Attributable, Legible, Contemporaneous, Original, Accurate
    D. Attributable, Legible, Clear, Original, Accessible

8. Match the following types of CGMP documents to their purpose:
    A. Standard operating procedure (SOP) ____
    B. Batch production record ____
    C. Validation report ____
    D. Material release specification ____

    1. Provides step-by-step instructions for tasks
    2. Captures results of executed validation activities
    3. Serves as input for evaluating material quality
    4. Provides instructions and records details of batch-specific activities

9. True or false? Because electronic batch records used in CGMP may help to minimize the occurrence of documentation errors, they are not subject to the ALCOA+ principles to ensure data integrity.

10. What is the purpose of cleaning validation in biopharmaceutical manufacturing?
    A. Reducing manufacturing costs
    B. Verifying that equipment functions properly
    C. Ensuring no product or other process residues remain on equipment
    D. Simplifying batch release procedures

11. Choose the pair of words that correctly fill the blanks in the following statement:
    Process validation is a documented program that provides a _____ that a biopharmaceutical process will _____ produce a product of appropriate quality.
    A. hope, most of the time
    B. high degree of assurance, sometimes
    C. high degree of assurance, always
    D. high degree of assurance, consistently

12. A deviation report is submitted with the following sections included: a full description of the event, immediate assessment and containment, root cause analysis, and corrective/preventive actions. Which important section is missing?

13. The agitator on a solution prep vessel malfunctioned while preparing a 0.5 M NaOH solution. Classify the actions below as a correction (or containment action), corrective action, or preventive action.
    A. Replaced faulty part on all similar agitators in facility, and placed on PM program
    B. Fixed the faulty agitator and placed it on a preventive maintenance program
    C. Disposed of 0.5M NaOH due to lack of proper mixing

14. True or false? As long as a batch of drug product for an FDA-approved biopharmaceutical meets all its analytical testing requirements, it can be dispositioned as released for use by patients.

Chapter 4
Upstream operations

The first step in a biopharmaceutical production process is the generation of product. Often, this is achieved by cultivation of an engineered cell line at large volumes. During this cultivation, growth substrates (i.e., nutrients that feed the cells) are consumed by the cells and converted into desired product via a combination of native and recombinant metabolic pathways. Once the upstream process is complete, broth is harvested that contains the cells used as a catalyst for production, spent medium, and the desired product. This chapter covers upstream production of biopharmaceuticals by answering the following questions:

- How do we manipulate engineered cells into producing our product?
- How do we operate the process to maximize productivity and product quality?
- How are bioreactor vessels designed to support a cell growth process?
- How does current good manufacturing practice (CGMP) influence bioreactor design, process development, and execution of an upstream process?
- What are the process parameters that influence cell behavior and process performance?
- What process attributes are used to assess the upstream process?
- What are the major considerations during process development, and how are bioreactor operations scaled up?

Growing cells are the catalyst for the upstream process. As such, their behavior determines the behavior of the overall process. Accordingly, our discussion begins with information on kinetics of cell growth, which refers to the growth of an individual cell and an increase in the number of cells, and the factors that influence it.

4.1 Describing cell growth

A variety of different cells are used as factories for production of biopharmaceuticals. The host is selected based on a variety of criteria. Characteristics of the various cell types used in biopharmaceutical manufacturing are shown in Table 4.1. The organism must be generally recognized as safe (GRAS), a designation granted by regulators that a substance is recognized as safe by qualified experts. Cost of cultivation is another main consideration that can vary widely from cell type to cell type. Growth rate is similarly important because a faster growing host cell results in a shorter process time, which leads to higher productivity. A cell line that is tolerant of a range of process conditions may be desirable as it simplifies processing. Many therapeutic proteins require modifications post-translation in order to be active. Commonly, glycosylation is an essential modification for protein function. If a glycosylated final product

https://doi.org/10.1515/9783111112459-004

is required, a cell type capable of performing that modification must be chosen. Finally, secretion of proteins to the extracellular medium may simplify and reduce the cost of downstream processing. Commonly used cell lines are selected for their ability to meet the above criteria: *E. coli* (bacterial), *S. cerevisiae* (yeast), *Spodoptera frugiperda* (insect), Chinese hamster ovary (CHO, mammalian), and human embryonic kidney (HEK, human). Table 4.1 highlights some of the attributes of each type.

Table 4.1: Characteristics of various cell types used in biopharmaceutical processing.

	Bacteria	Yeast	Insect	Mammalian (nonhuman)	Human
Medium cost	Low	Low	Medium	High	High
Batch time	Hours–days	Days–weeks	Days–weeks	Weeks–months	Weeks–months
Expression level	High	High	Low–high	Low–high	Low–medium
Process ranges	Broad	Broad	Medium	Narrow	Narrow
Secretion	No	Yes	Yes	Yes	Yes
Glycosylation	None	Simple	Simple	Complex	Human

Using living cells as a catalyst offers a variety of benefits that make them suitable for biopharmaceutical manufacturing. Metabolic processes are specific (catalyzing only a single reaction) and selective (producing only a single stereoisomer product in most cases). Furthermore, complex polymerization processes such as DNA replication and RNA translation are carried out with a fidelity and efficiency that is impossible with traditional chemical synthesis. These useful processes are all performed during the course of the cell's main objective: proliferation. In an upstream process we aim to facilitate cell growth while simultaneously appropriating these processes to our own ends: synthesis of a biopharmaceutical. In many cases, the rate of a process, such as protein or DNA production, is directly proportional to the rate of cell growth [115–117]. As such, understanding the process of cell growth provides insight into how we might optimize a production process.

The growth of industrially relevant cell lines can be described by four distinct phases of behavior: the lag, the exponential, the stationary, and the death phase. A typical growth curve representing each of these phases is shown in Figure 4.1. It should be noted that the distinction between fermentation and cell culture is defined in earlier chapters. In this chapter, we also use the term culture to refer to either the process of cultivating growing organisms or the actual process fluid that results from culturing cells and that contains a population of growing cells of any type.

The lag phase can be described as a period during which the cells are acclimatizing to their new environment. In most cases, production is initiated by diluting a high-density culture into fresh medium. This represents a major environmental change, and the cells require time to adjust. Indeed, physiological changes occur during the lag phase as activities such as gene expression and regulation adjust. During this time, there is no appreciable change in the cell density (i.e., the number of cells per unit volume of broth).

Next is the exponential phase, during which cells begin to grow. Industrial cell lines proliferate by division of a single cell into two cells. Often called daughter cells, these two products of cell division are genetically identical (barring any random mutation events). This process goes by different names, depending on the type of cell: binary fission for bacteria like *E. coli*, budding for yeast like *Saccharomyces cerevisiae*, or mitosis for mammalian cells like CHO cells. Regardless of the nomenclature, the cells undergo the process of replicating all genetic material as well as cell components such as organelles and membrane components before segregating the copies and dividing into two cells.

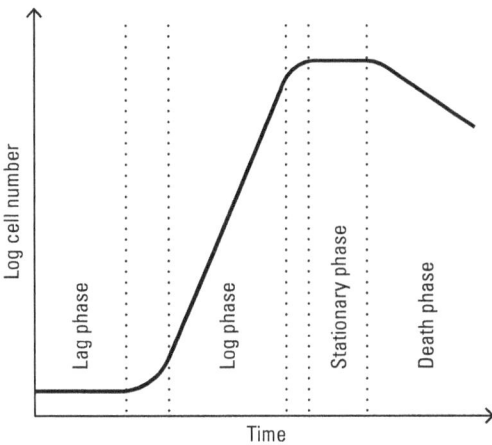

Figure 4.1: Typical growth curve of an industrial cell line. Image © NC State University; reprinted with permission.

Regardless of the specific type of cell line used, the ultimate outcome of cell growth and division is the doubling of cells. One cell will become two cells; those two cells will divide to form four cells; those four will become eight, and so on. This exponential growth is perhaps most easily described by the Monod model of cell growth, which assumes the growth rate, dX/dt, is proportional to the concentration of cells:

$$\frac{dX}{dt} = \mu X \qquad (4.1)$$

where X is the cell concentration, t the time, and μ the specific cell growth rate [118]. This simple model plainly states that the rate of increase in cell density is proportional to some growth constant multiplied by the current cell density. Upon integration of the differential equation, we see that cell density follows an exponential trend with respect to time:

$$X = X_0 e^{\mu t} \qquad (4.2)$$

where X_0 is the initial cell concentration. Upon further examination, it is clear that the rate of cell growth is entirely determined by the numerical value of the growth rate, μ. Further, growth rate varies as a function of process parameters such as temperature, pH, and media composition, to name a few [119–121]. This exponential growth rate continues throughout the exponential phase.

If a nutrient is depleted or byproducts accumulate to toxic levels, growth enters the stationary phase. During this phase there is little if any cell growth. Cell mass can be maintained for a short time as the cells scavenge the waste products they produced during growth on their preferred substrate.

Eventually the medium will be truly depleted of nutrients to the point it can no longer even sustain the culture that has grown. As cells experience prolonged stress created by this depletion, they begin to die. This death phase is characterized by a drop in the viable cell count in the bioreactor.

The overall dynamics of an upstream process are driven by the behavior of the cells. Cell proliferation is directly coupled to both consumption of nutrients (often called growth substrate(s)) and formation of desired product. Furthermore, because of its impact on the rate of cell growth, specific growth rate determines much of the overall performance of an upstream process.

First, we consider substrate consumption. In order for cells to divide, they must duplicate everything from DNA to proteins. The synthesis of these cellular components is supported by the growth medium. The medium must be able to provide the atoms required to generate all cellular components – carbon, nitrogen, phosphorus, sulfur, oxygen, and hydrogen. Generally, the first four are supplied by soluble media ingredients, while oxygen and hydrogen are provided by aeration and the water in which the ingredients are dissolved. Carbon is often provided by a carbohydrate such as glucose or glycerol. Nitrogen can be supplied by a variety of chemicals ranging from ammonium sulfate to less chemically defined components such as yeast extract or enzymatic digests of plant or animal proteins. Though media comprises many different components, its function is to provide essential nutrition to the growing cells. As such, the faster the cells grow, the faster the media components will be consumed. For example, the rate of consumption of a carbon source can be given by the following expression:

$$\frac{dS}{dt} = \frac{\mu X}{Y_{X|S}}$$ (4.3)

where S is the unconsumed substrate concentration, t the time, μ the specific cell growth rate, X is cell concentration, and $Y_{X|S}$ the yield of cells on substrate.

Here we define a yield of cells on substrate, which is simply the number of grams of cells produced by consuming a single gram of carbon source. The value of the yield is influenced by a variety of process factors, such as temperature, pH, carbon source, and oxygen availability to name a few. Theoretically, under constant conditions the value of this yield should be constant. This equation shows that not only is the rate of substrate consumption directly proportional to growth rate but also that substrate consumption increases as cell density increases. Cell growth continues until the growth substrate has been exhausted from the medium.

Next, we consider product formation. Biopharmaceutical products are formed as a result of the proliferation and metabolism of the host cell. During these processes, a portion of the substrate consumed is directed toward the synthesis pathway for the desired molecule, while the remaining substrate is consumed to produce components required for proliferation or waste products. Additionally, the product will constitute some fraction of the total cell mass generated. As such, there are two ways to describe product yield: on a substrate-consumption basis or on a biomass-generation basis. Yield of product on substrate, $Y_{p|s}$, is simply the number of grams of product formed when one gram of substrate is consumed. Yield of product on cells, $Y_{p|x}$, is the number of grams of product formed when one gram of biomass is generated. We can write expressions to describe product yield similar to those for substrate consumption in equation (4.3) using either one of these yields. As with substrate dynamics, we would see that product formation is proportional to growth rate and cell density. Because cell growth ceases upon exhaustion of substrate, so too will product formation.

Example: Calculation of the specific growth rate and $Y_{x|s}$ for E. coli BL21
Consider the following growth curve and corresponding data. Use this information to calculate the following: (a) the growth rate between hours 3 and 2.5 of the fermentation, (b) the doubling time between hours 2.5 and 3 of the fermentation and (c) the yield of cells on glucose, $Y_{x|s}$, assuming an initial glucose concentration of 1.5 g/L.

Growth of E. coli BL21

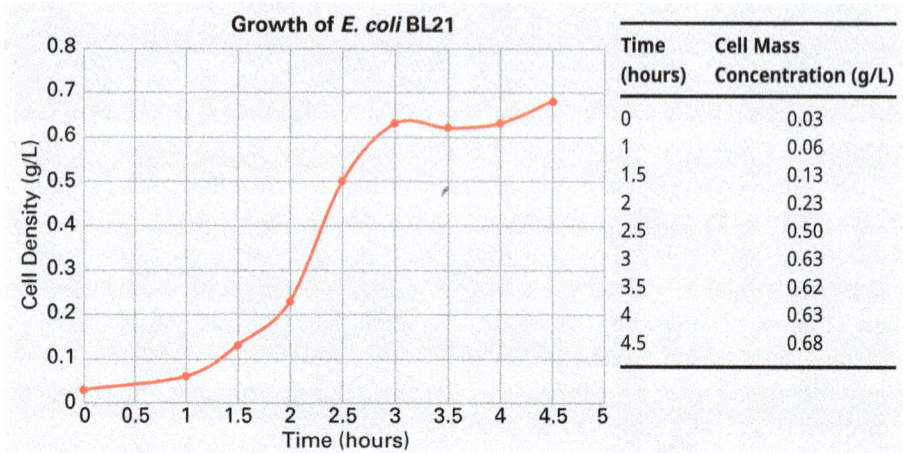

Time (hours)	Cell Mass Concentration (g/L)
0	0.03
1	0.06
1.5	0.13
2	0.23
2.5	0.50
3	0.63
3.5	0.62
4	0.63
4.5	0.68

Image © NC State University; reprinted with permission.

Solution

The growth rate between two time points within the exponential phase can be calculated by rearranging equation (4.2):

$$\ln\left(\frac{x_2}{x_1}\right) = \mu(t_2 - t_1)$$

(a) Plugging in the values from above, an expression for the growth rate between hours 3 and 2.5, or $\mu_{2.5,3}$ can be written:

$$\mu_{2.5,3} = \frac{\ln\left(\frac{0.63}{0.50}\right)}{3 - 2.5} = \frac{0.231}{0.5} = 0.462 \ h^{-1}$$

(b) By definition, a doubling time is the time it takes for cell density to double. In that case the natural log term would be equal to the natural log of 2, or 0.693. Solving for the time difference yields the doubling time:

$$\text{doubling time} = \frac{\ln(2)}{\mu} = \frac{0.693}{0.462 \ h^{-1}} = 1.5 \ h$$

This doubling time indicates that every 1.5 h the total amount of cells in the culture will double.

(c) Finally, the yield of cells on substrate over the course of the fermentation can be calculated as:

$$Y_{x|S} = \frac{\text{Cell Mass Generated}}{\text{Substrate Consumed}} = \frac{\Delta \text{Cell Mass}}{-\Delta \text{Substrate}}$$

Note that in the above equation, Δ refers to the quantity at a later time minus the quantity at an earlier time. In order to solve for yield with the data at hand, it can be assumed that all the glucose initially present was consumed coincidentally with the start of the stationary phase. As such:

$$Y_{x|S} = \frac{0.63 - 0.03}{-(0 - 1.5)} = 0.41$$

Every 1 g of glucose consumed will lead to generation of 0.41 g of cell mass.

Finally, it is important to understand that cells will consume substrate to produce things other than cells and product. This waste production is essential for balanced and efficient growth. The specific identity of waste products is often cell line and process specific, but there are a handful that are universally produced. First, carbon dioxide is a product of several of the steps of the tricarboxylic acid pathway that is used to produce energy as well as the carbon-containing precursors of amino and nucleic acids among other molecules. Additionally, organic acids such as lactic acid and acetic acid are common byproducts of sugar metabolism. These waste products are important to understand. Their accumulation often has inhibitory effects on cell growth. Additionally, in some cases cells can utilize waste products as a carbon source when the initial carbon source has been exhausted.

Ultimately, the dynamics of cell growth and the interaction of the living catalyst with its medium determine the operational details of a production-scale bioreactor for biopharmaceutical production.

4.2 Upstream processing and production modes: batch versus fed batch versus continuous

Production of a biopharmaceutical product begins with a vial from a working cell bank (WCB). For the purpose of this discussion, we assume the process relies on an engineered cell line to produce a protein therapeutic. Of course, a number of biopharmaceuticals are produced using cells that are not genetically engineered, as discussed in Chapter 2 (e.g., some flu vaccines); however, much of this discussion applies to that production scenario as well. The WCB comprises many identical vials of the cell line that has been grown to a desired point in the growth cycle and then frozen. This process involves the addition of a cryoprotectant, such as glycerol or dimethyl siloxane, before lowering the temperature below −70 °C. This combination of cryoprotectant and low temperatures captures the cells in a state of suspended animation so that when they are thawed, they are resuscitated to the state they were in prior to freezing.

The WCB is generated carefully from a master cell bank (MCB). The MCB contains cryogenically frozen vials of the host cell line for the process. The characterization requirements for the MCB are discussed later in this chapter. Generation of the WCB is initiated by thawing of a vial of the MCB and subsequently inoculating those cells into a flask containing sterile growth media. The cells are allowed to grow for a specific amount of time, corresponding to a specific part of the cell cycle, and prepared for freezing. As discussed above, this involves addition of a cryoprotectant and controlled freezing to temperatures at or below −70 °C. The size of a WCB is determined by the expected number of production runs.

As illustrated in Figure 4.2, a production run is initiated with a single vial from the WCB. Generally, a cell bank vial contains 1 mL of frozen cells. A production-scale bioreactor is typically operated at a volume ranging between 2,000 and 20,000 L. Inoculating such a large volume with a single mL of cells is problematic as seen from the growth equations (4.2) and (4.3): cell growth and product formation is proportional to cell density, and starting with a low cell density in the vessel means more time is required to reach a desired cell density and production rate. To speed up growth to the desired number of cells, a typical inoculum volume is 10 percent of the final working volume of the production bioreactor. Using this 1:10 ratio, a 20,000 L production bioreactor requires a 2,000 L inoculum. As such, culture expansion from 1 mL of WCB to 2,000 L of inoculum is required. As shown in Figure 4.2, this expansion is achieved by operation of a series of bioreactors of increasing size leading up to the production bioreactor, often referred to as a seed train.

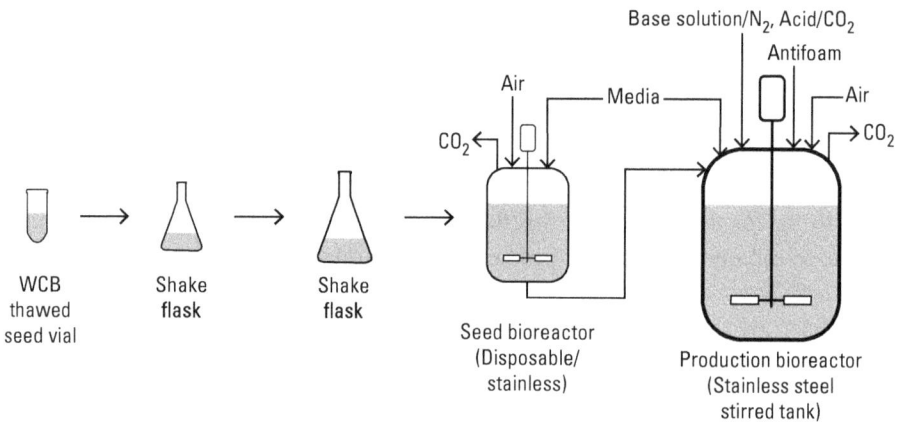

Figure 4.2: Expansion of a cell line from cell bank to production scale. Image © NC State University; reprinted with permission.

The first step in the seed train expansion typically involves inoculation of the working cell bank into between 500 mL and 10 L of fresh medium. This medium is specifically formulated to support desired growth rates and final cell densities as well as to establish a population of host cells that is ready for optimal performance in subsequent steps. This culture is incubated until the cells are at the desired cell density and stage of their life cycle. This culture is then used as the inoculum for the next bioreactor in the series. Each culture step increases the volume of culture and the total cell population by a factor of 10. This passaging is continued until the culture has been expanded to the volume necessary to inoculate the production bioreactor.

Example: Calculation of inoculum volume for a production bioreactor using CHO cells
The target starting cell density for a 4,000 L production bioreactor is 0.4 million cells/mL. After 10 days of cultivation, you measure the cell density in the seed bioreactor to be 5 million cells/mL. What volume of seed culture should be transferred to the production bioreactor as inoculum? What working volume, that is what total volume of culture, would you recommend for your seed vessel?

Solution
The process of inoculation is essentially a dilution of one culture with fresh medium. As such a dilution equation that equates the number of cells before and after dilution can be written as

$$C_1 V_1 = C_2 V_2 \tag{4.4}$$

Here, C and V are the concentration of cells and volume of culture, respectively. Subscripts 1 and 2 refer to before and after dilution, respectively. Applying the information we have, our initial concentration of cells is 5×10^6 cells/mL, and our final target concentration is 0.4×10^6 cells/mL at a volume of 4,000:

$$V_1 = \frac{0.4 \times 10^6 \frac{cells}{mL} \times 4{,}000 \text{ L}}{5 \times 10^6 \text{ cells/mL}} = 320 \text{ L}$$

It is good practice to cultivate a volume slightly in excess of what you need for inoculation, so in the example above, a culture of 350 to 400 L would be suitable.

Once the production bioreactor has been inoculated, the process of generating product begins. In the case of genetically engineered cells, providing the host cell population with the optimal environment for cell growth and product formation is required. This is achieved by careful formulation of the growth medium to provide all necessary nutrients as well as control of process parameters that are known to affect the behavior of the cell population. Once the production run is complete, the bioreactor contents are harvested and downstream processing begins.

Though the general process of seed expansion and production is similar across different types of cell lines and products, there are many process parameters that are manipulated differently for each particular process. In addition to controlling process parameters to a specific setpoint, which will be discussed in greater detail in subsequent sections, a ubiquitous strategy for maximizing the productivity – that is, the amount of product produced per unit volume of bioreactor per unit of time – is through the use of a particular mode of operation of the production bioreactor. Production modes fall into three different categories: batch, fed-batch, and continuous. Each mode has its own benefits and drawbacks to be considered before deciding on which to implement.

Batch production is the simplest mode. With this approach, the media is formulated and sterilized with the bioreactor itself. The bioreactor is then inoculated, and the process is allowed to proceed until the stationary phase is reached. Completion of the batch typically coincides with depletion of one or more nutrients from the medium, though it may also follow accumulation of an inhibitory substance to toxic levels. Alternatively, the batch may be ended at a specific time determined to be the most efficient or cost effective. These events correspond with the beginning of the stationary phase as previously discussed. Batch cultures are easy to perform. One major consideration in designing a batch culture process is that the cells see a constantly changing nutritional environment as substrates are consumed and products, including waste, are formed. This changing environment combined with the exponential dynamics of cell growth can lead cells to experience many different stresses that may have deleterious effects on growth and product formation. On the other hand, a contamination event is nearly impossible in this operation mode as the production system is completely closed upon inoculation. Additionally, executing an effective batch production requires very little process knowledge and generally requires relatively simple controls to maintain process setpoints. Batch mode is ideal for production of products that have solubility limits in the medium or are toxic to the host cell at lower concentrations. Batch processing is also very common in seed-train expansion steps.

Fed-batch production is simply a batch process that has one or more feeds introduced at some point before completion. This additional feeding has many advantages. It can be used to maintain a constant substrate concentration, leading to more consistent cell behavior. In some cases it can be used to switch growth substrates in the middle of the run, which may improve growth or product formation at a key time in the process. For example, a methanol feed can be introduced to initiate protein overexpression in the methylotrophic yeast *P. pastoris*. One of the biggest advantages of fed-batch operation is the potential to extend the exponential growth phase by feeding additional nutrients. The exponential nature of cell growth means that the final hour of a production run may be more productive than the previous 5 or 10 h. As such, extending the exponential phase by only a few hours may as much as double the final product titer. In some cases, extension of the exponential phase can be achieved by simply altering the initial media formulation to contain more substrate. Practically, however, these high initial substrate concentrations are often inhibitory to cell growth. Alternatively, by accounting for the rate at which cells are consuming substrate, a complementary feed rate can be set to ensure the substrate concentration never becomes prohibitive to cell growth, either too high or low.

Though fed-batch operation is an effective way to increase productivity, it does have limitations. First, despite substrate concentrations remaining stable, waste products, including toxic ones, accumulate. Furthermore, the more substrate that is fed, the more waste that is generated and the more severe the inhibitory effects of the waste products during the later stages of the process. In many cases, this inhibitory effect is what limits productivity. There are other engineering challenges as well. As cell density increases exponentially so does the demand for nutrients. At sufficiently high cell densities, it may be impossible to supply these nutrients in needed amounts due to limitations of substrate solubility, pump speed, air flow, or even total available volume in the bioreactor (i.e., the volume of nutrient solution required may cause the bioreactor to exceed its allowable working volume). Additionally, extremes of metabolic activity caused by increased cell density create challenges with respect to controlling parameters such as pH and temperature. Finally, design of an optimal feeding strategy requires a more detailed understanding of the system than batch operation. This understanding must be gained by analysis of data generated by process runs. Improper design or unexpected departure from expected behavior often leads to unexpected departure from control of process parameters, which is catastrophic in a fed-batch process. Despite these limitations, fed-batch operation is the most commonly used mode for biopharmaceutical production owing to increased productivity relative to batch operation with fairly straightforward feeding strategies.

Continuous production is a mode in which cells are grown in a bioreactor into which fresh media is being constantly fed at the same rate that spent medium is removed. This can be done with or without retention of the cells from the waste effluent stream. If cells are not retained within the process, it operates as a continuous stirred tank bioreactor. If cells are retained within the vessel, the process is referred to as perfusion. A perfusion process typically employs a microfiltration membrane as a cell retention device. Perfusion has been used for a number of years and is in use for approximately 20 biopharmaceutical products [122]. This mode of operation has the benefit of a substrate feed increasing the productivity of the system with the added benefit of removing toxic and inhibitory compounds in the effluent. Furthermore, in the case that the cell line secretes the product of interest, such as the production of monoclonal antibodies in CHO cells, this effluent will contain product. Typically, perfusion culture is only used for production of extracellular products. This operation mode benefits from a truly constant nutritional environment for the cell population, leading to much more consistent behavior throughout the run. This constant productivity means that utilization of the process equipment is high. This production mode, however, can be costly due to the need for large volumes of media. Constant feeding of medium also carries an increased risk of microbial contamination. In theory, continuous production can be carried out indefinitely because cells continue growing exponentially as long as continuous feed and harvest continue. Practically, however, there is a risk of a genetic mutation occurring during each cell division cycle. A mutation can potentially lead to reduced cell growth or product formation or, in the worst case, production of an altered product. Consequently, a limit on the number of generations in each run is typically defined. The duration of a generation is defined as the time between cell division events, or the doubling time. Doubling times range from 30 min for bacteria, to several hours for yeast, to a day or more for insect and mammalian cell lines.

Because each mode has its benefits and drawbacks, considerations related to cell physiology, engineering, regulatory, cost, and market considerations are carefully weighed when deciding on a production mode.

4.3 Bioreactor equipment design

Proper design of the bioreactor vessel is critical for proper execution of a fermentation or cell culture process from seed expansion to production. There are two main design criteria for any bioreactor: (1) it must enable aseptic operation so that the risk of contamination is minimized, and (2) it must be capable of providing an optimal environment for growth and product formation for each cell grown in the bioreactor. It should be noted that these two criteria are not of equal importance: the ability to maintain aseptic operation is paramount to CGMP manufacturing. Any design changes that improve feeding of media components or oxygen delivery, for example, must be done in a way that does not compromise the ability to prevent microbial or cross contamination in the process. With these guidelines in mind, we must also consider the material of construction and its interaction with the growing culture as well as the geometry and configuration of the vessel and its various attachments. These criteria, combined with the nature of the process fluid and requirements for CGMP manufacturing, lead to designs that fall into two categories: reusable or traditional bioreactors and single-use or disposable bioreactors.

Traditional bioreactors are also commonly referred to as stainless-steel bioreactors in reference to their material of construction. An example is shown in Figure 4.3. Stainless steel is used because it is tolerant to the wide range of conditions that a reusable bioreactor may be subject to. Because the bioreactor is reusable, it must tolerate the caustic and acidic cleaning solutions used during clean in place (CIP) without degradation or decomposition. Additionally, it should be inert to all the components of growth medium and not introduce any toxic components during the production process. Finally, it must be able to withstand the high temperatures of the sterilization process, which can exceed 121 °C. Stainless steel is the most practical material (from an availability and cost standpoint) that satisfies the above needs. In addition to the metal used for the vessel construction and the piping used to distribute gas and liquid components, soft parts are needed to form seals, piping junctions, and sites of flow regulation, such as harvest or sampling ports. These soft parts must be made of a material that satisfies the same criteria above. In this case, choices include Teflon, silicone, ethylene propylene diene monomer (EPDM) rubber, and fluororubber compounds. The American Society of Mechanical Engineers publishes an excellent standard describing requirements for equipment design that apply to bioreactors [69].

Single-use, or disposable, bioreactors have different material considerations. An example is shown in Figure 4.4. First, disposable bioreactors do not need to be cleaned because they are disposed after use. Further, they don't need to be sterilized by the user prior to use because they are typically sterilized, via gamma radiation, by the vendor. This eliminates the need for a temperature- and acid/base-resistant material. Typically, these bioreactors, sometimes called "bags," are made of polymeric materials, which are well suited to sterilization by gamma irradiation. Though sterilization

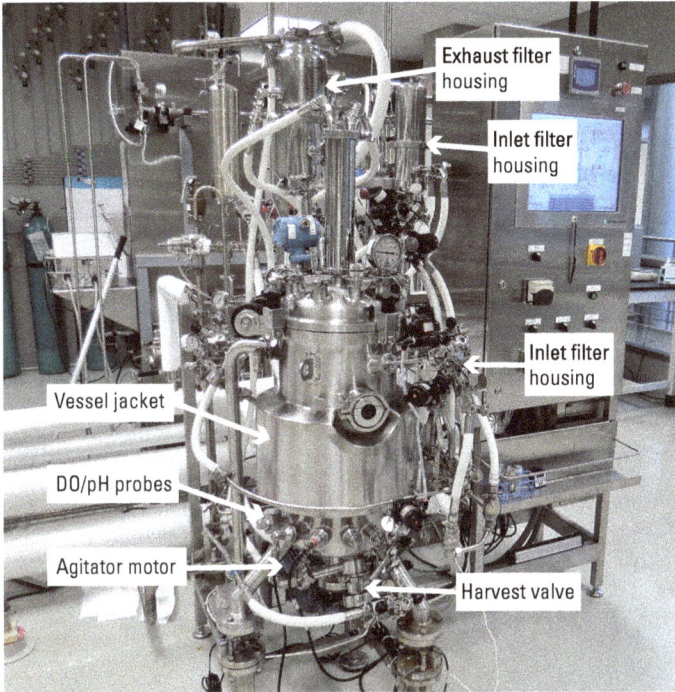

Figure 4.3: Example of a 30 L stainless steel bioreactor. Image © NC State University; reprinted with permission.

and cleaning have been eliminated, the materials chosen must be inert to the conditions present during cell cultivation while providing the structural integrity to contain the volume of liquid and achieve control over process conditions. In general, no single polymeric material has the appropriate combination of physical and chemical properties required. Because of this, the walls of the disposable vessel are made of laminated materials referred to as multi-layer films. The innermost layer that is exposed to the contents of the vessel is chosen for its chemical stability and reduced amount of compounds that can be leached by the culture medium. The outermost layer is chosen for its rigidity and mechanical strength to give the bag the ability to hold a shape. These layers are joined by an adhesive. Between the inert and the structural layer there may also be a layer of material that confers impermeability to moisture and gas to prevent losses through the film. Examples of commercially available single-use bioreactors are given in Table 4.2.

In addition to material considerations, the configuration of the bioreactor is critical to the ability to provide an optimal environment for the growing cells. The bioreactor must be configured such that mixing can be achieved, air can be introduced, heat can be added or removed to maintain temperature, and solutions can be added or removed to allow for pH control or a variety of feeding strategies. Probes and sen-

(a)

(b)

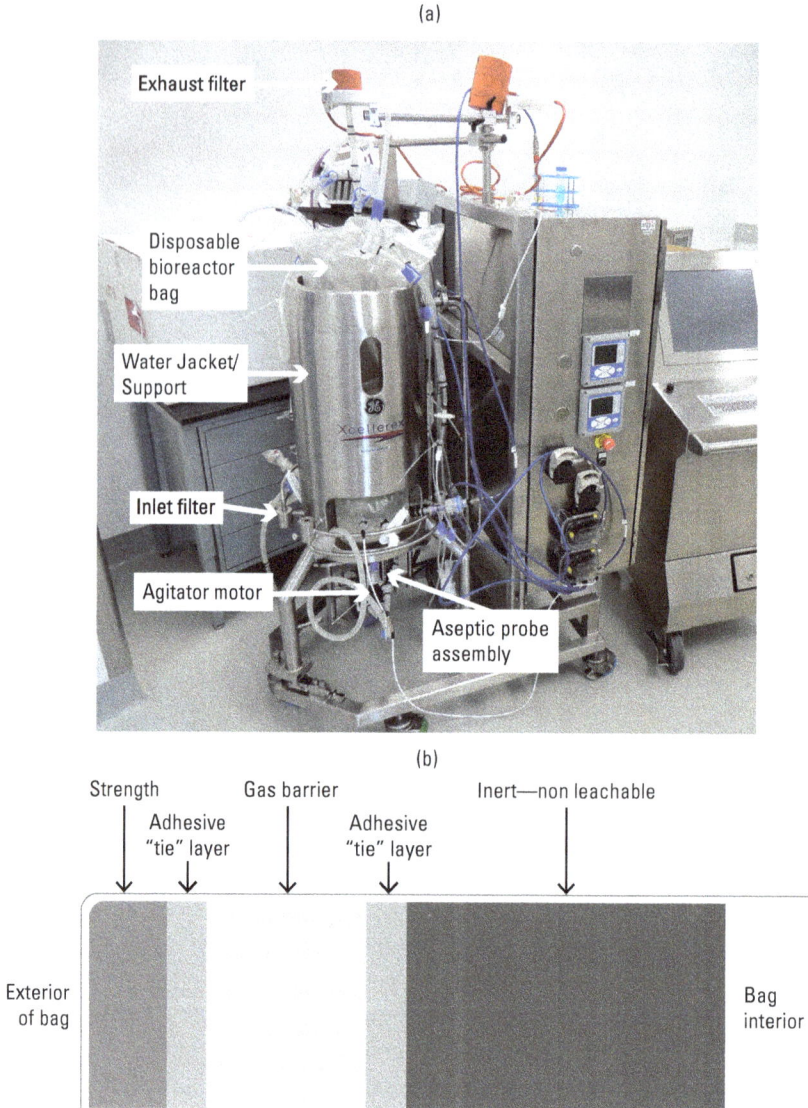

Figure 4.4: (a) An example of a 50 L disposable bioreactor and (b) an illustration of the typical construction of a single-use vessel. Image © NC State University; reprinted with permission.

sors for monitoring important process parameters must also be incorporated into the system in a way that useful measurements can be taken.

In stirred-tank bioreactors, whether reusable or single use, mixing is achieved by attaching impellers to a central agitator shaft, which is turned by a motor. Mixing serves to distribute nutrients to each and every growing cell within the reactor volume and to maintain uniformity of important parameters such as pH and tempera-

Table 4.2: Examples of commercially available single-use bioreactors.

Manufacturer	Trade name	Mixing type	Working volume
Cytiva	Xcellerex™ (various)	Stirred tank	4.5 L to 2,000 L
Cytiva	WAVE™ 25	Rocker bag	0.3 L to 25 L
MilliporeSigma	Mobius® Bioreactors	Stirred tank	1 L to 2,000 L
Sartorius	Univessel® SU	Stirred tank	0.6 L to 2 L
Sartorius	Ambr® 15 and 250	Stirred tank	10 mL to 250 mL
Sartorius	Biostat STR® Generation 3	Stirred tank	12.5 L to 2,000 L
Sartorius	Biostat® RM	Rocker bag	0.1 L to 100 L

ture. Impeller size and shape affect the physics of mixing and must be selected based on the mixing needs of the production system. Disposable bioreactors can also be configured so that mixing is achieved by a rocking motion. These aptly named "rocker bags" use a platform that tilts back and forth at a specified frequency to drive mixing of the liquid inside. This design does not use a shaft or any impellers and, consequently, delivers reduced mixing intensity and lower shear.

The two biggest considerations when selecting a mixing strategy and/or an impeller type are the fluid shear generated and the mixing paradigm. An impeller with flat and vertical blades, as shown in Figure 4.5(a), is often called a Rushton impeller. An impeller with curved and vertical blades, as shown in Figure 4.5(b), is referred to as a Smith turbine. Both push fluid radially outward as shown in Figure 4.5(c). This mixing is very efficient, but high shear rates are generated at the edges of the impeller. These high shear rates can lead to considerable cell lysis with certain cell lines. Reducing the shear generated can be achieved by pitching (i.e., angling) the impeller blades as shown in Figures 4.5(d) and 4.5(e), to create flow in the axial direction as shown in Figure 4.5(f). This axial flow causes pumping motion in the upward or downward direction. Other mixing design considerations are the number of impellers as well as their spacing. Multiple impellers on the agitator shaft should be distributed to ensure consistent mixing throughout the entire volume of the bioreactor vessel without creating dead zones. Generally, microbial cell lines can withstand much higher shear rates than those used in cell culture processes (insect and animal cells).

The location of the drive motor must also be carefully chosen. A bottom-mounted motor has good power efficiency, but the point of insertion of the agitator shaft will be below the surface of the liquid. This necessitates the design of a sanitary seal, to ensure the bioreactor contents do not make contact with the dirty motor, to maintain sterility. A failure of this seal can lead to either introduction of contaminants into the culture broth or loss of culture volume through the seal. While top-mounted drive motor configurations do not carry this risk, the power transfer is less efficient, mean-

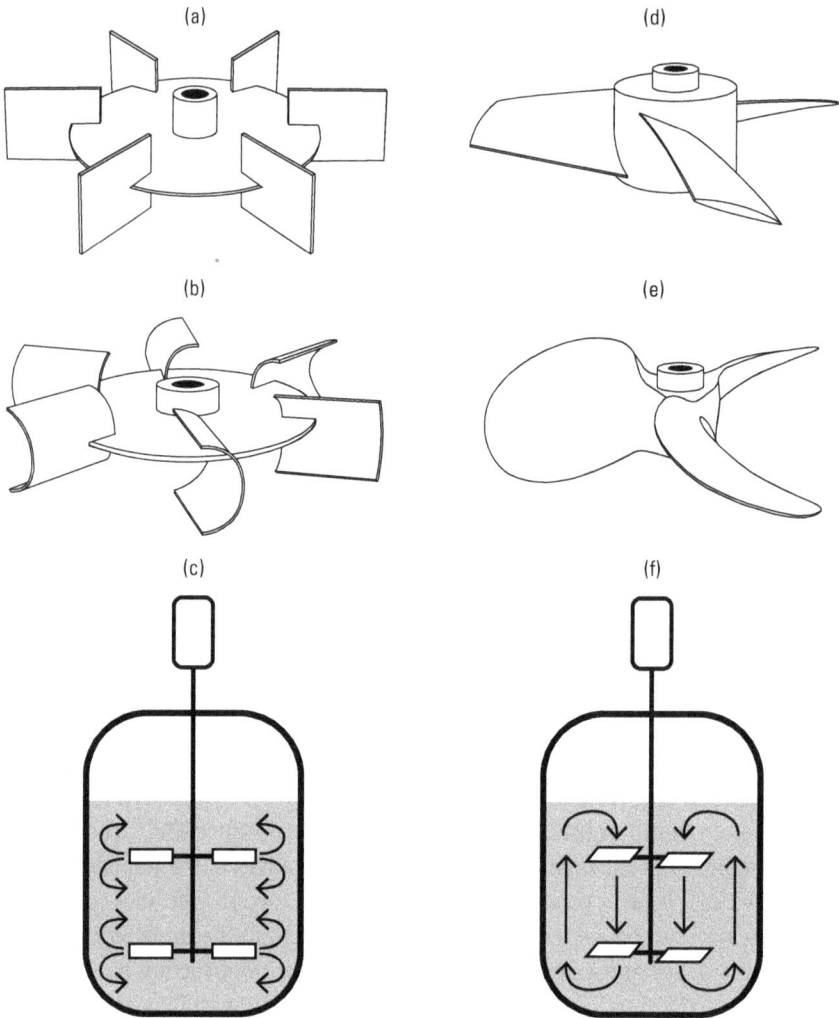

Figure 4.5: Examples of flat (a, d), curved (b, e), vertical (a, b), and pitched (d, e) impellers for use in bioreactor mixing. Illustrations of radial (c) and axial (f) flow patterns. Image © NC State University; reprinted with permission.

ing the same mixing performance will require a larger motor. Furthermore, larger bioreactors can be upward of 40 feet tall. Mounting a large motor over 40 feet off the ground comes with maintenance, safety, and infrastructure challenges.

Recall from previous discussion that gases, such as air and oxygen, must be fed to the bioreactor for cell growth. Gas feeds are achieved in one of two ways: below the liquid surface or into the vessel headspace. Subsurface gas addition is achieved by sparging a gas through the liquid. A sparger is a device that creates fine bubbles of a gas to be introduced into the bioreactor's liquid phase. Gases are pumped through a sparger, which is generally located below the bottom-most impeller to allow for optimal incorporation of gas bubbles into the liquid, thereby facilitating maximum mass transfer. Spargers can vary in shape, diameter, size and number of the openings through which gases flow, to name a few characteristics. These sparger design parameters are manipulated to provide the necessary flow rate and gas transfer rate as determined by process development studies. The second method for introducing gases into a bioreactor – directly into the headspace, or overlay – is also used. Compared to sparging, it can provide gas delivery without increasing foaming and bubble formation, both of which can negatively impact cell growth and product formation. Additionally, feeding gas into the headspace can aid in removing unwanted gases, namely carbon dioxide, from the culture medium. In any case, these gases must be sterile upon entry into the vessel. Sterilization is typically achieved by sterile a 0.22 μm filter.

A variety of liquids may need to be introduced during a run, including feeds and various solutions for pH adjustment to name a few. Liquid feeds are introduced through feed ports. These ports consist of a length of piping or tubing downstream of a control valve. These valves open to allow pumping of a solution into the vessel. These liquid feeds include solutions of carbon or nitrogen sources, acid or caustic for pH adjustment, or antifoaming agents and are used in all modes of operation. Inoculation is also carried out using these ports as well. Ports can be configured to introduce the feed either above or below the surface of the culture broth. This choice is determined by the nature of the feed, with substrates and inoculum typically being introduced in a subsurface fashion. These same ports can also be used for removal of culture broth during continuous operation. In this case, the piping must be below the surface of the broth.

4.4 CGMP considerations for upstream processing

In addition to the considerations made for facility design and environmental control discussed in Chapter 2, a major consideration for upstream processing in a CGMP environment is hygienic design of the equipment. This includes everything from the configuration of piping junctions, to valve architecture, to bioreactor vessel geometry to allow for adequate cleaning. The goal of hygienic design is to reduce the likelihood of contamination by carryover from previous processes, adventitious microorganisms, or other sources.

The need for adequate cleaning between runs drives much of the design of stainless steel bioreactors. Cleaning is achieved using a CIP system as described in Chapter 2. Cleaning the interior of a bioreactor vessel is facilitated by a sprayball device, which uniformly sprays the interior surfaces of the vessel with cleaning and rinse solutions. Additionally, flow paths (e.g., piping) are cleaned by turbulent flow of cleaning and rinse solutions across all surfaces that may potentially come into contact with process fluid. To facilitate this process, the bioreactor must be configured to ensure that no surfaces are obstructed from flow of the cleaning solutions or shadowed by other parts of the bioreactor, which may reduce or deter proper contact with cleaning solutions. Flat surfaces, including horizontal runs of piping, should be eliminated, and the vessel should be capable of fully draining by gravity to ensure that wet spots where microbes can settle and proliferate are eliminated. Good cleaning by flow also requires very smooth surfaces. Rougher surfaces have peaks and valleys that may be of the appropriate depth for microbes to attach and avoid removal by the fluid flow. It is also important that material junctions such as welds and clamped connections do not create deep pits or grooves where microbes can establish themselves. This effectively prohibits threaded connections from any area which sees process fluid. As an alternative, flanged connections are clamped together with a sanitary gasket forming a tight seal. The worst-case scenario with surface imperfections is the development of biofilms, which are notoriously difficult to remediate.

The orientation and design of flow paths to allow adequate cleaning is also very important. Because flowing liquids are used to clean surfaces, it is paramount that all surfaces are exposed to that flow. One of the biggest challenges in this area is the elimination of flow paths that cannot be configured to accept this flow during cleaning. These paths that cannot accommodate flow during cleaning are often called "dead legs." Without flow across these surfaces, they cannot be adequately cleaned of any residual cells, product, or media components between batches. It is desirable to eliminate any dead legs through system design. In the case of a tee that is perpendicular to a main run of pipe, as illustrated in Figure 4.6(a), a depth-to-width ratio for the tee (i.e., L/D in Figure 4.6(a)) of less than 2 is required to ensure that the tee is adequately exposed to cleaning solution. A typical double block-and-bleed valve configuration, shown in Figure 4.6(c), is an excellent way to avoid dead legs while allowing repeated sterilization of piping paths. This configuration arranges four valves (two block and two bleed valves) such that steam can be directed in one of two flow paths. These valves can then be manipulated to isolate the region that has been sterilized to maintain sterility.

Valves and pumps must also be designed so that the surfaces that come into contact with process fluid can be completely cleaned. Diaphragm valves are commonly used to satisfy this requirement. Diaphragm valves work by actuating a soft diaphragm up and down against a weir to permit or stop flow as shown in Figure 4.6(b). This diaphragm is the only moving part that is exposed to process fluid and is completely cleanable. A variety of hygienic pump designs exist and should be selected based on flow needs such as viscosity, pressure, and flow rate. Perhaps the most commonly used pump is a peristaltic type where a soft tubing can be placed in the pump head, and the flow is controlled by the rate of rotation of rollers that compress the soft tubing and force liquid to be pumped through the tube. In this case, the pump never needs to be cleaned because none of the pump parts ever contact process fluid (and the soft tubing placed in the pump head is disposed after each use).

As discussed in Chapter 3, validation of cleaning procedures is required for reusable equipment (either dedicated to one product or used for multiple products) to

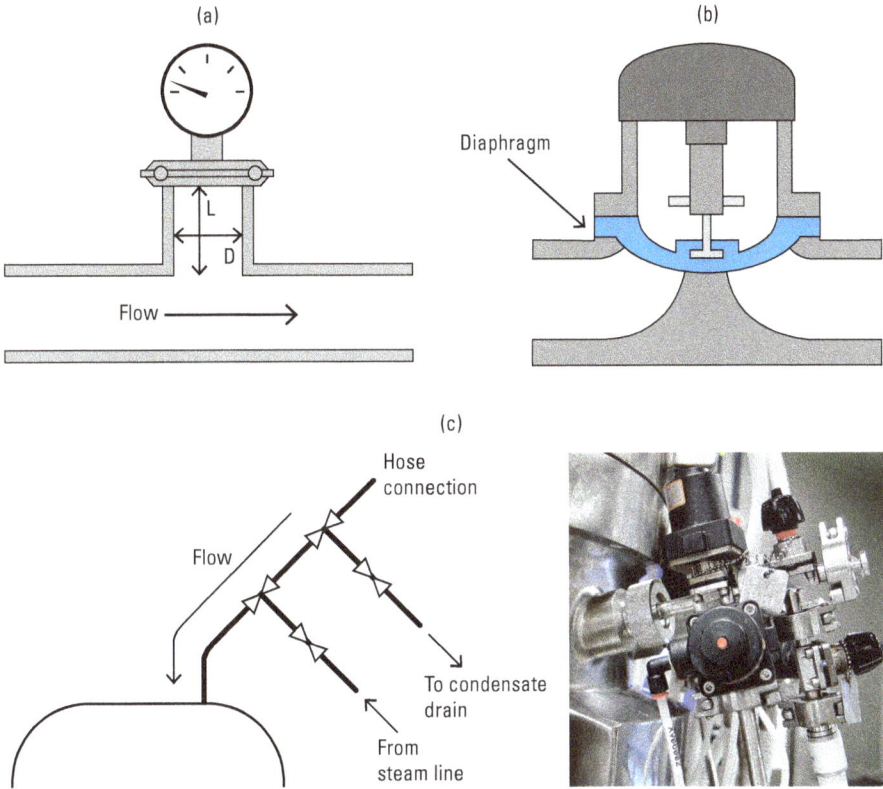

Figure 4.6: (a) An example of a dead-leg pathway. (b) An illustration of a typical diaphragm valve configuration. (c) An illustration and a photograph of a typical double block-and-bleed valve assembly. Image and photo © NC State University; reprinted with permission.

demonstrate that a cleaning process is consistently able to remove process soils and cleaning agents to acceptable levels between uses. This includes equipment used in CGMP fermentation and cell culture steps. As previously mentioned, bioreactor cleaning is accomplished by pumping the cleaning solutions to a spray device located inside the bioreactor. The spray device sprays the interior components of the bioreactor vessel with water and other cleaning solutions. Consistent performance of CIP procedures is achieved through careful control of pressure of the cleaning solution (at the spray ball), temperature of cleaning solution, contact time, and concentrations of cleaning agents. This process must be validated for the removal of cells and other components of the process between runs as well as the removal of any residual cleaning agents by final rinsing steps. Cleaning performance is measured by testing rinse water and swab samples, as described in Chapter 3, and comparing results to allowable limits. Swabbing locations are often chosen as areas within a bioreactor that are judged particularly difficult to clean. These challenging locations may include the underside of impellers or sparge rings, probe insertion sites, or baffle/wall junctions to name a few. Typically, a minimum of three consecutive successful cleanings – that is, cleanings that meet all acceptance criteria – are required to call a cleaning procedure validated. Once validated, the extensive sampling and testing performed during the validation runs are no longer required.

In similar fashion, sterilization protocols must also be validated to ensure that contaminating organisms are not present at the start of a run. Sterilization is achieved by exposing all surfaces to steam to raise the temperature to a desired value for a defined time period. The temperature is selected to provide a necessary degree of cell and spore destruction, typically 121 °C. Validation of sterilization-in-place (SIP) serves to assure that no residual microbes or their spores have survived the process. If they do survive, they can potentially begin to proliferate in the growth medium, ultimately leading to process failure. Detailed guidelines for equipment and sterilization process design can be found in Technical Report 61 from the Parenteral Drug Association [123]. Validation of sterilization protocols involves executing the procedure and then aseptically introducing sterile growth medium into the sterilized vessel. The sterile growth medium is held for a period of time to allow any surviving organisms to grow. The process is monitored for signs of growth, such as pH and dissolved oxygen (DO) changes. Samples are also taken and tested for bioburden. The sterilization process is modified until all indications of cell growth during a sterile hold are eliminated. Once validated, evaluation of both cleaning and sterilization efficacy is performed periodically to guarantee their ongoing acceptable performance.

Use of disposable bioreactors eliminates many – though not all – of the hygienic design considerations given that cleaning is not required. Many probes and sensors are still without single-use options. In the case where a reusable probe must be used in a single-use bioreactor, a method for sterilization and insertion of the probe without compromising the sterility of the bioreactor is necessary. A variety of proprietary systems exist to meet this need. Additionally, the connection needed for feeding or harvest cannot be sterilized by steam as with a stainless steel system. In these cases a tubing welder is used to make an aseptic connection. As such, the destination vessel as well as the feed container must have the appropriate weldable tubing present to make the connection. One final consideration for single-use systems is chemicals from the polymeric material leaching into the process fluid. Though complete elimination of leachables may not be possible, identification and characterization of potential chemicals is necessary as part of process validation (as described in Chapter 3). This characterization should be performed under conditions comparable to those present throughout the bioprocess. These conditions may include media composition, temperature, pH, and duration of exposure.

As discussed previously, the upstream process begins with expansion of cells from a WCB. There are a number of regulatory considerations related to the generation and maintenance of both MCBs and WCBs. The process for generating both is shown in Figure 4.7. The initiation of a process from a WCB was described previously.

That working bank is itself generated from an MCB. This MCB is generated from a single clone of the cell line in use for a process. Use of a working cell bank generated from a master cell bank ensures that each new production batch starts with cells from a population that is genetically identical. Once a clone is selected as part of cell line development activities, it is cultivated in flasks until it reaches a desired density and volume. This culture is then harvested, aliquoted into vials, and carefully frozen. Controlling freezing rate and incorporation of a cryoprotectant help prevent damage to cells during the freezing and subsequent thawing process. A typical storage temperature is −70 °C for microbial and −130 °C for animal cell lines. Each and every batch for production of a particular drug must be initiated from WCBs generated from a single MCB. As such, it is important to identify an appropriate size for the bank based on expected usage.

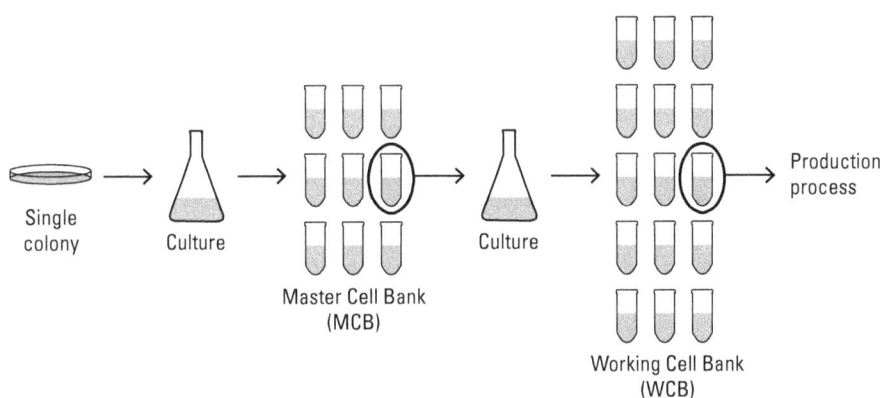

Figure 4.7: Typical process for generations of MCB and WCB. Image © NC State University; reprinted with permission.

Once generated, extensive testing is performed on the MCB. Specific guidelines can be found in FDA and EMA guidance documents [124, 125]. In short, three criteria are assessed: purity, identity, and stability. Purity is confirmed by testing for the presence of potential contaminating species such as microbes or viruses. DNA sequencing can be performed to confirm the species and the clonality of the cell stocks (i.e., the identity). Finally, the bank is grown and cells compared to a similar sample taken from the production bioreactor, in which cells have been passaged multiple times, upon the completion of a batch. This testing is described in Table 3.5. This comparison confirms that the cell line is suitable for stable cultivation over the course of a production run.

Working cell banks are generated from the master bank as described previously. Much of the same testing is performed on working cell banks to confirm purity, identity, and stability. Generally WCB characterization is less stringent than MCB generation with the assumption that, barring a contamination or mutation event, the WCB will be of suitable quality when generated with fidelity from the master bank. Additionally, purity and identity of cells are monitored throughout the seed train to ensure the same cells are being cultivated all the way through production [124]. In addition to their generation, storage of banks is important. The temperature of freezers must be carefully controlled and monitored to ensure their proper preservation. Generally, redundant and distinct storage locations are maintained, and access by personnel is strictly regulated.

Overall, a typical upstream production campaign follows a fairly consistent series of steps. The initial shake flask is sterilized and filled with fresh sterile medium. The WCB vial is carefully thawed and used to inoculate the first seed vial. During this cultivation, the next vessel (assuming it is reusable) in the seed train is being cleaned, sterilized, and filled with sterile culture medium. A sterile hold for at least 24 h is performed to test for latent microbial contamination. The preparation of the next vessel should be scheduled so that completion coincides with the completion of the previous cultivation step. A sterile "before-inoculation" sample is taken and sent to QC for sterility testing, and the vessel is inoculated from the previous seed-train step. End-point samples for each step can also be collected for analysis, to include potential contaminants and comparison of cell behavior to a validated process. This process is repeated at increasing scale until the vessel being inoculated is the production vessel. This vessel is then operated under the appropriate conditions until completion. At this time a final sample is collected for analysis, and the contents of the vessel are harvested for subsequent downstream processing.

4.5 Upstream process parameters and input material attributes

The performance of the cell drives the performance of an upstream process. Because of this, process parameters for an upstream process are those that affect the physiology, metabolism, and ultimately the growth rate of the host cell line. The foundation for cell growth is the nutrients provided by the growth medium. As mentioned earlier in the chapter, certain elements must be available for the cells to assimilate and convert into energy and compounds necessary for proliferation and product formation. Of particular importance are the carbon and nitrogen sources. These elements comprise the bulk of the mass of both cells and products. The specific carbon source significantly affects energy production as well as the waste products produced. Glucose, for example, is an ideal carbon source because it is highly oxidized and able to produce large amounts of adenosine triphosphate (ATP) and nicotinamide adenine dinucleotide (NADH) needed to produce the energy required for cell metabolism. Glycerol, on the other hand, can be catabolized to produce a similar amount of energy as glucose while producing more reducing equivalents [126]. This increased reducing power may be advantageous in production of products that require excessive reducing power in their metabolic production pathways. In some cases, this benefit comes at the cost of reduced carbon consumption and lower cell growth rates.

The specific nitrogen source can significantly affect product formation. Defined sources such as ammonium salts provide pure, soluble nitrogen for cell metabolism. This nitrogen must combine with carbon-containing molecules to form amino and nucleic acids before being incorporated into proteins or DNA, however. In some cases, pure amino acids can be used instead. Cell strains with a deficiency in production of a certain essential metabolite, or auxotroph strains that require a nutrient that the normal strain does not, often require supplementation of one or more specific amino acids. Often, use of pure amino acids can be cost prohibitive, and other nitrogen sources are used. Complex nitrogen sources such as proteolytic extracts of bulk protein from varied sources such as soy, animal tissue, milk, or yeast contain a mixture of the proteinogenic amino acids and nucleic acids along with other vitamins and minerals. While these complex sources may be more readily metabolized, they come with an increased lot-to-lot variability. In many cases, a medium will be formulated from multiple nitrogen sources.

While the identity of media components is known to impact cell growth and product formation, so too can the relative amounts of different components. The carbon-to-nitrogen ratio, for example, affects regulation of a variety of metabolic pathways, which will affect the proliferation of cells and product synthesis. The ratio of ammonium to free amino acids can affect the uptake of amino acids and formation of proteins. Other important components of media include buffering agents, divalent cations, and necessary vitamins and cofactors to name a few. Initial media formulation and feeding strategies (i.e., fed-batch or continuous operation) are carefully balanced and designed to optimize growth rate and product formation with respect to total process time, product quality, formation of impurities, and other downstream processing considerations.

Oxygen is a critically important nutrient that cannot be supplied in excess as a media component due to its low liquid solubility. As such, it must be continuously supplied from a gas phase in contact with the liquid. Oxygen in the bioreactor is controlled by achieving a specified DO concentration, an extremely important process parameter. In the presence of oxygen, cells generate large amounts of energy in the form of adenosine triphosphate (ATP) as well as important cofactors necessary for other metabolic processes [127]. When oxygen concentrations drop below a certain threshold, cell metabolism changes, and growth and production suffer. On the other hand, excessively high oxygen concentrations can lead to oxidative stress from formation of reactive oxygen species. Agitation, air flow rate to a sparger, oxygen enrichment of the inlet gas, and vessel pressure are process parameters that are controlled to maintain the optimal DO for the specific process by ensuring the rate of oxygen supply to the liquid matches the rate at which it is being consumed by the cells.

Another environmental process parameter that affects cell physiology and metabolism is extracellular pH. Each cell regulates the pH within its cytosol and various organelles. This is achieved by the pumping of protons, or hydrogen ions, across the various membranes [128–130]. These proton pumps are evolved to function within a narrow range of surrounding pH values. Any excursions from this pH range mean regulation is reduced, and a host of cell stresses occur due to protonation or deprotonation of metabolites and enzymes. Extracellular pH is an important process parameter and control strategies are designed to maintain its value. In some cases, this can be achieved by including simple buffering components in the medium, but in many cases acid and/or base addition may be necessary throughout a run. The specific optimal pH value can vary from as low as 4 to as high as 8 depending on the cell line and the product of interest.

Temperature affects many of the processes occurring at a cellular level in an upstream process. The activity of metabolic enzymes is precariously dependent on temperature as are many physiological factors, such as membrane and cytosol fluidity and regulation of gene expression [131, 132]. Changes in temperature, therefore, lead to changes in metabolism that in turn affect growth rate, product formation, and waste generation. Temperature also affects the properties of the medium. Decreasing temperature lowers the solubility of media components and increases the solubility of gases, notably carbon dioxide and oxygen. Temperature also affects pH.

4.6 Upstream performance parameters

Though a process is generally designed with a target product titer in mind, other parameters, such as those that describe cell growth and metabolism dynamics, also give insight into process performance. Parameters that describe the rates of cell growth or product formation, for example, capture the kinetics of the cell-based processes. Other parameters describe the stoichiometry of the relevant reaction, i.e., substrate to product, and inform on the metabolic and physiological state of the cell catalyst. Consistency in all of these parameters from run to run gives assurance that the cells are behaving in the desired way and, by extension, consistently producing the expected product at the expected quantity and quality.

Lag time, growth rate (μ as defined in equation (4.1)), and total process time are all performance parameters that may be monitored and characterize the kinetics of the cell catalyst. The lag time, or duration of the lag phase defined previously, is affected by the state of the cells in the inoculum as well as the conditions in the bioreactor at the time of inoculation. A departure of the lag time from expected values may indicate that an error occurred during the seed-train process, including variability in the composition of the growth medium. As discussed previously in this chapter, cell growth varies with a variety of process parameters. Excursions in pH, temperature, and substrate feed composition, for example, often manifest as a change in the observed growth rate. Related to growth rate is the total process time. As mentioned above, a process may be terminated based on the time of substrate exhaustion or accumulation of inhibitory compounds. Though this time will certainly vary with growth rate, it may also be affected by variation in substrate availability or marked changes in the metabolism of the host cell independent of growth rate. Consistency from run to run in lag time, growth rate, and total process time indicate that fermentation or cell culture step is producing the expected product at the expected quantity and quality.

In-process measurements such as final total cell mass concentration and final product titer provide abundant information about process performance as well. Consistency in these values indicates that the cell catalyst is converting substrate into more cells and desired product in the expected proportions. This consistency in stoichiometry strongly suggests cellular processes are generating product in the expected way, implying fidelity of critical quality attributes. Another parameter that suggests the process is performing in a way to produce product of the intended quality is the concentrations of select waste products. Though direct measurements of waste products give useful information, many other parameters indirectly inform on the behavior of the system. Because agitation is used to control DO, time trends for agitation act as a history of the oxygen demand, which is directly related to metabolic activity. Similar trends can be collected for pH control pumps or exhaust gas carbon dioxide concentrations, for example.

It should be noted that specific processes may include additional performance parameters or even omit any of the above. Indeed, the parameters that may influence performance are diverse and complex. Simply put, however, changes in these performance parameters can typically be directly attributed to the cell-based catalyst's significant departure from expected behavior. This departure strongly suggests that the physiology and metabolism have diverged from that which was validated. This divergence can have impacts on process productivity and product quality and can often lead to batch failure.

4.7 Upstream process development

Process development for fermentation or cell culture requires the methodical exploration of the effects of the numerous process parameters that can be expected to impact various process outputs. Though mathematical models can be used to predict process performance with changes in process conditions, the overall complexity and variation in a living biological system makes it difficult, if not impossible, to completely capture the performance of bioreactor operations in a mechanistic mathematical model (i.e., the use of mathematical expressions that govern the underlying biological processes).

As such, development of bioreactor steps is largely an experimental exercise necessary to generate the understanding required for process optimization. The necessary studies are typically conducted using bench scale (or miniaturized) systems to keep material requirements low. Once the process is defined, it is scaled up to production.

A major endeavor during upstream process development is determination of an optimal media formulation. As discussed at length in this chapter, the components of the media and their concentrations have a tremendous effect on process performance. Generally, industrial cell lines have a set of fairly standardized media compositions such as a Luria-Bertani broth for *E. coli* or Dulbecco's modified Eagle medium for CHO cells. These general recipes provide a good starting point for media formulation development, but it is often necessary to modify these formulae to get optimal performance from the system. A common approach is to thoroughly analyze the spent medium after a typical run to determine which nutrients may be limiting or are poorly utilized for that particular cell line. Knowledge of the underlying biochemistry of the organism can also be combined with this information to determine other supplements that may need to be added.

Design-of-experiment (DOE) approaches should be employed to ensure that effects of process variables – as well as interactions between different process variables – on the desired outputs of the process are captured with appropriate resolution. It should be noted that the impact of different process variables may depend on media formulation. Therefore, relevant variations on media recipes should be included at this stage. The overall design process is outlined in Chapter 2 as an example for a CHO cell culture process. Identification of an optimization goal, as well as process limitations, is an important part of this analysis. An obvious goal is maximization of final product titer while producing product of the intended quality, as defined by CQAs. This is the approach described in the example in Chapter 2, but many other criteria are typically worth considering. When comparing the performance of different batch or fed-batch production strategies, it can be helpful to optimize another performance parameter: productivity. As mentioned previously, it is defined as the ratio of product titer to process time. This parameter enables direct comparison on both a titer and a time basis. Other pertinent criteria to consider as part of the design process for a fermentation or cell culture step may include cost and availability of raw materials, feasibility of control strategies, or impacts on cost and efficacy of downstream steps. Finally, the selected conditions must meet the necessary targets for critical quality attributes of the product. Once complete, process development studies should provide the design space for the upstream process.

4.8 Scaling up fermentation and cell culture

As mentioned in Chapter 2, process development studies are typically performed at small scale with volumes not exceeding two liters. Execution at small-scale facilitates the completion of the high number of necessary runs in a cost- and time-effective way. Though the information gleaned at development scale is useful, it is unwise to assume that cell growth and the production process will behave identically at production scale, which typically ranges from 2,000 to 20,000 L or more. As such, care must be taken when scaling up to ensure that the necessary consistency and performance defined during process development can be achieved at production scale. Scale-up methods for bioreactors and challenges are presented here.

This chapter has discussed the importance of environmental conditions (e.g., medium pH and temperature) at length. A cell's environment, in this case, can be described by the conditions only nearby, within microns of the cell surface. As such, heterogeneity of relevant process conditions within the bulk volume of the bioreactor leads to different portions of the cell population experiencing different environments. This discrepancy can lead to varied behavior of different cells within the same bioreactor. This inconsistency, in turn, leads to variability in process performance.

The bioreactor system relies on efficient and thorough mixing to create a system that is spatially homogenous in process parameters such as dissolved oxygen and medium components, among others. The effect of mixing on oxygen delivery is discussed later in this chapter. Many of the problems that arise when scaling up can be attributed to differences in mixing dynamics at different scales. A bioreactor system is a complex one. All three phases of matter are present, and their behavior is meaningful to the process. Furthermore, process parameters change across a wide range of time and length scales: micrometer-sized cells reside in a vessel potentially several meters wide growing over hours while oxygen transfer takes place on the order of seconds. Achieving consistency throughout the bioreactor becomes more challenging as liquid volumes increase. More power is required to generate the same sort of turbulent flow because bubbles, solutes, and solids must travel farther to be evenly distributed, and gravity and pressure effects become more significant.

Because mixing is so critical, scale-up approaches generally aim to keep certain parameters related to mixing constant as culture volumes increase. Many parameters can be used to describe mixing: power/volume, Reynolds number, or mixing time, for example, capture the momentum transfer in the system. Unfortunately, these different parameters scale differently with volume, and holding one value fixed (e.g., power/volume) as scale increases means others (e.g., Reynolds number and mixing time) will vary. If those parameters that vary upon scale-up have a significant impact on mixing dynamics, performance may begin to decrease with increasing volume. Reynolds number, for example, increases with the square of the impeller diameter, while the power/volume ratio increases with the fifth power of impeller diameter. So if bioreactor scale-up is performed by holding the Reynolds number constant, the calculated impeller diameter will result in a power/volume ratio that is lower for the production bioreactor than the smaller scale unit. From this example, it is clear that the decision on which parameter to design around is, therefore, critically important and somewhat perilous.

Let's consider another example to illustrate how holding different mixing parameters constant on scale up affects the Reynolds number around the impeller. The Reynolds number is a dimensionless number that captures the ratio of inertial forces of fluid flow (i.e., the input forces) to the resistive forces from fluid viscosity. At high Reynolds numbers, above 10,000, the flow resulting from the impeller is turbulent, which is desirable for effective mixing. For an impeller, it can be written as follows:

$$Re = \frac{D_i^2 n \rho}{\mu} \tag{4.5}$$

where Re is the Reynolds number, Di the impeller diameter, n the rotational speed (in revolutions per minute), ρ the fluid density, and μ the fluid viscosity.

Consider the case of scaling up a bioreactor by a factor of 1,000 (e.g., scaling up from 2 L to 2,000 L), so that $V_2 = 1,000 \times V_1$ (where subscripts 1 and 2 refer to small and large scales, respectively). To scale up, the following geometric parameters will be held constant:

– the aspect ratio, which is the ratio of the vessel height to diameter, at a value of 3:1;
– the ratio of impeller diameter to tank diameter.

The volume of each vessel is approximated as the volume of a cylinder and can be written as follows for each:

$$V_1 = \pi h_1 \left(\frac{D_{T,1}}{2}\right)^2 ; V_2 = \pi h_2 \left(\frac{D_{T,2}}{2}\right)^2 \tag{4.6}$$

where $D_{T,1}$ is the diameter of tank 1, $D_{T,2}$ the diameter of tank 2, h_1 the height of tank 1, and h_2 the height of tank 2.

Using a constant aspect ratio of 3:1, equation (4.6) becomes:

$$V_1 = 3\pi D_{T,1} \left(\frac{D_{T,1}}{2}\right)^2 ; V_2 = 3\pi D_{T,2} \left(\frac{D_{T,2}}{2}\right)^2 \tag{4.7}$$

Because the ratio of V_2 to V_1 is equal to 1,000, the ratio of tank diameters can be written as

$$\frac{V_2}{V_1} = 1,000 = \frac{\frac{3}{4}\pi D_{T,2}^3}{\frac{3}{4}\pi D_{T,1}^3} \tag{4.8}$$

Simplifying the expression by cancelling and taking the cube root of both sides leads to

$$D_{T,2} = 10 \times D_{T,1} \tag{4.9}$$

Because we assume the ratio of tank to impeller diameter is constant upon scale-up, then

$$D_{i,2} = 10 \times D_{i,1} \tag{4.10}$$

where $D_{i,1}$ is the diameter of impeller on tank 1 and $D_{i,2}$ the diameter of impeller on tank 2.

With the relationship between impeller diameters at each scale established, we consider the value for Reynolds number, Re, at each scale. Assuming that the density and viscosity of the fluid in the bioreactor are the same at each scale leads to the following equation:

$$\frac{Re_2}{Re_1} = \frac{(D_2)^2 n_2}{(D_1)^2 n_1} \tag{4.11}$$

Next, we consider holding different mixing parameters constant at scale-up and then assessing the impact that each scenario has on the Reynolds number ratio in equation (4.11). In the first scenario, rotational speed (rpm) is held constant at each scale. In the second scenario, the linear velocity of the impeller tip is held constant at each scale. If the same rotational speed is used at each scale (i.e., $n_1 = n_2$), the ratio of Reynolds numbers at each scale becomes:

$$\frac{Re_2}{Re_1} = \frac{(10 * D_{i,1})^2}{(D_{i,1})^2} = 100 \tag{4.12}$$

If the linear velocity of the impeller tip is held constant between scales (note that the linear velocity of the impeller is proportional to the impeller diameter, so that $n_2 = 1/10 \times n_1$), the following relationship results:

$$\frac{Re_2}{Re_1} = \frac{(10 * D_{i,1})^2 \frac{1}{10} n_1}{(D_{i,1})^2 n_1} = 10 \tag{4.13}$$

This analysis demonstrates how keeping certain mixing parameters constant – in this example, either rotational speed or linear velocity of the impeller – can result in significantly different values for the impeller Reynolds number, an important measure of mixing performance. Finding the parameter that best captures the behavior of your system is therefore critical for successful scale-up.

Aeration is another important consideration during scale-up. Delivery of oxygen is often the limit on cell growth and is affected by process conditions and bioreactor architecture. Because of this, it is important that aeration efficiency is maintained as an optimized process is scaled up. Simple parameters describing aeration, such as vessel volumes per minute (vvm), superficial gas velocity (flow rate/cross sectional area of bioreactor), or gas residence time, are all straightforward and are easily scaled. Furthermore, if a consistent bioreactor aspect ratio is maintained, these will all scale linearly with volume. It should be noted that airflow to the overlay does not contribute as efficiently to oxygen transfer compared to sparging.

Example: Scaling air flow rates and gas velocities
Consider a pilot-scale bioreactor with total liquid volume of 2,000 L and a height-to-diameter ratio of 3:1. The setpoint for airflow in this process is 1,250 L of air per minute. If the final production reactor is to be 15,000 L with the same aspect ratio, what gas flow rate is necessary to match the vessel volumes per minute used at pilot scale?

Solution
First, the flow rate in vvm is calculated by simply taking the ratio of gas flow rate to liquid volume

$$vvm = \frac{\text{gas flow rate (L/min)}}{\text{liquid volume (L)}} = \frac{GFR}{V}$$

$$vvm_{Pilot} = \frac{GFR_{Pilot}}{V_{Pilot}} = \frac{1,250 \text{ L/min}}{2,000 \text{ L}} = 0.625 \text{ vvm}$$

Keeping vvm constant allows the following expression:

$$vvm_{Production} = 0.625 = \frac{GFR_{Production}}{V_{Production}} = \frac{GFR_{Production}}{15,000 \text{ L}}$$

$$0.625 \text{ vvm} * 15,000 \text{ L} = GFR_{Production} = 9,375 \text{ L/min}$$

Another parameter used to characterize aeration is the overall oxygen mass transfer coefficient, often denoted as $k_L a$. The transfer of oxygen from the gas into the liquid is driven by the concentration difference between the surface of the bubble and the bulk of fluid as illustrated in Figure 4.8. This gradient exists across a finite distance known as the boundary layer. $k_L a$ is a lumped parameter – the product of k_L, the specific mass transfer coefficient, and a, the total interfacial surface area between the gas and liquid phases. k_L is a function of the thickness of the boundary layer (i.e., the stagnant liquid layer represented by the outer layer in tan in Figure 4.8) that surrounds a gas bubble and through which oxygen must diffuse to be distributed throughout the bioreactor. Boundary layer thickness is affected by the fluid properties of the liquid to a large degree, and the intensity of mixing to a lesser degree. The interfacial surface area is affected by the total volume of air flow as well as the size of the air bubbles. Smaller bubbles have a higher surface-area-to-volume ratio compared to larger bubbles, and increasing mixing efficiency helps to create smaller bubbles. It should be noted that various physical limits, namely surface tension, put a practical lower limit on bubble diameter. When lumped, $k_L a$ is fairly easily determined for a system based on vessel geometry, total gas flow, and agitation.

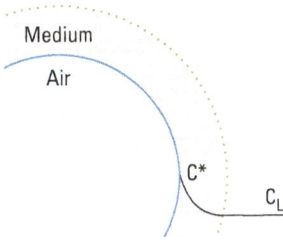

Medium

Air

C^*

C_L

Figure 4.8: Diffusion of oxygen from a gas bubble into the bulk liquid for consumption by growing cells. C^* represents the concentration of oxygen in the liquid medium in equilibrium with the gas-phase oxygen. C_L represents the concentration of oxygen in the bulk liquid medium. Image © NC State University; reprinted with permission.

It should be noted that mixing and aeration within the system are likely to lead to the formation of entrained foam at the top of the liquid, and increases in agitation and gas flow will increase foam formation. Foam is problematic for a few reasons. First, it will entrain cells within the bubble film and create an environment that is not able to be replenished with nutrients as readily as the bulk liquid. Second, foam occupies significant bioreactor volume if not controlled and can actually cause the total volume of the contents to exceed the available volume in the bioreactor and result in what is referred to as "foaming out." As such, delivering similar process performance across scales without causing excessive foaming is an ongoing challenge with upstream process development and scale-up.

One final consideration for bioreactor scale-up is temperature control. As mentioned previously, temperature affects cell growth as well as a variety of other process parameters, which themselves affect cell growth. Metabolically active cells produce heat as a waste product, and this heat must be removed in order to control the temperature. This is accomplished by circulating a cooling fluid through the jacket of the bioreactor vessel to transfer heat away from the vessel. Bioreactor vessels are typically cylindrical and, as such, as they get larger the surface-area-to-volume ratio decreases. This decrease in surface area to volume can lead to a scenario in which there is not enough surface area for heat to be removed as quickly as it can be generated. It should be noted that this limitation is more commonly encountered during microbial fermentation than animal cell culture. This can be overcome by either using cooling fluid whose temperature can be reduced beyond that of water (water can be cooled to no lower than its freezing point of 0 °C) or by increasing cooling surface area by adding coils to the interior of the vessel.

There are many different approaches to discern parameters that are important for describing scale-up effects. First, development of a computational fluid dynamic model for a bioreactor can help predict how mixing changes as volume increases. While a computational approach only provides an approximation, it provides insight into which parameters are useful for capturing changes with scale. Ultimately, it can point to which parameters should be explored experimentally.

Often, performance studies are conducted at pilot scale – that is, a scale between development and production. The choice of the size of this pilot bioreactor depends on the ultimate final production scale. At pilot scale, performance is compared to bench-scale and an assessment made of which mixing and aeration parameters explain departures in performance. This assessment then informs on which parameters to hold constant and which to change on scale-up to the final desired volume. This process is more practical at pilot scale than final production scale for cost reasons (e.g., material costs for the studies will be less than at production scale). In addition, the magnitude of change is smaller and the source of variation is easier to discern.

Another common approach is to use a scale-down model. In this method, you characterize your production-scale bioreactor and then try to impose similar behavior at the bench-scale that matches the expected behavior of your final production bioreactor. Limits on parameters such as agitation speed or airflow that exist in the production bioreactor can be designed into the development bioreactor to restrict values for characteristic parameters like $k_L a$ or Reynolds number. This approach requires knowledge of the characteristics of the final production bioreactor, which may not have been built or even designed in the earliest stages of process development. However, heuristics can be applied that make this approach more certain. Often, scale-down is applied during troubleshooting of an established process. In this case, the production-scale bioreactor is scaled down to bench scale to explore causes of problems at production scale in a fast and inexpensive way.

Scaling up of upstream operations is challenging. Each particular process will have its own criteria, limitations, constraints, and challenges that can be hard to accurately predict or explain even with ample data. In some cases where scale-up is particularly problematic and the desired performance cannot be reached, scale-out is another option for increasing process capacity. Scale-out increases capacity by simply operating a larger number of smaller bioreactors that are able to meet necessary desired performance parameters. While this strategy eliminates the need for scale-up studies, it inherently increases variability in performance owing to each vessel being a separate production run. This inconsistency comes with other regulatory and processing challenges. Furthermore, multiple small vessels have a higher capital cost than one larger vessel.

4.9 Summary

The chemical complexity of biopharmaceuticals necessitates growing cells for their production. A variety of cell lines are used – from simple bacteria like *E. coli* to mammalian (non-human) cell lines such as CHO to human cell lines such as HEK293 cells –

to generate products such as recombinant proteins, viral vectors for gene therapy, or DNA. Though the specifics of the host cell and desired product may vary from process to process, the principle is the same: cells consume nutrients and perform a complex series of metabolic reactions to produce an incredibly complex and highly valuable product. This chapter describes a general process for taking a cell from a frozen stock (i.e., working cell bank) to product-containing broth ready for harvest. Key considerations in producing a biopharmaceutical product in the required quantity and with the intended quality include the following:

– *Bioreactor design,* which ensures our ability to fastidiously cultivate cells at large scale. Bioreactor vessels must be designed to deliver an optimal environment for cell growth and product formation while reducing the risk of contamination. Material of construction is an important consideration. Additionally the architecture of the vessels must be carefully designed to allow hygienic and aseptic operation in order to eliminate both microbial and cross-contamination. Single-use polymeric bioreactors and stainless steel options are both suitable options for upstream processing, though each have their benefits and disadvantages.

– *Adherence to regulatory requirements,* which also drives equipment design considerations. Effective procedures for cleaning and sterilization must be developed and validated to applicable standards. Additionally, cell banks used to initiate production runs must be thoroughly characterized to ensure that the desired cell line and only the desired cell line is being grown for product formation.

– *Control of cell growth,* which is the main tool for influencing process behavior. Cells rely on growth media for the nutrients necessary for growth. The specific medium recipe for each process must provide all the components that cells require to proliferate and produce the desired product. Simple components such as sugars and amino acids provide a majority of these needs. More complex components such as growth factors, vitamins, or proteins like insulin may be necessary for more challenging host cells or products. In addition to nutrients, cells require other conditions such as pH, DO, and temperature to be carefully controlled. All of these factors affect the metabolism and physiology of the growing cells, which, in turn, affects the biological processes they use to produce the desired product. Maintaining consistent cell behavior is therefore crucial for maintaining consistent product quality. Finally, different feeding strategies can be employed to improve the productivity and reproducibility of a process.

Finally, approaches and considerations for process development and scale-up were presented. The goal of process development is twofold: generate process understanding and define the optimal operating conditions to maximize profitability. This process involves identifying those process parameters that impact productivity and product quality and testing different combinations until a set of conditions that lead to desired process outputs is determined. Most commonly, details around medium formulation and process parameters such as temperature, pH, and DO are identified. Process outputs for optimization may include product titer, volumetric productivity, and product purity. Scale-up is complementary to process development with the goal of applying the optimal conditions identified during development to a production process at a larger scale in order to meet market demands. Challenges with scale-up center around changing mixing dynamics with increasing volumes. Achieving homogeneity of process conditions throughout the entire volume becomes increasingly difficult as volume increases. The heterogeneity that results creates heterogeneity in the behavior of the host cells and, ultimately, product formation. Different approaches to scale-up were presented.

4.10 Review questions

1. Consider the growth curve in example 4.1. Identify the lag, exponential, and stationary phases.

2. You are trying to optimize a medium for production of a protein by fermentation. After performing several experiments, you determine the following:
    – Complete exhaustion of 20 g/L of glucose yields 4 g/L biomass in an otherwise nutrient-rich medium.
    – Repeating the same fermentation with induction of protein expression yields 3.6 g/L of biomass and 0.65 g/L of protein product.
    – The biomass is 60% carbon and 20% nitrogen by mass.
    – The protein product contains 15% nitrogen by mass.

    Based on the above, answer the following questions:
    (a) What are the values for $Y_{X/S}, Y_{P/S}$, and $Y_{P/X}$?
    (b) How many grams of nitrogen must be available to the cells to achieve the biomass and protein yields obtained in experiment 2 above?

    You decide to use ammonium chloride as a nitrogen source in your medium.
    (c) What biomass and protein titers can you expect if you grow this strain and induce protein expression in a medium that contains 20 g/L glucose and 2 g/L ammonium chloride?
    (d) What will the residual glucose and ammonium chloride concentrations be?

3. Explain how the following would impact each of the performance parameters: final total cell mass, growth rate, and final product titer.
    (a) Nitrogen content of the medium was less than specified in the formulation
    (b) The solution record for a glucose feed had an extra zero, leading to a 10-fold excess of glucose in the feed

(c) Sterilization was not completely effective and an unknown microbe survived the process

(d) Double the correct amount of inoculum was added to initiate the run.

4. You are comparing data from three candidate runs. The first is a batch process that produces 2 g/L product in 10 h. The second is a fed-batch process that produces 4 g/L in 24 h. The third is a continuous process that produces a harvest stream with 1 g/L of product at a flow rate of 15 L/h. Each has a working volume of 100 L.
   (a) Calculate the productivity for each process.
   (b) Why might you recommend each over the other options?

5. This number is used to describe how efficiently a stirrer transfers its energy into movement of the fluid. It can be described as the power consumed by the fluid as it resists the motion of the impeller. This power is subsequently used to mix the liquid contents. Consider the example given in 4.8 regarding Reynolds number at different scale. Perform a similar analysis to determine how power-to-volume ratio changes with scale:

$$N_p = \frac{P}{\rho n^3 D^5}$$

where N_p is the power number, P the power applied to the fluid, ρ the fluid density, n the rotational speed, and D the impeller diameter.

6. Derive an expression for the surface-area-to-volume ratio for a sphere. Assuming bubbles in a bioreactor to be spherical, describe how smaller bubbles leads to increased surface area available for gas transfer.

7. The transfer of oxygen from the gas phase to the liquid phase can be described by the following equation:

$$V\frac{dC_L}{dt} = k_L a(C^* - C_L)V$$

where V is the culture volume, CL the oxygen concentration in the bulk fluid, C^* the oxygen concentration in the liquid at the bubble surface, and $k_L a$ the overall mass transfer coefficient.

You decide to try and quantify the $k_L a$ at a specific agitation and airflow. You've calibrated your probe to 100% saturation of the liquid with oxygen. You turn on the agitation and airflow and record the reading every 10 min for 2 h. Use this data to estimate $k_L a$.

Time (min)	Probe reading (%)
0	10
10	21
20	31
30	38
40	48
50	54
60	60
70	63
80	70
90	73
100	75
110	79
120	82

Chapter 5
Harvest operations, part 1: cell lysis

Once the product has been generated, it must be separated from the production system (e.g., cells), and all solids must be removed to create a clarified liquid in preparation for subsequent chromatography steps, which may clog if solids are present in the feed. The steps involved comprise the harvest stage and are the subject of the next three chapters.

Most biopharmaceuticals are produced using cells, so feed to the harvest stage is typically a broth – the cells used to produce the biopharmaceutical suspended in the spent medium used to support cell growth. For those products that are extracelluar – that is, products that reside in the liquid medium surrounding the cell after production – cells are removed from the broth so that the medium can be processed further. We discuss methods for separating solids, like cells, from liquid in the next two chapters. For those products that are intracellular, cell lysis, or the breaking open of cells, is required to release the biopharmaceutical product into the surrounding liquid phase. Separation of the resulting cell debris from the surrounding liquid is required afterwards. This chapter covers cell lysis for biopharmaceutical production by addressing the following questions:

- What is the structure of the cells typically used in biopharmaceutical production, and if a lysis step is required, how does the cell structure impact the choice of lysis technique?
- What are examples of intracellular and extracellular biopharmaceutical products?
- In addition to cell lysis, what other processing steps are used in the product harvest stage for intracellular products?
- What are the different methods used for cell lysis? Which are scalable and which are not? And, in addition to cell type, what other factors impact the choice of method?
- How does one of the most common cell lysis unit operations – high-pressure homogenization – work?
- What are the process parameters that impact performance of a high-pressure homogenizer, and how are ranges for these parameters determined?
- What is a typical operating procedure for a high-pressure homogenizer used for current good manufacturing practice (CGMP) production of biopharmaceuticals?

We start the discussion with information on basic cell structure, focusing on the outer layers of cells that must be ruptured to release intracellular product. This background information helps to put subsequent discussion on lysis methods in context.

https://doi.org/10.1515/9783111112459-005

5.1 Structure of cells commonly used in biopharmaceutical production

Generally speaking, cells can be classified as prokaryotic or eukaryotic. Eukaryotes are organisms made up of one or more complex cells and include animals, plants, fungi, and protists. The eukaryotic cell, as shown in Figure 5.1(a), contains a membrane-bound nucleus; numerous membrane-bound organelles, such as the endoplasmic reticulum, Golgi apparatus, and mitochondria, each with a specialized function; and rod-shaped chromosomes that carry genetic information for each cell. Eukaryotic cells may be bound by only a cell membrane (also referred to as the cytoplasmic membrane or plasma membrane) or by both a cell membrane *and* rigid cell wall, which provide a barrier between the cell cytoplasm (i.e., the liquid that fills the inside of a cell) and fluid surrounding the cell. Examples of eukaryotic cells commonly used in biopharmaceu-

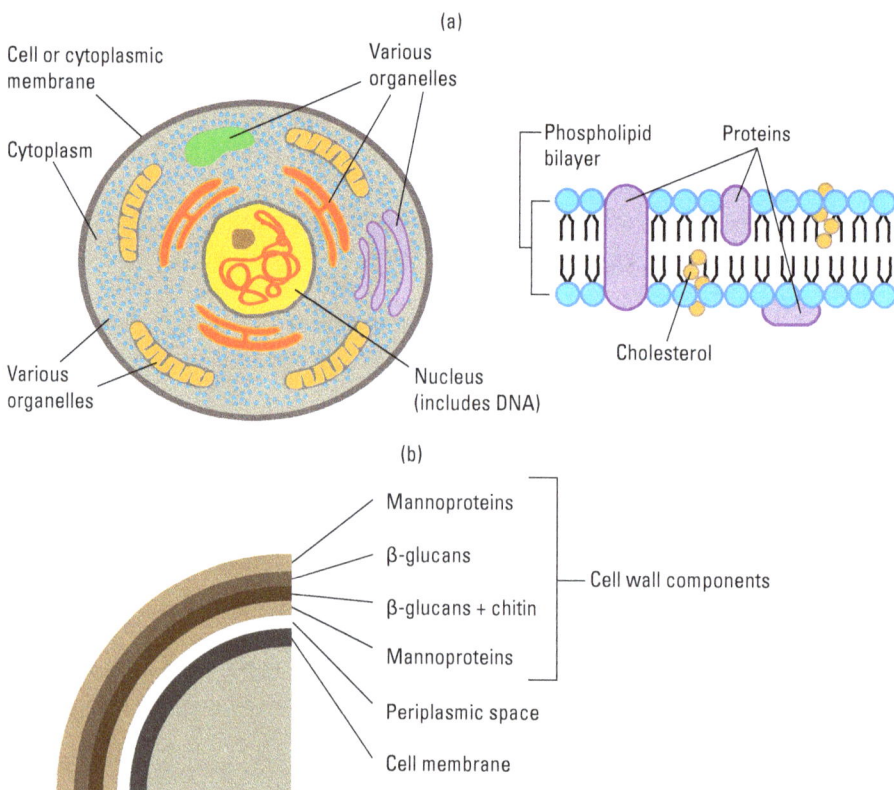

Figure 5.1: Basic structure of eukaryotic cells: (a) illustration of an animal cell (such as a CHO cell), which has no cell wall but does have a lipid bilayer membrane surrounding the cell, which is shown to the right; (b) illustration of the thick cell wall of the eukaryote *Saccharomyces cerevisiae*, a yeast. Image © NC State University; reprinted with permission.

tical production and discussed previously include Chinese hamster ovary (CHO) cells and *Saccharomyces cerevisiae*, a yeast. CHO cells grown in suspension are spherical and have a diameter of approximately 15 μm [133]. *Saccharomyces cerevisiae* is egg-shaped and has a length of 5–10 μm [134]. One big difference between animal cells and yeast is that animal cells like CHO are enclosed only in a cell membrane primarily comprised of a lipid bilayer with embedded proteins, while yeast such as *Saccharomyces cerevisiae* have a thick cell wall surrounding its cell membrane. The cell wall for *Saccharomyces cerevisiae* is made of β-glucans (polysaccharides), proteins, and chitin, as shown in Figure 5.1(b), and has a thickness of 110–200 nm [135]. The thick cell wall of yeasts make them more difficult to lyse compared to animal cells bound only by a membrane.

Prokaryotes are less structurally complex single-celled organisms. All bacteria, including *E. coli*, are prokaryotes. Prokaryotic cells do not have a membrane-enclosed nucleus or membrane-bound organelles like eukaryotic cells. In essence, most prokaryotes are molecules surrounded by a membrane and a cell wall that provide a barrier between the cell cytoplasm and surrounding fluid. An illustration of bacteria cells, highlighting the cell envelope for both gram-positive and gram-negative bacteria, is shown in Figure 5.2. Note that *E. coli* is rod shaped with dimensions of approximately 1×3 μm [136] and as such is smaller than either CHO or *Saccharomyces cerevisiae*.

Because the layers surrounding the cell must be disrupted during lysis, more detail on the constituents of those layers is presented in Table 5.1. Give this table a careful read, as subsequent sections of this chapter rely on the information found here.

The gram-positive and gram-negative designation of bacteria is based on a staining method known as the Gram stain test. A violet dye is applied to cells, followed by a decolorizing agent, followed by a second dye that is red. Bacteria with a thick peptidoglycan layer retain the violet dye used in the Gram stain procedure and are referred to as gram positive; bacteria with a thinner peptidoglycan layer lose the initial violet dye during decolorization and are stained pink by the red dye. These bacteria are referred to as gram negative. Also worth noting is that the outer membrane of gram-negative cells such as *E. coli* is a source of endotoxins in biopharmaceutical processes. Endotoxin specifically means toxin "from within." Endotoxins can cause fever, trigger the coagulation cascade, and cause endotoxin shock among other adverse effects; therefore, they must be removed in the downstream process to ensure safe drug product [140]. The source of endotoxins is the lipopolysaccharides that make up the outer membrane. Lipopolysaccharides consist of both lipids and sugars. One of the lipids in the outer membrane of gram negative cells is lipid A, the toxic portion of the lipopolysaccharide. When gram-negative bacterial cells are lysed, lipopolysaccharides (i.e., the endotoxins) are released. Consequently, endotoxin levels in *E. coli*-based processes are particularly high. While endotoxins are a particular concern in an *E. coli*-based biopharmaceutical processes, steps for removal of endotoxins are designed into processes that rely on other host systems as well, such as CHO, as contamination by endotoxin-producing gram-negative bacteria that enter the process stream adventi-

(a)

(b)

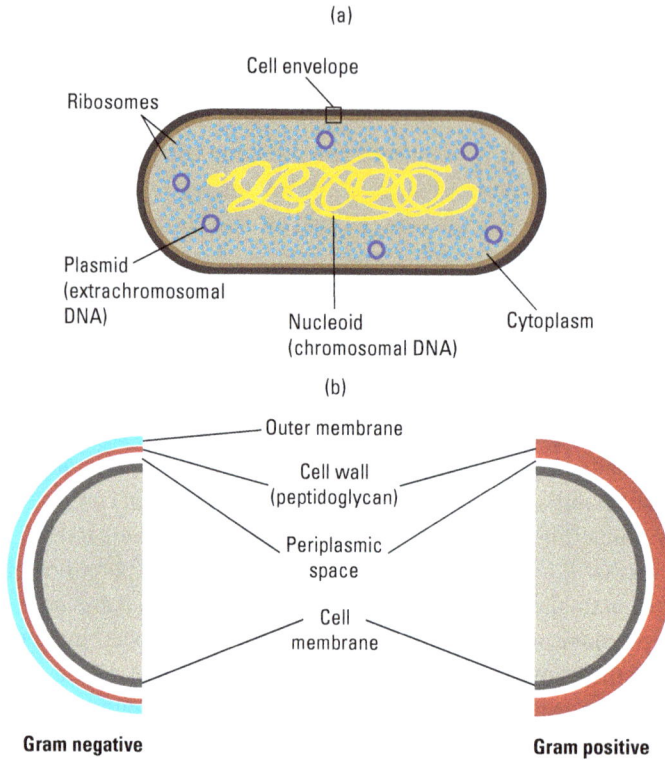

Figure 5.2: (a) Structure of bacteria, a prokaryote. Note that the inner layer of the cell envelope represents the cell membrane, and the out layer represents other envelope components. (b) Close-up of the cell envelopes of gram-negative and gram-positive bacteria. Note that, in this book, the term cell wall for bacteria refers specifically to the layer containing peptidoglycan, which is a component of the cell envelope. Image © NC State University; reprinted with permission.

Table 5.1: Description of the outer layers surrounding animal cells, bacteria, and yeast.

Cell type	Outer layers of cell looking from inside out	Composition
Animal	Cell membrane[a]	A lipid bilayer with embedded proteins. See Figure 5.1(a). Cell membranes generally have a thickness of 5–10 nm [137].
Yeast	Cell membrane[a]	A lipid bilayer with embedded proteins, 7.5 nm wide in *S. cerevisiae* [138]. See Figure 5.1(a).
	Periplasmic space	Thin region external to the cell membrane comprised mainly of secreted proteins that are unable to permeate the cell wall. See Figure 5.1(b).

Table 5.1 (continued)

Cell type	Outer layers of cell looking from inside out	Composition
	Cell wall	A thick layer (110–200 nm in *S. cerevisiae*) made of polysaccharides (β1,3- and β1,6-linked glucans), mannoproteins, and chitin. Chitin is a polymer of *N*-acetylglucosamine [135]. See Figure 5.1(b).
Bacteria (gram negative)	Cell membrane[a]	A lipid bilayer with embedded proteins, similar to that shown for animal cells in Figure 5.1(a). Approximately the same thickness as the peptidoglycan layer in *E. coli* [139] and part of the bacterial cell envelope.
	Periplasmic space	Thin region external to the cell membrane that contains the peptidoglycan layer and proteins. Part of the cell envelope as drawn in Figure 5.2(b).
	Peptidoglycan layer	See Figure 5.2(b). Peptidoglycan is a polymer consisting of alternating residues of two carbohydrates: β-(1, 4) linked *N*-acetylglucosamine (NAG) and *N*-acetylmuramic acid (NAM), along with attached peptides. It is a relatively thin layer, only about 6–7 nm thick in *E. coli* [139]. It's a main component of the cell wall in Figure 5.2(b) and is part of the bacterial cell envelope.
	Outer membrane	A lipid bilayer. The outer surface of the bilayer is made up of lipopolysaccharide and protein. Lipopolysaccharides are large molecules consisting of a lipid and polysaccharide. Lipopolysaccharides are endotoxins, a common impurity in biopharmaceutical processes, at particularly high levels in *E. coli*-based processes. Part of the cell envelope as drawn in Figure 5.2(b).
Bacteria (gram positive)	Cell membrane[a]	A lipid bilayer with embedded proteins, similar to that shown for animal cells in Figure 5.1(a).
	Periplasmic space	Thin region external to cell membrane that can be slightly thicker than periplasmic space in gram-negative bacteria, although there is some debate about whether a periplasmic space exists in gram-positive bacteria.
	Cell wall	See Figure 5.2(b). The cell wall is made up of peptidoglycan, described above. In gram-positive bacteria, the cell wall is thick relative to that in gram-negative bacteria [139].

[a]Also referred to as the cytoplasmic membrane and plasma membrane.

tiously is a possibility. Endotoxins along with other impurities are discussed in greater detail in Chapter 8.

Generally speaking, bacteria and yeast, with their tough cell walls, are more difficult to lyse by mechanical means than animal cells. Among bacteria, gram-positive cells are more difficult to lyse by mechanical methods than gram-negative cells due to the presence of a thicker peptidoglycan layer in gram-positive bacteria.

5.2 The need for cell lysis: intracellular versus extracellular products

Once produced, the biopharmaceutical product may remain inside the cell as an intracellular product or be released to the medium surrounding the cells as an extracellular product. Whether a cell produces intracellular or extracellular products depends on the host system. For example, products produced using *E. coli* are intracellular. Processes for intracellular products require a cell lysis step for release into the liquid surrounding the cell. Depending on the process design, this liquid may be the aqueous spent medium from the bioreactor or a buffer solution used to suspend whole cells recovered from the bioreactor. Processes for extracellular products do not require a lysis step because the product is secreted out of the cell. Further explanation of intracellular and extracellular products is provided below:

– Intracellular products reside within the cell cytoplasm or the cell periplasm. Biopharmaceutical products produced in *E. coli*, such as insulin and human growth hormone, are examples. The antigen produced for Recombivax HB® hepatitis B vaccine, which is produced in *S. cerevisiae*, is another [34].

– Recombinant intracellular products can be soluble, insoluble, or some combination of both. Insoluble products often form inclusion bodies – micron-sized solid aggregates that consist of relatively pure (typically greater than 80%), but inactive product [141, 142]. Insoluble production as inclusion bodies can lead to high purity going into the purification stage of a process because of the high purity of the inclusion bodies; however, because the protein that makes up the inclusion body is misfolded, the inclusion body must be solubilized by unfolding the protein with a denaturant and then refolding the protein to create a bioactive, soluble biopharmaceutical product. The recovery of product during this refolding step is typically very low due to a significant amount of product precipitation that occurs [143]. Therefore, soluble production is generally desirable over insoluble because it results in good overall process yields.

– For intracellular product that resides within the cell periplasm, yields tend to be low and recovery of the periplasmic protein can be poor. However, an advantage of periplasmic product is the possibility of designing a cell lysis step that selectively releases the product contained within the periplasm without releasing the contents of the cytoplasm. This selective release would result in a higher product purity going into the purification stage of the process. However, methods for selective release of periplasmic product can be challenging to implement [144].

– Extracellular products are released into the medium used to grow the cells. A common example is monoclonal antibodies produced in CHO cells, which are capable of secreting proteins [145]. Extracellular products have the advantage of requiring no lysis step that, in addition to releasing product, releases undesirable cell components into the process stream; thus, extracellular product typically results in fewer soluble impurities entering the purification stage. As mentioned previously, *E. coli* does not typically secrete recombinant proteins, although work is being done to engineer *E. coli* strains to do just that [146].

– Some processes produce significant proportions of both intracellular and extracellular product. An example is the production of adeno-associated virus (AAV) using human embryonic kidney 293 (HEK293) cells, as shown in Figure 5.3 [147]. AAV viral capsids are commonly used as vectors for gene therapy, as discussed in Chapter 1. From Figure 5.3, certain AAV serotypes, such as AAV5 and AAV9, are produced as both intracellular and extracellular capsids using HEK293 cells. Other serotypes, such as AAV2, are primarily intracellular, while most other serotypes (AAV1, AAV6, AAV7, and AAV8) are primarily extracellular. AAV2 is primarily intracellular in a HEK293 production system due to its affinity to heparin, which is present on the HEK293 cell surface [147].

Whether product is intracellular, extracellular, or some combination impacts the design of a production process. Examples of two different harvest process designs for intracellular products are shown in Figure 5.4 along with the design for an extracellular product for comparison. Each process starts with broth from a bioreactor and ends with a clarified process intermediate, prepared for loading to a chromatography column in the purification stage.

In the first process shown, cells from the bioreactor are recovered by centrifugation, then suspended in a buffered solution that is appropriate for the product and for the first chromatography step in the purification stage. The resulting cell suspension is homogenized to create the lysate, also referred to as the *homogenate* when using a homogenizer, then filtered to produce clarified lysate. In the second design in Figure 5.4, product is lysed directly in the bioreactor using a detergent. The resulting lysate is passed through a depth filter, or multiple depth-filter stages, for clarification followed by an ultrafiltration step for product concentration and buffer exchange. This latter step ensures that product is in the correct buffer system for the first chromatography column in purification stage. This second design is particularly useful when a significant portion of product is both intracellular and extracellular, as is the case with certain viral vectors used for gene therapy, because all bioreactor contents move forward in the process. By contrast, in the first process, because only the cells from the bioreactor move forward to the next step, any product that resides in the spent medium would be lost. However, the first process design offers the advantage of not requiring a buffer exchange step (ultrafiltration) prior to the

(a)

(b)

AAV2

20–25 nm

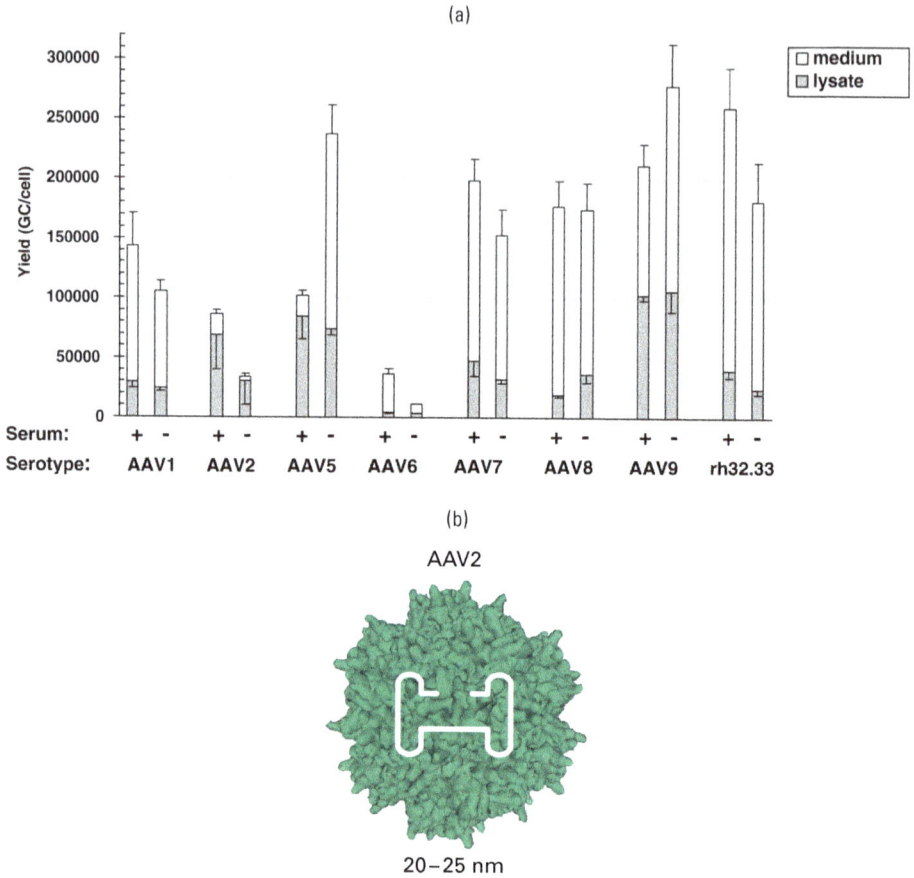

Figure 5.3: Distribution of intracellular (recovered after lysis, in the lysate) and extracellular (in the cell culture medium) AAV capsids produced in HEK293 cells. The y-axis shows the number of genome copies (i.e., the number of capsids that contain the gene of interest) produced per HEK293 cell (GC/cell). The x-axis shows the AAV serotype. The + and – in the figure indicate whether or not cells were incubated in fetal bovine serum following the transfection step [147]. Below the graph is an illustration of an AAV2 capsid. The graph is reprinted with permission from "Efficient Serotype-Dependent Release of Functional Vector into the Culture Medium During Adeno-Associated Virus Manufacturing," by Luk H. Vandenberghe, Ru Xiao, Martin Lock, Jianping Lin, Michael Korn, and James M. Wilson, 2010, Human Gene Therapy, 21(10), 1253. ©Mary Ann Liebert, Inc [147]. The AAV2 capsid is adapted from PDB ID: 1LP3, *The Atomic Structure of Adeno-Associated Virus (AAV-2), a Vector for Human Gene Therapy*, DOI: 10.2210/pdb1LP3/pdb, RCSB Protein Data Bank, rcsb.org, 2002 [148].

purification stage. Suspending cells in the buffered solution required for the first column chromatography step eliminates this need. The topic of liquid–solid separation steps (i.e., centrifugation and filtration) that follow cell lysis is covered in greater detail in the next two chapters.

Harvest process #1 for intracellular product:

Broth from bioreactor

↓

Centrifugation

Recover cells

↓

Cell suspension (in a buffer)

Slurry cells to flow through homogenizer; put product in correct buffer for purification

↓

Homogenization

Lyse cells

↓

Tangential-flow microfiltration

Remove cell debris created during lysis (clarification)

↓

Clarified liquid to purification

Harvest process #2 for intracellular product:

Broth from bioreactor

↓

Incubation with detergent (directly in bioreactor)

Lyse cells

↓

Depth filtration

Remove cell debris created during lysis (clarification)

↓

Ultrafiltration

Concentrate product and exchange buffer to put product into correct matrix for purification

↓

Clarified liquid to purification

Harvest process for extracellular product:

Broth from bioreactor

↓

Centrifugation

Remove particulate (clarification)

↓

Depth filtration

Remove residual particulate (because centrifugation alone is unlikely to remove all particulate)

↓

Clarified liquid to purification

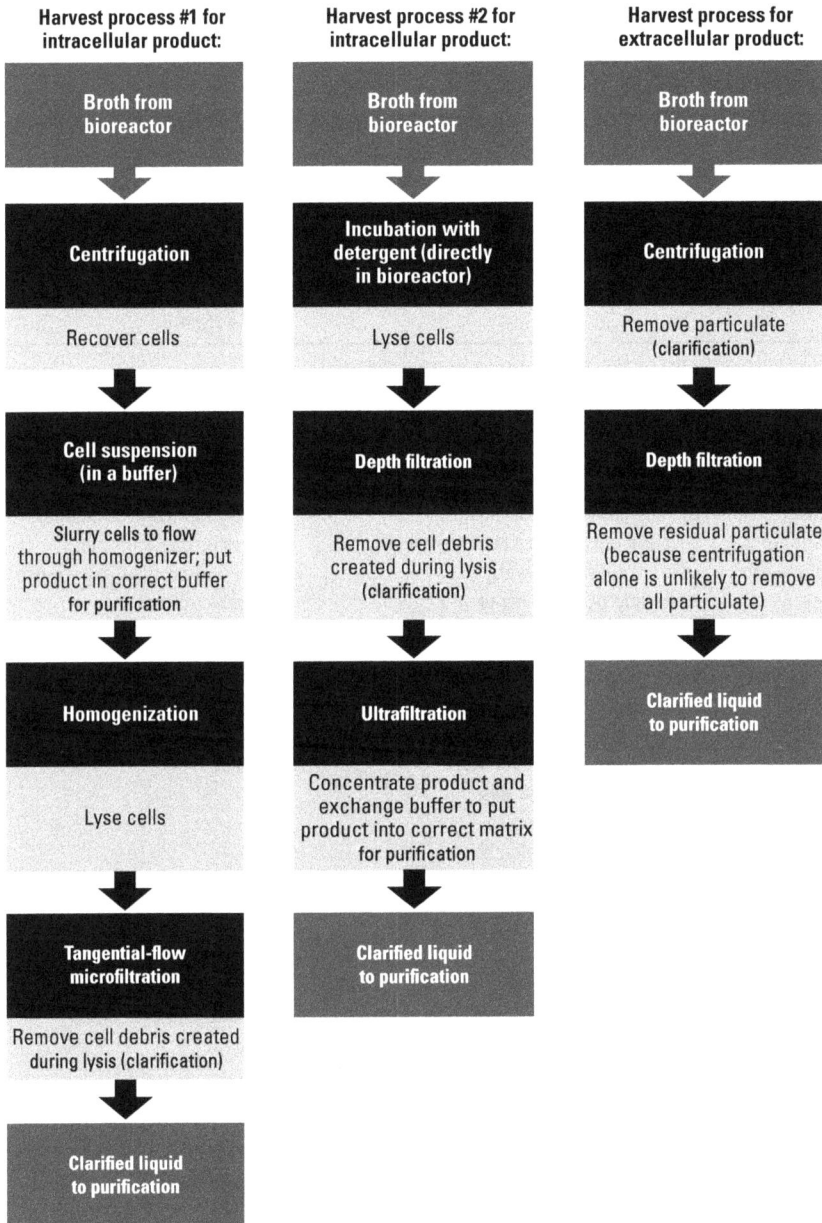

Figure 5.4: Examples of three different harvest processes: two different processes for intracellular products and one for an extracellular product. Note whether or not a UF step is needed in process #2 depends on the requirements of the next step, usually chromatography. It is possible that a UF step may be required after depth filtration in the process for extracellular product as well. Image © NC State University; reprinted with permission.

5.3 Cell lysis methods

There are a number of lysis methods in use to recover intracellular biopharmaceutical products. Table 5.2 provides a summary and classifies them as either nonmechanical or mechanical.

Table 5.2: Common cell lysis methods.

Method	How it works	Scalable?	Common cell types suitable for the method
Nonmechanical (includes both chemical and physical disruption methods)			
Treatment with nonionic surfactant (e.g., Tween™ 20)	Surfactant solubilizes the cell membrane.	✓	Animal cells, such as CHO and HEK293
Treatment with organic solvent (e.g., ethanol, isobutanol, and toluene)	Organic solvent extracts lipids from the cell membrane of animal cells or cell walls, thereby increasing permeability of intracellular components.	✓	Typically not used for biopharma processing due to risk of denaturing proteins in presence of organic solvent
Enzymatic treatment (e.g., lysozyme)	Enzyme digests bacterial or yeast cell walls. For example, lysozyme is a well-known enzyme that digests the peptidoglycan layer of bacteria [149]. To use this method, however, a gram-negative cell would first be treated with a detergent to remove the outer membrane [149].	✓	Gram-negative bacteria, such as *E. coli*, gram-positive bacteria, yeast such as *Saccharomyces cerevisiae*
Alkaline treatment	A lysis solution containing sodium hydroxide and a surfactant, such as sodium dodecyl sulfate, are used to break down the cell wall and cell membrane of bacteria.	✓	Commonly used for isolating plasmid DNA from bacteria
Osmotic shock	Osmolality in the surrounding fluid is lowered to drive water into cells, breaking the cell membrane.	✓	Primarily animal cells; the force is not great enough to lyse tougher bacterial and yeast cell walls
Freeze/thaw	Suspension to be lysed is exposed to multiple freeze-thaw cycles; the resulting ice crystals will disrupt the integrity of cell membranes.	X	Bench-scale method primarily animal cells, but may find use for bacteria

Table 5.2 (continued)

Method	How it works	Scalable?	Common cell types suitable for the method
Mechanical			
High-pressure homogenization	In this book, we use the term *high-pressure homogenization* to refer specifically to forcing a cell slurry at high pressure through the small orifice created by a homogenizing valve resulting in lysis. Additional detail is provided in the text.	✓	Variety of cells, including both eukaryotic and prokaryotic
Lysis using a Microfluidizer®	A different type of high-pressure homogenization in which a cell slurry is forced through microchannels at a high pressure; high fluid pressure is converted to shear force, which results in cell lysis. In this book, we refer to this type of high-pressure homogenizer as a *Microfluidizer®*.	✓	Variety of cells, including both eukaryotic and prokaryotic
Bead milling	Cells are mixed with beads made from glass or stainless steel and agitated vigorously, which promotes an abrasive action that disrupts cells.	✓	Commonly used to disrupt yeast
Sonication	Electronic generator sends waves of high-frequency energy through a metallic tip. Vibration in the cell slurry results in cavitation, which disrupts cells.	X	Not commonly used in biopharma production; may not be suitable for lysis of tough cells like yeast

The choice of lysis method depends on a number of factors, notably the ease with which cells are disrupted so that product can be released (to enable high product recovery), the ability to maintain the integrity of the target molecule (i.e., avoid denaturation), and the scalability of the method. Other factors to consider include the ease of removal of the cell debris produced (generally, the smaller the debris, the more difficult it is to clarify the lysate), processing time, and, of course, cost.

Among the nonmechanical methods shown in Table 5.2 are chemical treatments used to "permeabilize" the membrane of animal cells, which allows for release of intracellular product to the fluid surrounding the cell. Nonionic surfactants such as Tween™ 20 are used for this purpose, as is treating cells with organic solvents like toluene. However, treatment with organic solvent is not commonly used in production of biopharmaceuticals given the risk of denaturing the product.

Table 5.2 also shows that enzymatic treatment can be used to digest cell walls in bacteria, such as *E. coli*, and yeast, such as *Saccharomyces cerevisiae*, which leads to cell rupture. For example, the enzyme lysozyme breaks the cell wall of gram-negative bacteria by catalyzing the splitting of the polysaccharide chains in the peptidoglycan layer [149]. However, in gram-negative bacteria, access to the peptidoglycan is limited by the outer membrane, which may inhibit effectiveness of the enzymatic treatment [149]. Therefore, it may be necessary to expose the cells to a surfactant to solubilize the outer cell membrane. Generally speaking, a disadvantage to the use of enzymatic treatment for lysis of bacteria or yeasts is the high cost and possible lack of availability of the required enzymes [150].

Osmotic shock can be used to lyse animal cells by creating an osmotic pressure gradient between the liquid surrounding the cell and the interior of the cell. Osmotic pressure is the hydrostatic pressure produced by a difference in concentration of solutes between a semipermeable membrane. When the extracellular solute concentration becomes small, the osmotic gradient causes water to flow into the cell, which eventually leads to rupture. The last nonmechanical method listed in Table 5.2 is the use of freeze-thaw cycles, particularly for animal cells. Ice crystals that form during the process disrupt the integrity of cell membranes. While commonly used in a bench-scale lab environment, this method is not typically used in a manufacturing environment due to lack of scalability.

Numerous mechanical methods are also available for cell lysis. The four listed in Table 5.2 are among the most common. High-pressure homogenization is discussed separately in the next section, as homogenizers are the most common devices for cell disruption in biopharmaceutical production [151]. We use this term to specifically refer to forcing a cell slurry at a high pressure through a small orifice created by a valve.

A Microfluidizer®, a different type of high-pressure homogenizer, pumps the cell slurry to an interaction chamber at high pressures – up to 40,000 psi. The proprietary interaction chamber contains microchannels in which pressure is converted to kinetic energy, and the resulting shear and impact forces produce cell lysis. Bead mills consist of a cylinder (grinding chamber) typically filled with glass or stainless steel beads. The beads and cells are agitated using agitator disks on a shaft inside a grinding chamber. Cells are lysed due to grinding between and collisions with beads. They are often used for tough-to-lyse cells like yeast. Ultrasonic disruption (i.e., sonication) relies on high-frequency sound that is transported through a metallic tip to a cell suspension. The sound creates cavitation – the formation, growth, and collapse of vapor-filled bubbles, which leads to shearing and ultimately cell lysis. While effective, this method is often limited to lab-scale applications.

The mechanical and nonmechanical cell lysis techniques have relative advantages and disadvantages, which are summarized below.

– Mechanical methods are generally effective at lysing tough microbial cells; however, they tend to break cells completely, which means that all intracellular materials are released. This results in relatively high amounts of impurities going to the purification stage. The nonmechanical methods may produce gentler lysis and therefore less release of unwanted intracellular impurities.

– Mechanical methods tend to create smaller particles than nonmechanical, which can make subsequent clarification of the lysate by centrifugation and/or filtration difficult.

– Mechanical methods generate more heat than nonmechanical methods, which can adversely impact the activity of the product. At manufacturing scale, design of the lysis step must include ways to minimize the temperature increase. Equipment for most mechanical methods is equipped with a heat exchanger to cool down product post lysis. In addition, it is good practice to cool the process feed to these units (e.g., to 2–8 °C) so that even with heat generation, the lysate temperature remains low enough to avoid denaturation of the biopharmaceutical product. For many proteins, thermal denaturation begins at 40 °C as explained in Chapter 1. If the feed is not chilled, temperatures approaching this level are possible in homogenizer lysate.

– Chemical (nonmechanical) methods require chemical additives that must be removed in subsequent downstream processing steps. Mechanical methods require no additives; therefore, there is no concern over the removal of additional process-related impurities.

5.4 High-pressure homogenizers

Because high-pressure homogenization is commonly used for cell disruption in processing biopharmaceuticals, greater detail is provided in this section. Lysis in a homogenizer relies on pumping a cell slurry through a restricted orifice valve. An illustration of the flow of the slurry through a GEA Niro Soavi Type NS3006H unit is shown in the flow diagram in Figure 5.5(a). A photo of the unit is shown in Figure 5.5(b).

Cells, suspended in the spent medium from the bioreactor step or in a buffer, are fed through the product supply line to the piston pump. This flow path is highlighted in green. A rotary lobe pump in the product supply line keeps the piston pump flooded with cell slurry and is referred to as a stuffing pump. The piston pump moves the cell slurry to the homogenizer valve at pressures up to 1,000 bar (≈ 14,500 psi) for this particular unit. The homogenizer valve is discussed in greater detail shortly. Once cells are lysed, the lysate stream flows to a heat exchanger, in place to cool the lysate as a result of the heat generated during lysis. A temperature increase of approximately 2.4 °C per 100 bar of operating pressure can be expected [152]. Once the lysate

(a)

(b)

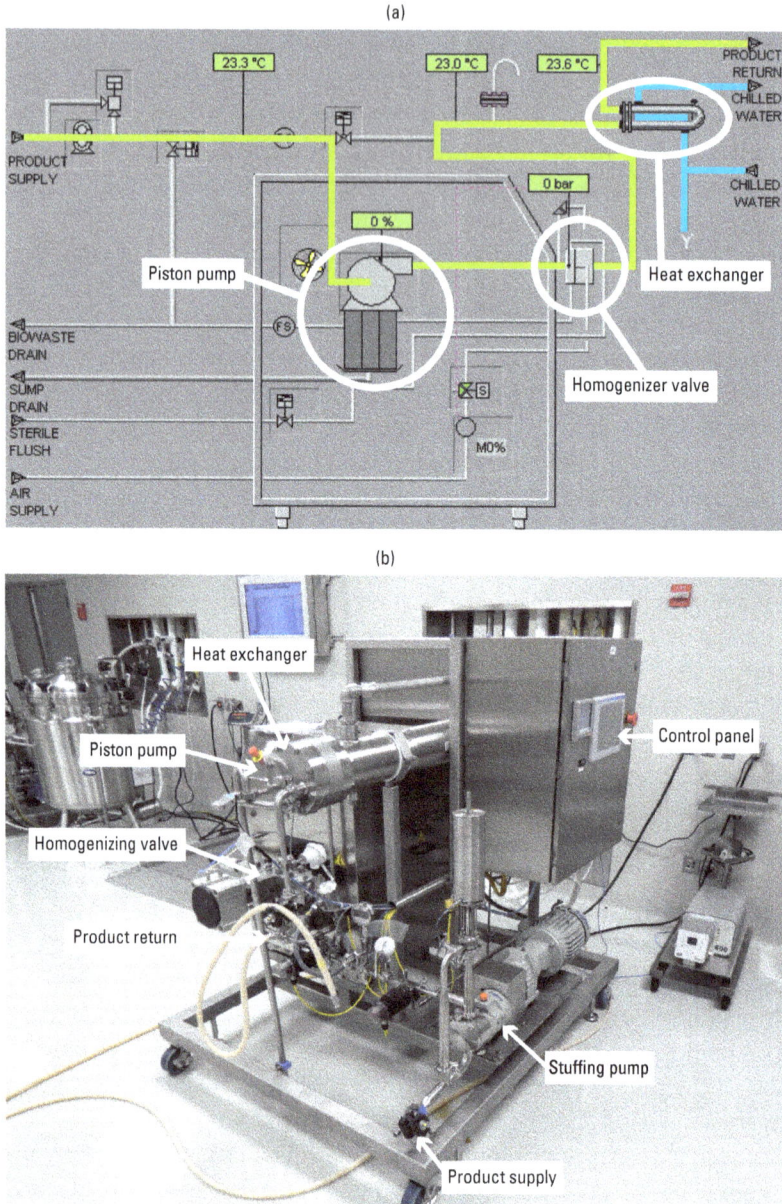

Figure 5.5: (a) Flow diagram for a Niro Soavi Type NS3006H homogenizer. The path of flow for product is highlighted in green. Major components are circled. The blue path shows chilled water entering and exiting the shell-and-tube heat exchanger located at the outlet of the homogenizing valve. Note that variables shown in the green boxes were measured when the system was not running. Image adapted from the NCSU MMI Manual and used with permission, © GEA Mechanical Equipment US, Inc.; (b) Photo of a Niro Soavi Type NS3006H high-pressure homogenizer with various components identified. Photo © NC State University; reprinted with permission.

passes through the heat exchanger, it exits the homogenizer through the product return line. The lysate can be sent to the next processing step or recycled to the homogenizer for another pass. The number of times the slurry is fed to the homogenizer is referred to as the number of passes, and it is often more than one for a homogenization step. Operating with multiple discrete passes requires at least two vessels: one that contains the initial feed slurry and a second that collects lysate from the first pass. To execute a second pass, the lysate collection vessel for the first pass becomes the feed vessel for the second and the original feed vessel is used to collect the second-pass lysate. As mentioned previously, it is good practice to cool the feed to the homogenizer (e.g., to 2–8 °C) to ensure that the temperature of the lysate just after the homogenizing valve but before the heat exchanger does not result in denaturation of product. Thus, it is desirable that both process vessels used be jacketed to allow flow of cooling water.

Figure 5.6 shows a photo of the homogenizer valve assembly and an illustration of how the slurry flows through it. The three pieces shown on the left side of the figure are the key components. From right to left they are the impact head (that is, the valve), the impact ring, and the passage head, also referred to as the valve seat. The image on the right shows how the three components come together to create the homogenizing valve assembly and the flow through it. The slurry passes through the passage head (seat) and flows radially through the opening, which is only a few millimeters. The slurry "hits" the impact ring, then exits the valve assembly as lysate. On one side of the valve, the pressure is thousands of pounds per square inch as mentioned previously, and it is controlled by adjusting the distance between the valve and the seat. Once the cells pass through the small orifice, the pressure drops to approximately atmospheric.

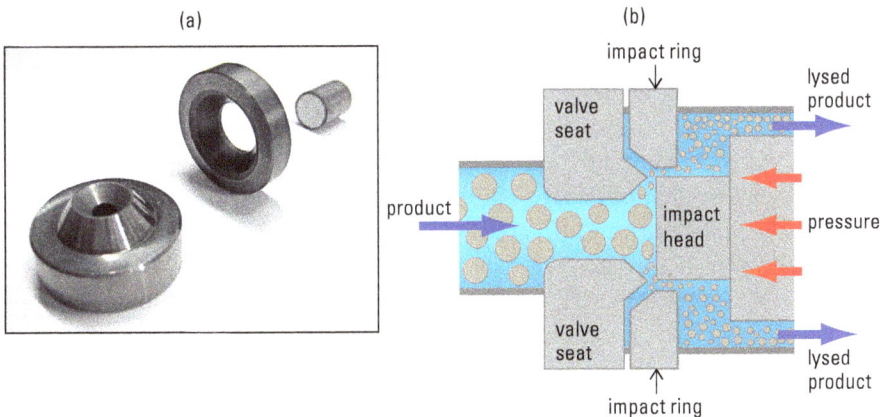

Figure 5.6: The homogenizing valve assembly. (a) The three components that make up the valve assembly, from right to left: impact head (the valve), impact ring, and the passage head or seat. (b) The path of fluid through the homogenizing valve assembly. Note that the valve design shown in this figure is referred to as a sharp-edge or knife-edge valve and is common for cell disruption. Other valve designs exist. Photo and image © NC State University; reprinted with permission.

Several mechanisms likely contribute to cell lysis in a high-pressure homogenizer, including: (1) shear forces in the liquid, (2) impingement of solids against the impact ring, (3) turbulence and turbulent eddies, (4) cavitation forces, and (5) the pressure drop across the valve [150, 152].

5.5 Homogenizer process and performance parameters

Let's start our discussion of parameters that impact homogenizer performance by considering a model presented by Hetherington et al. [153] for the homogenization of yeast. It assumes that the lysis process follows first-order kinetics – meaning that the rate of product released from a cell is proportional to the amount of releasable product. From this assumption, they went on to derive the following expression, which fit their data well:

$$\log_{10} \frac{R_{\mathrm{mp}}}{\left(R_{\mathrm{mp}} - R_p\right)} = K N_{\mathrm{passes}} P^{\alpha} \tag{5.1}$$

where R_p is the product released per unit cell mass, R_{mp} is the maximum amount of product that can be released per unit cell mass, K is the empirical constant that is a function of temperature, N_{passes} is the number of times suspension passes through the homogenizer, P is the pressure, and α is the empirically determined constant.

We will use equation (5.1) as a guide to understanding the parameters that impact lysis in a high-pressure homogenizer. From equation (5.1), pressure and number of passes impact the amount of product released (a performance parameter) during a homogenizer run; specifically as the pressure and/or number of passes increase, the amount of protein released, R_p, increases. These two process parameters are directly controllable and would require a range (for pressure) or target (for passes) to be specified in a batch production record for a CGMP homogenization step.

Data showing the impact of pressure and passes on the homogenization of E. coli genetically modified to produce green fluorescent protein (GFP), in this case used as a surrogate for a protein therapeutic, are shown in Figure 5.7. The data were generated from a bench-top high-pressure homogenizer, in which five passes were run at 300 bar (=4,350 psi) and at 900 bar (=13,050 psi). Lysate from each pass was centrifuged, and the concentration of GFP in the resulting supernatant, reported in units of mg GFP/mL of solution, was measured. As equation (5.1) would predict, a higher number of passes and higher pressure lead to more released product. At 900 bar, the concentration of GFP in the liquid levels off at approximately 2.66 mg/mL, which represents the maximum amount of GFP to be extracted from the E. coli cells used in this study. Note that data for the E. coli/GFP system (not shown here) demonstrate that pressures above 900 bar do not result in release of additional product.

In Figure 5.7(b), the data shown in Figure 5.7(a) are plotted in a slightly different way, as the extent of lysis for each pass. The extent of lysis is a performance parameter that measures the percent of the maximum concentration of product that has been released at any pass and is calculated by the following equation:

$$\text{Extent of lysis } (\%) = \frac{C - C_o}{C_{max} - C_o} \times 100 \tag{5.2}$$

where C is the concentration of the target component (product), C_o is the starting concentration of the target, and C_{max} is the maximum concentration of target after infinite time (assuming no denaturation of this compound).

The data in Figure 5.7 can be used to set a range on homogenizer pressure and target value for the number of passes. Specifically, three passes at 900 bar release most product from the cells; therefore, for CGMP manufacturing, batch record instructions would specify using three passes at, say 850–950 bar with a target or set point of 900 bar. No range on passes (e.g., 2–4 passes) is needed as executing three discrete passes is straightforward. Most production homogenizers are equipped with automated pressure control, so the range around the 900 bar set point depends on the ability of the homogenizer to hold the set point. Note that operating with more than three passes not only offers no benefit to product recovery but would have undesirable consequences such as increasing the process time and decreasing the size of the cell debris, making subsequent clarification steps more challenging.

Using the data shown in Figure 5.7 alone may be insufficient to set a range on operating pressure and a target on the number of passes. Additional data showing the impact of passes and pressure on the activity of the biopharmaceutical – that is, measuring activity as a performance parameter in addition to product concentration in the liquid – would more completely characterize the homogenization step and allow for a choice of process parameter values that ensures no negative impact to product quality.

In addition to pressure and the number of passes, other parameters affect homogenizer performance. These are summarized below:

– Feed temperature. The empirical parameter K in equation (5.1) increases with temperature. Consequently, the extent of lysis can generally be improved by increasing the temperature of the slurry fed to the homogenizer [152]; however, doing so poses a risk to the product given that increasing the temperature of the homogenizer feed would result in an even higher temperature after the slurry has passed through the homogenizing valve, which could lead to product denaturation.

– Cell type. The parameters a and K, and therefore the extent of lysis, depend on the type of cell. As discussed previously, yeast, with their thick cell wall, are more difficult to lyse than gram-negative bacteria such as E. coli, which has a thinner cell wall. In a biopharmaceutical process, the cell line being used is set as part of process design and is not directly controllable.

(a)

(b)

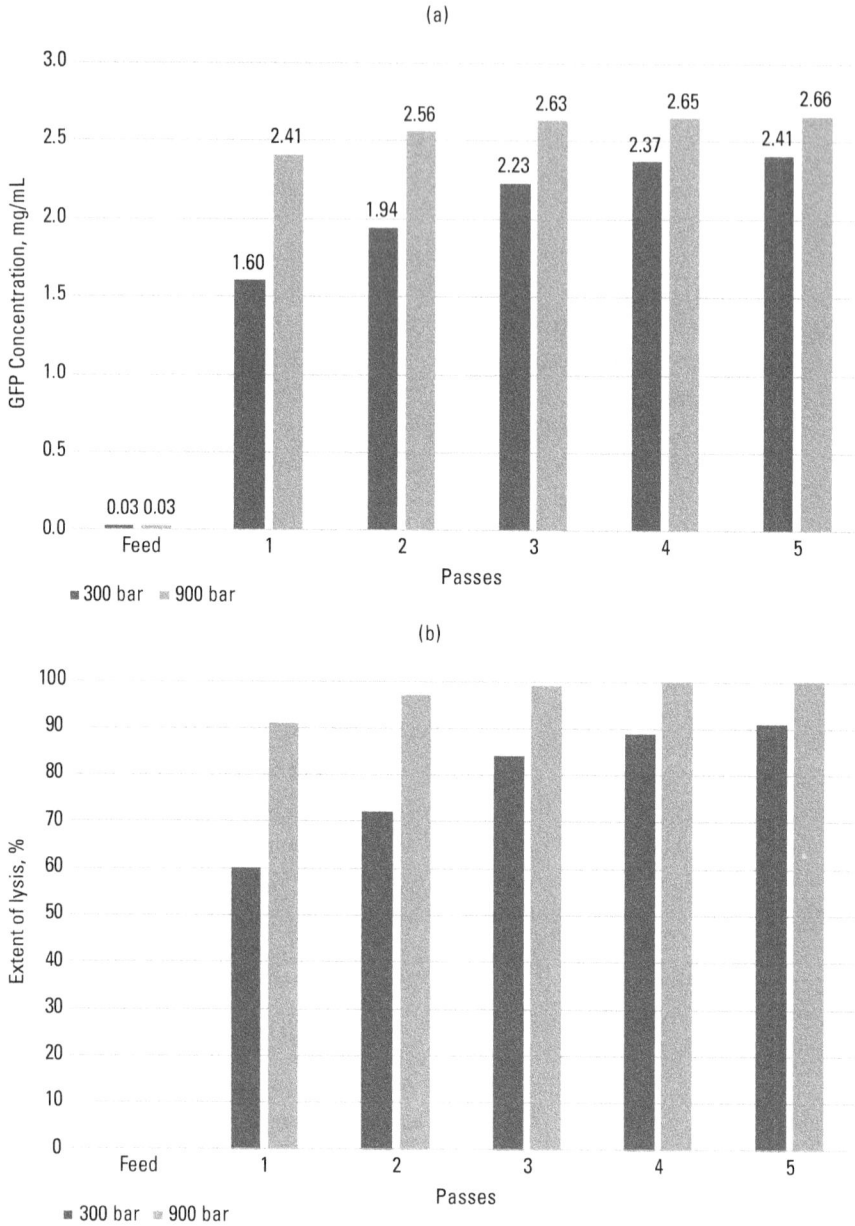

Figure 5.7: Impact of operating pressure and number of passes on release of GFP from *E. coli* using a GEA Niro Soavi Type NS1001 homogenizer. The *E. coli* cell line used is BL21(DE3)pET-17B GFPuv. (a) The concentration of GFP in the liquid phase of the lysates generated. (b) Data from (a) is converted to the extent of lysis calculated by equation (5.2). The calculation was performed assuming that the average of the GFP concentration values from passes three through five at 900 bar represent the maximum concentration of product that could be obtained (C_{max}). Images © NC State University; reprinted with permission.

- Numerous others. The growth phase of the cell at the time of bioreactor harvest, the growth medium composition, and the design of the homogenizer and the homogenizing valve all impact the extent of lysis in a single pass [152].

In the case of E. coli broth, disruption efficiency has been found to decrease with increasing cell concentration, but not to the extent that diluting the broth before homogenization is necessary [154].

The performance of a homogenizer can be quantified by either indirect or direct methods. We've already mentioned the amount of product released and the extent of lysis as performance parameters. Measuring either would be an indirect method for measuring homogenizer performance. In contrast, counting the number of cells destroyed would be considered a direct method. Quantifying the intracellular product released is the method used to generate the data shown in Figure 5.7. Although classified as an indirect method, it offers the advantage of providing a direct answer to the question, "How much biopharmaceutical product can be recovered from the cells?" – a question that needs to be answered for the sake of process design.

The time required to complete a homogenization step, $t_{process}$, can be calculated by the ratio of the total volume fed to the unit, which must account for the possibility of more than a single pass, to the volumetric flow rate or

$$t_{process} = \frac{V_{feed}N_{passes}}{Q} \tag{5.3}$$

where V_{feed} is the volume (e.g., units of L) of slurry fed in a single homogenizer pass, N_{passes} is the number of passes, and Q is the volumetric flow rate (e.g., units of L/h) through the homogenizer. V_{feed} is determined by process requirements. The number of passes needed is determined through a study, like the one previously described, to determine the number of passes necessary for maximum product recovery. Q depends on the operating pressure and size (e.g., model) of high-pressure homogenizer that you are working with. The supplier should be able to provide data showing the dependence of flow rate on operating pressure for a given homogenizer model [155].

Example: Calculating the homogenizer flow rate required to achieve a given processing time
You are developing a high-pressure homogenization step to process 300 L of suspended E. coli cells. Development studies show that two passes at 900 bar are required for acceptable lysis. The homogenizer supplier wants to know what flow rate is required, so they can recommend a model. What is your response?

Solution

Start by solving equation (5.3) for Q:

$$Q = \frac{V_{feed}N_{passes}}{t_{process}}$$

V_{feed} and N_{passes} are known, but a suitable process time must be chosen. Three hours seems reasonable given that you want to complete the homogenization step in a single 8-h shift and allow enough time for setup and cleanup. Therefore

$$Q = \frac{300 \frac{L}{pass} \times 2 \text{ passes}}{3 \text{ h}}$$

$$= 200 \text{ L/h}$$

5.6 Procedure for a high-pressure homogenization step

The procedure that we use for a (simulated) CGMP high-pressure homogenization step using the equipment in Figure 5.5 is depicted in Figure 5.8. After the homogenizer is set up, slurried cells are fed to the unit. Multiple passes may be required to completely lyse cells as described previously. The high-pressure homogenizer is then flushed with a buffer, and recovered buffer is added back to the final lysate to capture any residual product remaining in the hold-up volume of the unit. The system is then immediately cleaned and sanitized using a validated method. Vendors can make recommendations on appropriate cleaning solutions to use for their specific equipment – 0.5 M NaOH is common. Recall that an open downstream process such as that being described is expected to be under bioburden control (rather than sterile), so the high-pressure homogenizer is typically sanitized rather than sterilized between runs. The homogenizer is then rinsed with high purity water to remove residual cleaning/sanitizing agents and stored dry.

5.7 Summary

For biopharmaceutical processes involving intracellular products, cell lysis is essential to release the product from within the cell into the surrounding liquid phase. Because a variety of cells are used to produce biopharmaceutical products, there is variation in the outer layers that must be disrupted to extract the target molecule from the cell. For example, animal cells are enveloped with a membrane composed of a lipid bilayer but lack a cell wall. Prokaryotic cells such as bacteria, other eukaryotic cells like fungi, including yeast such as *Saccharomyces cerevisiae*, and plant cells all have both a cell membrane and a thicker cell wall, which make them more difficult to lyse by mechanical means than animal cells. The composition of these outer layers

```
┌─────────────────────────────┐
│   Suspend cell paste in     │
│     appropriate buffer      │
└─────────────────────────────┘
              ▼
┌─────────────────────────────┐
│      Set up homogenizer     │
└─────────────────────────────┘
              ▼
┌─────────────────────────────┐
│ Feed cell slurry to homogenizer │
└─────────────────────────────┘
              ▼
┌─────────────────────────────┐
│ Feed lysate from previous pass if │
│    multiple passes needed   │
└─────────────────────────────┘
              ▼
┌─────────────────────────────┐
│   Perform buffer flush for  │
│      product recovery       │
└─────────────────────────────┘
              ▼
┌─────────────────────────────┐
│  Clean/sanitize system with │
│         0.5M NaOH           │
└─────────────────────────────┘
              ▼
┌─────────────────────────────┐
│      Flush system with      │
│   high purity water (HPW)   │
└─────────────────────────────┘
              ▼
┌─────────────────────────────┐
│    Store homogenizer dry    │
└─────────────────────────────┘
```

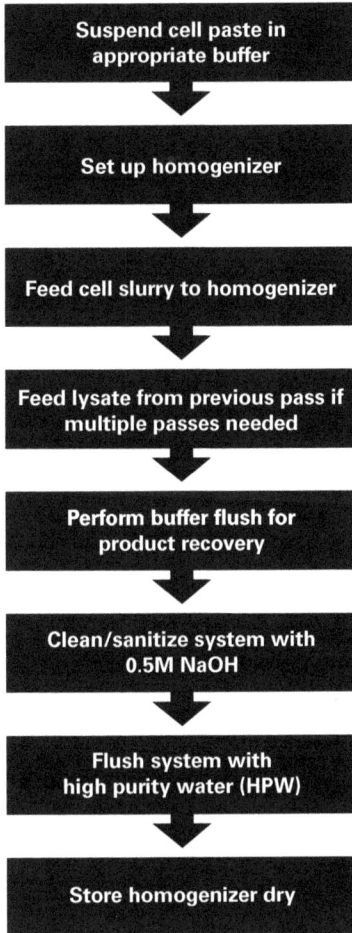

Figure 5.8: A typical procedure for recovery of intracellular biopharmaceutical products using a high-pressure homogenizer. Note that these steps are based on the first process design option presented in Figure 5.4. Image © NC State University; reprinted with permission.

surrounding the cell impacts choice of lysis method, and there are a variety to choose from. The choice of method depends on a number of factors, notably the ease with which cells are disrupted so that product can be released and the ability to maintain the integrity of the target molecule (i.e., avoid denaturation). Other factors to consider include the ease of removal of the cell debris produced – generally, the smaller the debris, the more difficult it is to clarify the lysate – level of soluble impurities released, processing time, and, of course, cost. Examples of nonmechanical methods that are scalable and that can be used in biopharmaceutical applications include:

– treatment with nonionic surfactant to solubilize cell membranes, which can be used to lyse animal cells;
– enzymatic treatment to digest cell walls of bacteria and yeast;
– osmotic shock to drive water into cells thereby breaking the cell membrane.

Examples of scalable mechanical methods that can be used in biopharmaceutical applications include high-pressure homogenizers, which send a slurry at high pressure through a small orifice created by a homogenizing valve, and bead mills that rupture cells by grinding with glass or stainless steel beads. Generally, mechanical methods are effective at lysing tough microbial cells; however, they tend to break cells completely, which means that all intracellular materials are released. They also create smaller particles relative to the nonmechanical methods, which can make subsequent clarification difficult. And they generate heat, which can adversely impact product activity. Nonmechanical chemical methods, on the other hand, require additives that must be removed in subsequent downstream processing steps.

High-pressure homogenization is a common lysis method used in biopharmaceutical production. Development studies should focus on specifying ranges for the operating pressure and number of passes that ensure maximum recovery without denaturing the product or creating cell debris that cannot be removed in subsequent clarification steps. Allowable ranges for each would be included in a batch production record.

5.8 Review questions

1. D.V. Yates, an operator in the large-scale recovery suite, has frequently failed to follow instructions in the batch record for the high-pressure homogenization of *E. coli* used to produce his company's blockbuster protein therapeutic. You are the process engineer assigned to this product and are responsible for writing the product quality impact assessment for these deviation reports. Explain whether or not the following deviation scenarios impact product quality.
   (a) The homogenization pressure is above the range written in the batch record.
   (b) Fewer than the number of passes required by the batch record are used.
   (c) The homogenization pressure is below the range written in the batch record.
   (d) Cooling water was not fed to the jacketed vessels used for the initial homogenizer feed or the lysate collection as directed by the batch record.

2. Lysis of cells using a high-pressure homogenizer usually requires multiple passes to achieve the desired extent of lysis. In addition to implementing discrete passes as described in the chapter, describe two other means to achieve multiple passes. What are the advantages/disadvantages of each?

3. You are developing a high-pressure homogenization step to process 1,000 L of suspended *E. coli* cells. Your development studies show that four passes at 700 bar are required. (a) Among the units listed in the table below, which will you pick for the large-scale process and why? (b) For the model chosen, what will the actual processing time be?

Table 5.3: Flow rates in L/h as a function of pressure for various high-pressure homogenizer models.

Model	Pressure (bar)					
	100	200	300	400	700	1,000
A	10	10	10	10	10	10
B	1,000	700	500	350	120	120
C	7,200	3,500	2,100	1,500	700	550
D	12,000	6,000	4,000	3,000	1,700	1,200
E	22,000	18,000	12,000	8,500	4,200	2,600

4. Using a high-pressure homogenizer to lyse *E. coli*, you use three passes at 300 bar. Estimate the extent of lysis, assuming the first-order kinetic expression in equation (5.1) applies, with constants of $K = 6.5 \times 10^{-4}$ and $\alpha = 1.71$, for a pressure in MPa [152].

5. 400 mL of *E. coli* cell paste, generated by removing the spent medium from a fermentation broth by centrifugation, was suspended in 4,000 mL of tris buffer to conduct range-finding studies for a homogenization step. The resulting cell suspension was homogenized to completion, using five passes at 900 bar, and the concentration of the protein therapeutic in the clarified lysate was 2.5 mg/mL after the fifth pass. The cells came from a 25-L fermentation with a solids concentration of 8% by volume. Assuming the homogenizer run extracted all of the intracellular protein therapeutic, calculate the fermentation titer.

6. Your course instructor gives the following quiz question: Which cell lysis method is likely to lead to a simpler purification "process" – high-pressure homogenization or treatment with a non-ionic surfactant? You circle treatment with a non-ionic surfactant. Your instructor says you are wrong, that high-pressure homogenization results in simpler purification because there are no chemicals added that must be subsequently removed. You still think your answer is correct. Why?

7. You are working on the design of an animal cell-based process for viral vector manufacture for a gene therapy product, and your company has set a goal to use only single-use equipment to minimize capital expenditures. How would you design your lysis step to meet this goal? In answering this question, address the following: (a) what method/equipment would you use for the lysis step, (b) where in the process would you implement the lysis step, and (c) what process and performance parameters are important to consider in process development studies?

Chapter 6
Harvest operations, part 2: solid-liquid separation by centrifugation

In this chapter we turn our attention to solid-liquid separation, which is required throughout a biopharmaceutical process and, in particular, in the harvest stage. As we have discussed previously, regardless of whether the biopharmaceutical product is intracellular or extracellular, the product intermediate from the harvest stage needs to be clarified – that is, made free of solids. More specifically, that intermediate typically comprises the active biopharmaceutical dissolved in a clarified liquid, such as spent media or a buffer solution, and is ready to be loaded to a chromatography column in the purification stage. The intermediate must be clarified because the first chromatography step typically involves a packed bed that is susceptible to clogging if solid particles are present in the feed. And solids from bioreactor (i.e., the host cells) are not part of the final drug product, unless, of course, the product is a cell therapy. In addition to this application, solid-liquid separation steps find other important uses in biopharmaceutical processes, and these are discussed shortly.

In this book, we focus on those techniques that are well suited to separating a relatively large amount of solid from a slurry as is required in the harvest stage of a biopharmaceutical process. This chapter is the first of two that considers these unit operations. In it, we focus on one of the most common methods – centrifugation – while the next chapter covers two common filtration methods – depth filtration and tangential-flow microfiltration. This chapter opens with a brief general discussion on solid-liquid separation in biopharmaceutical processes before turning its attention to centrifugation. Questions addressed include the following:

– What are examples of solid-liquid separations performed in biopharmaceutical processes? And, generally speaking, what unit operations are used to carry out these steps?
– What are the principles underlying centrifugation for separation of solids from liquids?
– How does a disc-stack centrifuge compare to a gravity settling tank for continuous liquid-solid separation?
– In addition to disc-stack centrifuges, what other types of centrifuges are used in biopharmaceutical processes?
– What are the process and performance parameters for disc-stack centrifugation steps?
– What is sigma analysis and how is it used in centrifuge scale-up?

https://doi.org/10.1515/9783111112459-006

6.1 Solid-liquid separations in biopharmaceutical processes

Previously Figure 2.3 showed a process flow diagram depicting monoclonal antibody (mAb) production using Chinese hamster ovary (CHO) cells that secrete product to the cell culture medium; that is, the product is extracellular. From that diagram, it is clear that steps for solid-liquid separation are required for:

1. Removing cells and cell debris from the cell culture broth, leaving a product that is dissolved in the spent clarified liquid medium, and that this separation can be carried out by centrifugation and depth filtration.

There are also a number of other uses for solid-liquid separation steps in biopharmaceutical processes, including:

2. Recovering cells from the bioreactor for further processing of intracellular product as depicted in Figure 5.4.
3. Clarifying the product following a cell lysis step, also shown in Figure 5.4.
4. Recovering inclusion bodies for processes in which the host system (e.g., *E. coli*) produces the product (a protein) in this form. Inclusion bodies were discussed in Chapter 5.
5. Recovering precipitated product or removing precipitated impurities in processes that use precipitation steps for purification. Note that the term *precipitation* refers to a process step in which a dissolved target component – either the product or impurities – is separated from other components also in solution by changing a property, such as temperature or pH, or adding a precipitant that decreases the solubility of the target and causes it to fall out of solution as a solid.
6. Removing microorganisms to produce a bioburden-free or sterile process stream.

Table 6.1 summarizes these scenarios and lists techniques that can be used to perform the separation.

Table 6.1: Solid-liquid separation examples common to biopharmaceutical processes and the unit operations used to perform the separation.

Solid–liquid separation scenario (refer to the text above)	Commonly used techniques
1	Centrifugation, depth filtration, tangential-flow microfiltration, flocculation followed by centrifugation or depth filtration, and expanded bed chromatography
2	Centrifugation and tangential-flow microfiltration
3	Tangential-flow microfiltration, depth filtration, flocculation followed by centrifugation or depth filtration, and expanded bed chromatography

Table 6.1 (continued)

Solid–liquid separation scenario (refer to the text above)	Commonly used techniques
4	Centrifugation and tangential-flow microfiltration
5	Centrifugation, depth filtration, and tangential-flow microfiltration
6	Surface (membrane) filtration

In the next chapter, we discuss the choice of a solid-liquid separation method in greater detail. Generally speaking, however, the method chosen and designed should (1) provide effective separation of the solid from the liquid, (2) produce a good product step yield, (3) be scalable, (4) be rapid, and (5) be cost-effective. Centrifugation typically meets these criteria and is the topic that we now turn to.

6.2 Centrifugation principles

Centrifugation is a unit operation that uses centrifugal force to separate components based on differences in density and size. In biopharmaceutical manufacturing, centrifugation is typically used to separate denser solids from less dense surrounding liquid in which they are suspended. We begin the discussion by considering centrifugal force. It is an apparent force that acts on an object moving in a circular (or curved) path and creates movement outward from the center of rotation. There are numerous examples of centrifugal force in our everyday lives. It is the reason that clothes in a washing machine get pushed to the wall of the machine drum during the spin cycle. The greater the speed of rotation and distance from the center of the motion, the greater the apparent force.

Centrifuges separate liquid from solids by rotating – that is, generating centrifugal force – which causes particles dispersed in a liquid to move away from the axis of rotation of the centrifuge as long as the density of the particles is greater than the density of the liquid in which they are suspended. Centrifuges have been used for many years. A relatively simple bench-top model is shown in Figure 6.1(a). Samples are put into small tubes, which are placed in the rotor. The rotor in the centrifuge in Figure 6.1(a) spins at a maximum rate of 14,000 revolutions per minute (rpm), and solids quickly settle to the bottom of the tubes as illustrated in Figure 6.1(b). To understand the magnitude of the centrifugal force that can be generated, it is useful to compare centrifugal force to gravitational force by defining a dimensionless acceleration, referred to as the relative centrifugal force (RCF), as follows:

$$\text{RCF} \equiv \frac{\omega^2 R}{g} \tag{6.1}$$

where ω is the angular velocity (in units of rad/s, for example), R is the distance from center of rotation, the product $\omega^2 R$ is the centrifugal acceleration, and g is the acceleration due to gravity (= 9.8 m/s^2). RCF provides a measure of the outward force generated by the rotation of the centrifuge relative to the force of gravity. In the Eppendorf centrifuge shown in Figure 6.1(a), the unit spinning at 14,000 rpm generates an RCF of 16,888 (calculation shown in Figure 6.1(c)), often written as 16,888 × g; thus, the force acting on the particles is 16,888 times the force of gravity, which results in much more rapid settling of particles in the fluid than what gravity could produce. This is an important point because fast particle settling leads to fast separation, which leads to short processing times as we will discuss shortly. (Note that the terms *settling* and *sedimentation* are used interchangeably in this book.)

(a) (b)

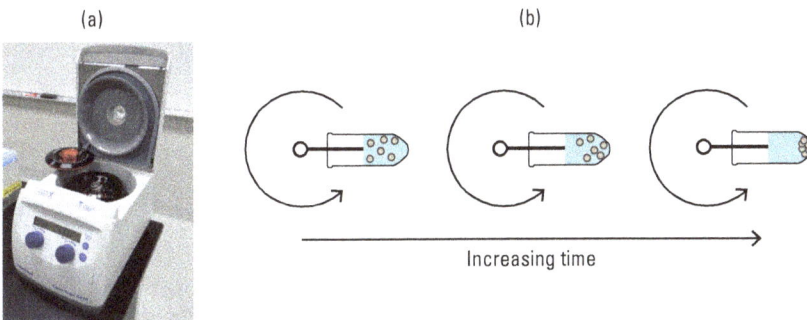

Increasing time

(c) RCF for the Eppendorf Centrifuge 5418 shown above:

RCF $\equiv \omega^2 \times R/g$

ω = 14,000 rev/min × 1 min/60 s × 2π radians/rev = 1,466/s

R = 7.7 cm (radius from center of axis of rotation to bottom of centrifuge tube)

RCF = $(1,466/s)^2$ × 7.7 cm/(980 cm/s^2) = 16,888 (often expressed as 16,888 × g)

Figure 6.1: A bench-top centrifuge. (a) A photo of the Eppendorf 5418, an example of a bench-top centrifuge. (b) The movement of solids in a spinning centrifuge tube with respect to time. (c) The calculation of RCF for the Eppendorf 5418. Note that samples in the centrifuge rotor are at an angle of 45° from vertical. Images © NC State University; reprinted with permission.

6.3 Solid-liquid separation by gravity

Despite the high RCF values generated by the centrifuge shown in Figure 6.1(a), it would clearly be difficult to scale up this system for use in a large-scale manufacturing process. To better understand how centrifugation is implemented at larger scale, we first consider a settling tank that relies on gravity to separate solids from liquid. Gravity separation of solids may seem like a digression from our main topic – centrifugation – but this discussion sets the foundation for production-scale centrifugation.

An example of an equipment setup for solid-liquid separation that relies on gravity for particle sedimentation is shown in Figure 6.2(a). Once particles settle to the bottom of the vessel, liquid can be pumped out and solids then removed. The time that it takes for the settling process to be completed, $t_{settling}$, is equal to the time required for a particle that starts at the top of the vessel to settle to the bottom and is given by the following equation:

$$t_{settling} = \frac{h}{v_g} \tag{6.2}$$

where h is the height of the liquid in the vessel and v_g is the settling velocity, which is simply the velocity at which the particle falls through the fluid.

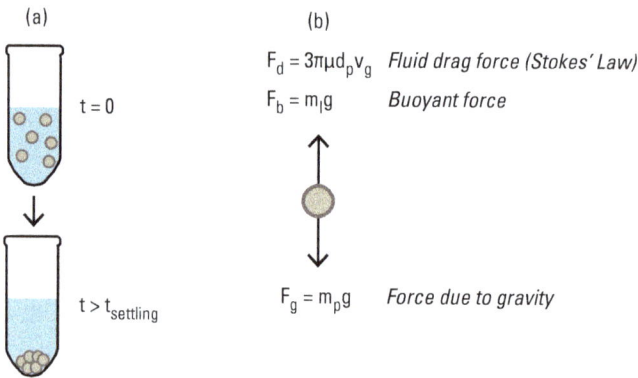

(a)

t = 0

t > $t_{settling}$

(b)

$F_d = 3\pi\mu d_p v_g$ Fluid drag force (Stokes' Law)

$F_b = m_l g$ Buoyant force

$F_g = m_p g$ Force due to gravity

(c)   A force balance on a particle settling under the force of gravity gives:

gravity = buoyancy + drag

$F_g = F_b + F_d$

$m_p g = m_l g + 3\pi\mu d_p v_g$

$\frac{4}{3}\pi d_p^3 \rho_p g = \frac{4}{3}\pi d_p^3 \rho_l g + 3\pi\mu d_p v_g$

Solving for the settling velocity, v_g:

$v_g = \frac{d_p^2(\rho_p - \rho_l)g}{18\mu}$

where:

m_p = particle mass
m_l = mass of liquid displaced
g = acceleration due to gravity
μ = liquid viscosity
d_p = particle diameter
ρ_p = particle density
ρ_l = liquid density

Figure 6.2: Particles settling in a vessel under the force of gravity. (a) Solid-liquid separation by gravity in a batch settling vessel. (b) Forces acting on a single particle settling in a fluid under the force of gravity. (c) A derivation of the settling velocity for a particle from a force balance. Image © NC State University; reprinted with permission.

When a particle moves through a fluid under the force of gravity, it quickly comes to a constant velocity, referred to as a settling velocity, sedimentation velocity, or terminal velocity. At steady state (i.e., constant settling velocity), gravity acting downward on a particle is balanced by the buoyant force and the Stokes' drag force acting upward on the particle, shown in Figure 6.2(b). Drag is the force of friction on a particle that resists its motion through a fluid. Buoyancy is the tendency of an object to float in a fluid. Equating the downward acting force of gravity to the upward acting drag and buoyancy forces results in the following expression for v_g:

$$v_g = \frac{d_p^2 \left(\rho_p - \rho_l \right) g}{18\mu} \tag{6.3}$$

where d_p is the particle diameter, ρ_p is the particle density, ρ_l is the liquid density, g is the acceleration due to gravity, and μ is the viscosity of the liquid through which the particle is settling. The derivation is shown in Figure 6.2(c). From equation (6.3), it is clear that a difference in density between the particle and fluid must exist for particle settling and that the larger the difference, the faster the settling velocity. Equation (6.3) also shows that large particles settle faster than small, and that particle settling becomes slower as fluid viscosity increases. The latter relationship, in particular, should be clear as a highly viscous fluid like motor oil impedes the movement of a particle to a much greater extent than a less viscous fluid like water. Note equation (6.3) also indicates that if the particle density is less than that of the liquid, the density difference is negative and the particle moves opposite the direction that gravity acts. This is what happens when an ice cube settles at the surface of water in a glass rather than sinking to the bottom of the glass.

We expand this analysis by considering gravity sedimentation with the continuous flow unit in Figure 6.3. This analysis, adapted from a publication [156] from Alfa Laval, a leading centrifuge manufacturer, provides insight into the principles underlying production centrifuges, and, in particular, the disc-stack centrifuge that is commonly used in the biopharmaceutical industry [157]. The illustration shows a slurry, such as a fermentation or cell culture broth, being fed to a vessel at a volumetric flow rate, Q. We assume that all particles in the feed are exactly the same size and that the inlet and outlet flow rates are the same so that the liquid volume in the tank does not change with time. The goal is to produce an effluent that is free of particles, which requires that all the solid particles that are introduced at the inlet have ample time to settle to the bottom of the tank before being pushed out of the tank by the liquid effluent. From a processing perspective, it is worth considering the maximum volumetric flow rate, Q_{max}, that will allow for a particle-free outlet stream, noting that an inlet flow rate that is too high will result in a liquid residence time within the unit that is too short to allow for complete sedimentation of particles (i.e., the particles will flow out of the tank with the liquid). Incomplete sedimentation would result in an unclarified effluent.

To derive an expression for Q_{max}, both the particle settling time and the residence time of the fluid in the tank must be considered. The time required for a par-

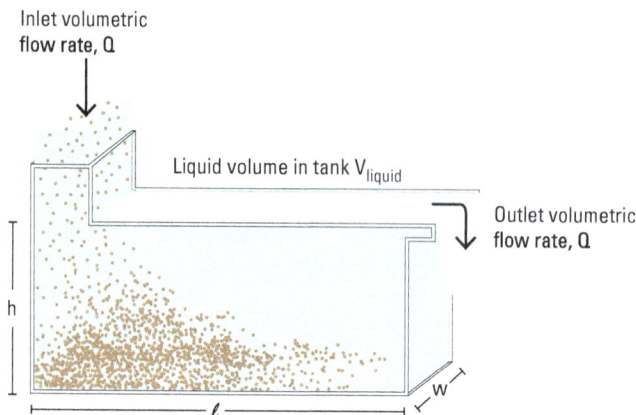

Figure 6.3: Continuous separation of solids from liquid using a gravity settling tank. Image courtesy of Alfa Laval and adapted with permission.

ticle to settle to the bottom of the unit is given by equation (6.2), just as it was for the batch settling vessel shown in Figure 6.2(a). The residence time of the liquid in the tank, $t_{residence,liquid}$, is the liquid volume in the tank, V_{liquid}, divided by the volumetric flow rate, Q:

$$t_{residence, liquid} = \frac{V_{liquid}}{Q} = \frac{lwh}{Q} \tag{6.4}$$

where l, w, and h are the tank length, width, and liquid height, respectively. To produce a particle-free effluent, the fluid must reside in the tank at least as long as is required for particles to settle to the tank bottom (i.e., $t_{residence,liquid} \geq t_{settling}$); otherwise, particles will be swept out of the vessel by the liquid effluent. The maximum inlet volumetric flow rate, Q_{max}, that allows for a particle-free outlet stream corresponds to the scenario in which $t_{residence,liquid}$ time, given by equation (6.4), is equal to $t_{settling}$, given by equation (6.2):

$$\frac{lwh}{Q_{max}} = \frac{h}{v_g}, \tag{6.5}$$

or

$$Q_{max} = v_g lw = v_g A_{settling} \tag{6.6}$$

where $A_{settling}$ (=$l \times w$) is the area for settling. Equation (6.6) shows that Q_{max} depends only on settling velocity and settling area. Interestingly, the height of the liquid does not impact Q_{max}. This results because even though an increase in height would increase the time required for particle settling, the residence time of the liquid for a fixed Q_{max} likewise increases and offsets the longer settling time.

We have established that the maximum volumetric flow rate for a gravity settling tank that produces a particle-free effluent is proportional to the settling velocity of the particle and the area available for settling. However, as mentioned previously, settling time under the force of gravity is relatively long because v_g is small. How small? Table 6.2 shows settling velocity values for *E. coli*, *Saccharomyces cerevisiae*, and CHO cells along with the parameters used to calculate the values according to equation (6.3). As you can see from the v_g values presented, cells settle slowly under the force of gravity; for example, *E. coli* settles only slightly faster than 0.10 cm/h! As a consequence, solid-liquid separation by gravity is, simply put, slow and not often used in biopharmaceutical processes. In general, the time to complete a process step, $t_{process}$, is the ratio of the total volume fed to a process step, V_{feed}, and the volumetric flow rate, Q. Applying this to the settling tank in Figure 6.3 leads to:

$$t_{process} = \frac{V_{feed}}{Q} = \frac{V_{feed}}{v_g A_{settling}} \qquad (6.7)$$

Clearly, a small particle settling velocity results in a long (and undesirable) process time. As an example, complete settling of *E. coli* cells from a 2,000 L fermentation under gravity in a vessel with 100 ft^2 (= 92,903 cm^2) of settling area would require 135 h (= 2,000,000 cm^3/(0.16 cm/h × 92,903 cm^2). So, small settling velocities indeed translate to undesirably long process times.

Table 6.2: Gravitational settling velocities calculated for cell types commonly used in biopharmaceutical processing.

Cell type	Diameter (µm)	Density (g/cm^3) [158]	Settling velocity (cm/h)[a]
E. coli	1 × 3 [136]	1.09	0.16
Saccharomyces cerevisiae	5–10 [134]	1.11	2.16
CHO	15 [133]	1.06	2.64

[a]The liquid density and viscosity are assumed to be that of water – 1 g/mL and 1 cp, respectively. We have also used the largest length dimension shown.

6.4 Production centrifuges

Equation (6.7) suggests that to reduce the processing time required by the unit in Figure 6.3 for a given V_{feed}, the volumetric flow rate Q should be increased. As equation (6.6) suggests, increasing settling area, settling velocity, or both will increase the maximum value for Q that will still allow for a clarified effluent from the tank. An increase in settling area can be accomplished by placing plates or discs into the settling tank. To increase settling velocity, equation (6.3) suggests that increasing the acceleration may be most effective because the other equation variables relate to particle and fluid properties that may not be adjustable in a process (e.g., their values are dependent on the cells used and bioreactor conditions). If we replace gravitational acceleration, g, in equation (6.3) with centrifugal acceleration, $\omega^2 R$, the particle settling velocity can be expressed as follows:

$$v_c = \frac{d_p^2 \left(\rho_p - \rho_l \right) \omega^2 R}{18\mu} = v_g \times \text{RCF} \tag{6.8}$$

We showed previously that RCF values for a centrifuge are in the thousands; thus, $v_c \gg v_g$ and much faster processing times result when separation is performed by centrifugation. This discussion points to the rationale for the design of the commonly used disc-stack centrifuge, which is the focus of much of this section.

6.4.1 Disc-stack centrifugation: an introduction

Imagine modifying the equipment for gravity separation in Figure 6.3 by adding settling plates, moving the feed closer to where the clarified liquid exits, turning the tank counterclockwise by 90 degrees, and rotating the resulting unit to generate centrifugal force. The result would resemble a disc-stack centrifuge, as shown in Figure 6.4. Making these changes would allow for an increased feed flow rate, Q, which would lead to shorter processing times.

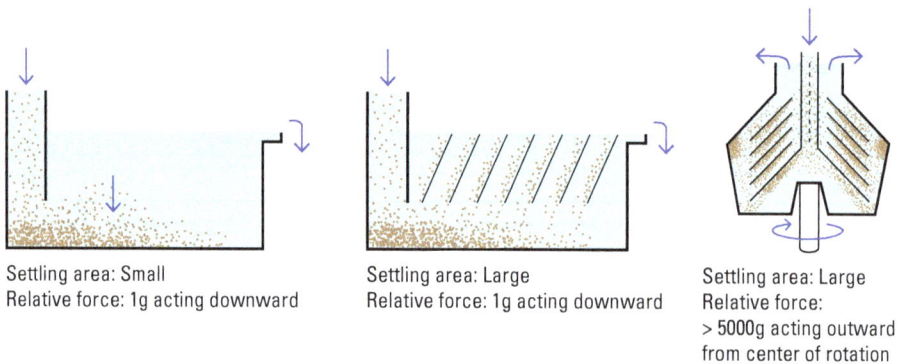

Settling area: Small
Relative force: 1g acting downward

Settling area: Large
Relative force: 1g acting downward

Settling area: Large
Relative force:
> 5000g acting outward
from center of rotation

Figure 6.4: Improvements to a continuous gravity settler lead to a disc-stack centrifuge. The disc-stack centrifuge improves upon the gravity tank by increasing both the settling area and the force acting on the particles (i.e., settling velocity). Image courtesy of Alfa Laval and adapted with permission.

Figure 6.5 shows the bowl and discs (inside the bowl) of a disc-stack centrifuge and flow of solid and liquid through the discs. The spacing between discs is small to create a short distance for particle settling. The feed – a slurry such as fermentation or cell culture broth – enters continuously at the top of the centrifuge through a stationary pipe and moves through the centrifuge bowl as it spins. The feed flow is generated by pressurizing the slurry feed vessel, in which case the flow rate is controlled by a valve in the feed line, or by pumping the slurry into the centrifuge inlet. Feed flows downward then upward in the space between the discs. As the centrifuge rotates, liquid continues to move between the discs and exits continuously as clarified liquid, also referred to as *centrate* or *supernatant*, near the top of the centrifuge. The movement of the liquid carries particles between the discs. But centrifugal force creates an outward force that results in particles settling on the underside of the discs and continuing to move to the bowl wall.

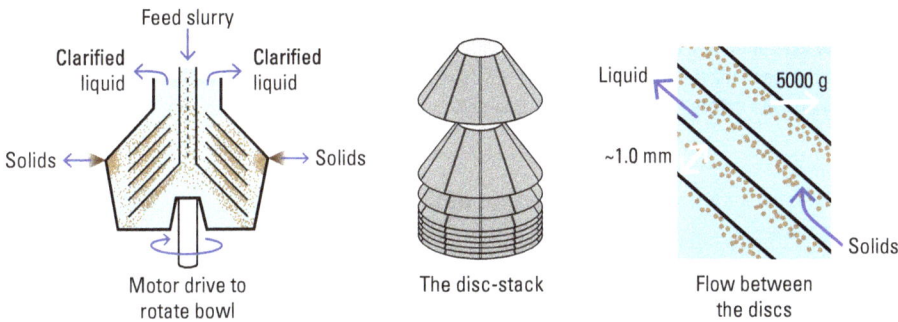

Figure 6.5: Schematic of the bowl (and flow through the bowl) of a disc-stack centrifuge. Image courtesy of Alfa Laval and adapted with permission.

As the run proceeds, the periphery of the bowl gradually fills with solids. The resulting solids "wall" grows radially inward towards the disc stack. These solids must be removed before they block the space between discs, which must remain open for fluid flow. Solids from disc-stack centrifuges can be removed manually, by disassembling the centrifuge and scraping solids out of the bowl, or automatically. Automatic removal may be intermittent or continuous. Intermittent removal is common and involves ejecting the accumulated solids from the centrifuge bowl at regular, predetermined intervals. The ejected solids from either intermittent or continuous removal are referred to as the *solids*, or perhaps more appropriately as the *concentrate* given that some liquid is ejected along with the solids. Note that either intermittent or continuous removal of solids provides for a continuous solid-liquid separation method unlike the bench-top centrifuge shown in Figure 6.1(a), which operates in batch mode. Note also that either the clarified liquid stream or the solids stream is readily collected; therefore, the biopharmaceutical product can be recovered regardless of whether it resides in the clarified liquid or solids stream. Figure 6.6 shows a disc-stack centrifuge at its different stages of operation.

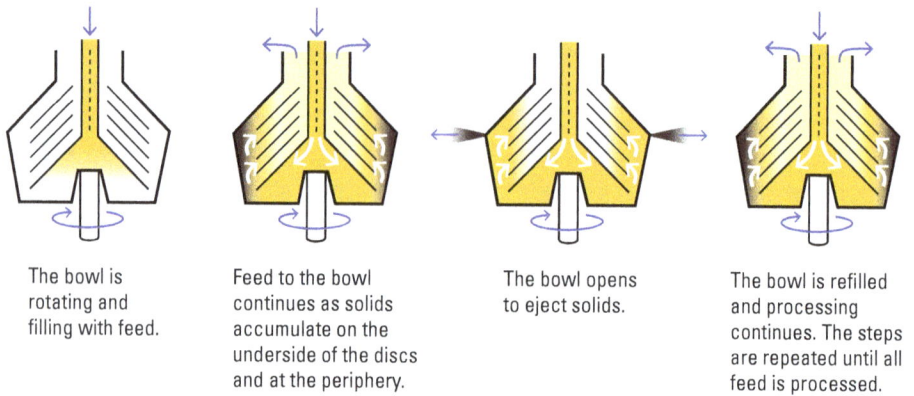

The bowl is rotating and filling with feed.

Feed to the bowl continues as solids accumulate on the underside of the discs and at the periphery.

The bowl opens to eject solids.

The bowl is refilled and processing continues. The steps are repeated until all feed is processed.

Figure 6.6: Operation of a disc-stack centrifuge with intermittent discharge. The bowl shown in these images discharges radially. Intermittent axially discharging bowls also existing in which solids flow downward (axially) before being discharged. Image © NC State University; reprinted with permission.

6.4.2 Disc-stack centrifuges: equipment details

Disc-stack centrifuges are mechanically complex, and covering the details of their design is beyond the scope of this book. However, we do want to highlight some features that will likely be of interest and directly impact processing considerations. The basis of this discussion is the Alfa Laval LAPX 404 SGP centrifuge, shown in Figure 6.7, which serves as an example of the type of disc-stack centrifuge commonly used for industrial applications.

Figure 6.8 shows three highlighted system lines that are central to the execution of a disc-stack centrifugation step: process liquid (i.e., feed) line 201, clarified liquid line 220, and concentrate line 222. A description of these lines, as well as other lines and important components of the centrifuge, follows:

– The centrifuge bowl, which is covered at the top by a stationary "hood." The volume of the bowl for the LAPX 404 centrifuge shown in Figure 6.7 is 2.2 L. Bowl rotation is driven by an electric motor via a belt and bowl spindle. The unit in Figure 6.7 rotates at a maximum speed of 9,600 rpm. The centrifuge is also equipped with a brake that can be used to stop bowl rotation once the power is off.
– Process liquid (feed) line 201. Feed slurry to be separated is fed through line 201. The temperature is measured by temperature transmitter TT201-1, and the feed flow rate by flow indicator transmitter FIT201-4. The flow control valve FCV201-2 is a diaphragm valve that controls the flow rate of the feed contained in a pressurized vessel.
– Clarified liquid line 220. The resulting clarified stream flows through line 220, where turbidity is measured with turbidimeter QIT220-4, pressure with pressure transmitter PT220-1, and temperature with temperature transmitter TT220-1. Pressure control valve PCV220-3 is in place to control back pressure to the centrifuge.

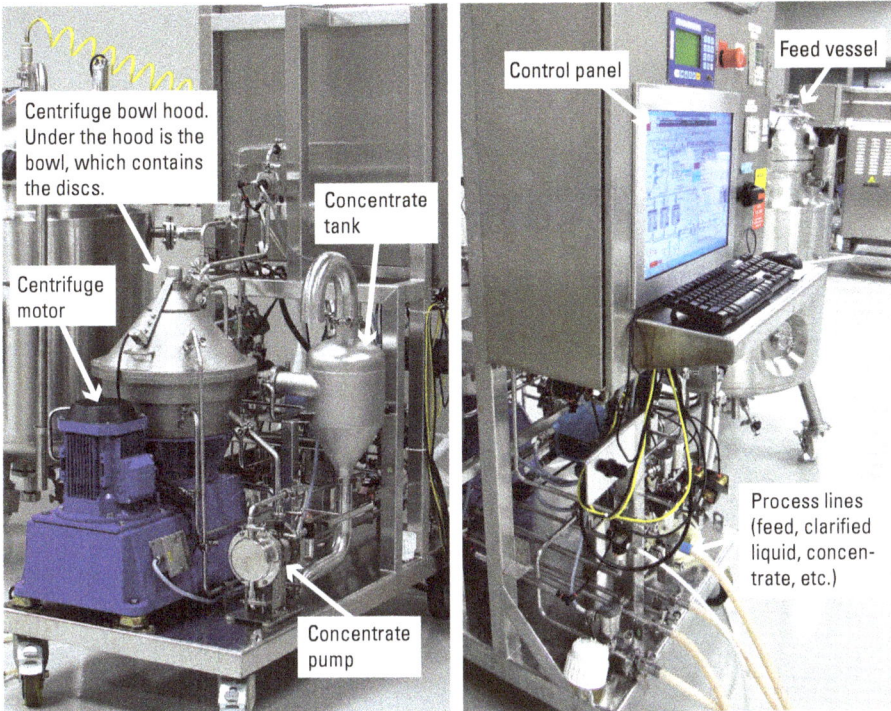

Figure 6.7: An Alfa Laval LAPX 404 SGP/TGP centrifuge. Image © NC State University; reprinted with permission.

- Concentrate (solids) line 222. Solids are discharged through ports in the bowl wall. For the centrifuge in Figure 6.7, these ports are covered between discharges by a sliding bowl bottom. Opening and closing of the bowl is controlled hydraulically, with line 370 feeding the necessary water. "Opening" water fed through the 372 branch of line 370 lowers the sliding bowl bottom, thereby exposing ports at the bowl wall. "Closing" water is supplied via the 376 branch of line 370 to the built-in operating water tank and pumped to the operating system through the bowl spindle. The closing water pushes the sliding bottom up, thereby closing the ports at the bowl wall. The ports on the bowl wall remain open for 1–2 s. The concentrate line 222 includes a small vessel for collecting solids as they make their way to an air-operated diaphragm pump (P222-1) that moves the solids through the concentrate line either for collection or disposal to waste.
- Flushing liquid line 301. This line allows cleaning agent to be sent to the bowl exterior and to the solids collection vessel, as shown in Figure 6.8. More detail on cleaning a disc-stack centrifuge is provided later in this chapter.
- Buffer liquid line 301a. This line can be used to feed a buffer (or other liquid) to the bowl if needed. This function is particularly useful when product is in the liquid

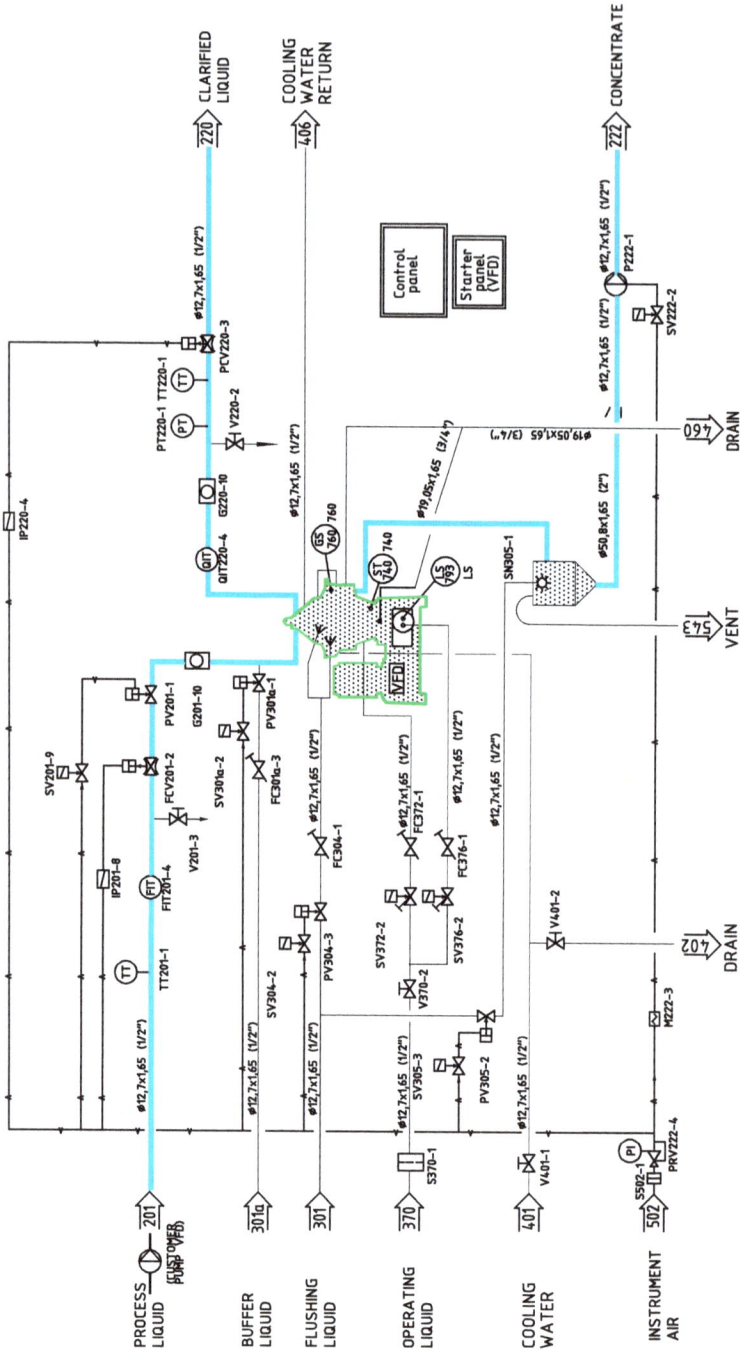

Figure 6.8: A piping and instrumentation diagram (P&ID) for the Alfa Laval LAPX 404 unit. Image courtesy of Alfa Laval and adapted with permission.

phase, as is the case with monoclonal antibodies produced using CHO cells that secrete the antibody. During bowl discharge, liquid, in addition to solid, may be ejected resulting in a yield loss for extracellular products since the discharge stream flowing through concentrate line 222 is typically directed to waste. To avoid product loss, the bowl can be flushed with a buffer through line 301a just before solids discharged – referred to as a predischarge flush – to push product-containing liquid within the bowl out of the system through line 220, thus enabling product recovery. Note that the bowl continues to spin during this step, so solids remain within the bowl while buffer flushes out held-up liquid. This step should be designed to minimize any disruption to solids in the bowl.

– Cooling water line 401 is used to feed water into the hood surrounding the bowl to reduce heating of the product as the centrifuge bowl rotates.

6.4.3 Other types of production centrifuges

While the disc-stack is the most commonly used type of centrifuge for biopharmaceutical applications such as cell recovery and cell removal, other types are available. For example, in biopharmaceutical processing, a tubular-bowl centrifuge is also common. Decanter and multi-chamber bowl centrifuges are also commonly used industrially, although less commonly for biopharmaceutical applications. A number of resources provide detailed coverage of the different types of centrifuges [159–161]. Here, we present only a summary of these and relevant characteristics of each in Table 6.3.

In addition to the centrifuges listed in Table 6.3, there are also a growing number of single-use centrifuges currently in use. These units offer many of the advantages described previously in Chapter 2 for single-use equipment, including elimination of cleaning and sanitization steps, which reduces the costs of water, cleaning agents, and cleaning validation. Examples of single-use units include the Unifuge® Pilot from CARR Biosystems and its scaled up version, U2k®, both single-use tubular devices. Feed is pumped to the single-use module. Cells settle to the periphery of the module and clarified liquid (supernatant) continuously flows. For the Unifuge®, once the bowl is filled with solids, feed is stopped, cell slurry is pumped out, and the process can start again. For the U2k®, the solids are continuously discharged. The Unifuge® operates at flow rates of 1–4 L/min and at RCF values of 300–4,000. The larger U2k® system operates at up to 20 L/min and RCF values from 300–4,000. Both units are recommended for clarification of mammalian cell cultures, such as CHO broths, and harvesting (including washing and concentrating) cells [164, 165].

Table 6.3: Commonly used centrifuges and important characteristics of each.

Centrifuge type	Illustration[a,b]	Feed rate (L/h) [159]	Typical solids content in feed (% by volume) [159]	Σ (m²) [159]	Other characteristics [159, 162]
Disc-stack		20–500,000	≤30	1,000–500,000	– RCF up to 15,000 × g – Continuous operation (i.e., solids discharge possible) – Bowl cooling – Solids are wet, paste contains liquid – Difficult to clean
Tubular		20–7,000	<1	1,400–4,500	– RCF up to 20,000 × g – Firm paste (i.e., good removal of liquid) – Easy to clean – Limited solids capacity – Recovery of solids typically manual and difficult
Decanter		300–200,000	5–50	400–25,000	– RCF up to 5,000 × g – Continuous solids discharge – Handles high feed solids concentration
Multi-chamber bowl		100–20,000	<5	–	– RCF up to 9,000 × g – Firm paste (i.e., good removal of liquid) – Large solids holding capacity – Bowl cooling – Recovery of solids typically manual and difficult – Difficult to clean

Note that the sigma (Σ) values shown in the table are discussed shortly, in Section 6.7.

[a] Arrows indicate path of liquid phase; dashed lines indicate solids accumulation.

[b] Images adapted from *Fermentation and Enzyme Technology*, p. 263, by D. I. C. Wang et al., New York: John Wiley & Sons, Inc. Copyright (1979) by John Wiley and Sons, Inc. [163].

6.5 Disc-stack centrifuge CGMP operating procedure

A typical operating procedure for an automated disc-stack centrifuge with intermittent discharge, such as the one shown in the photo in Figure 6.7, is outlined below:

– Set up the unit.
– Measure the % solids in the feed and calculate the feed interval.
– Input parameters (e.g., feed flow rate and feed interval) to the centrifuge controller.
– Begin feeding.
– Allow multiple feed/discharge cycles to execute until feed is completely processed.
– Upon completion of feeding, perform system flush with high purity water (HPW) or a buffer. Note this flush may be built into the clean-in-place (CIP) cycle.
– Clean and sanitize the machine.
– Perform final HPW rinse to remove cleaning/sanitizing agents. Note that this rinse may be built into the CIP cycle.
– Store system dry.

As we discussed in Chapter 3, current good manufacturing practice (CGMP) production of biopharmaceuticals requires that equipment cleaning be validated. Centrifuge cleaning and sanitization are often carried out using a CIP skid programmed to run an automated CIP cycle. Table 6.4 provides details of a CIP cycle that has been validated for the disc-stack centrifuge shown in Figure 6.7. As shown, the cycle comprises multiple phases, each with a distinct objective. To execute the CIP cycle, connections are made from the CIP skid, via a CIP station, to three inlets on the centrifuge. In the P&ID in Figure 6.8, they are the process liquid (i.e., feed) line 201, the buffer liquid line 301A, and the flushing liquid line 301. These connections enable all product-contact surfaces to be cleaned, and the cleaning fluids are cycled through each line throughout the CIP cycle. Spent cleaning fluid is returned to the automated CIP skid through the clarified liquid line 220 and the concentrate (i.e., solids) line 222. In addition, during CIP, the bowl is discharged at two-minute intervals to ensure adequate cleaning of the interior of the bowl. It is also worth noting that immediately after use, prior to execution of the automated CIP cycle, a "manual" rinse of the centrifuge is performed with 100 L of HPW. HPW is fed only through line 201 at flow rate of 500 L/h, and the bowl is discharged at two-minute intervals. The manual rinse is in place to enable extended time between processing and running the automated CIP cycle. This time is needed because the centrifuge step is often not completed until the end of a shift, making it necessary to delay execution of the automated CIP cycle until the next shift arrives more than 12 h later. (It is desirable to have staff present as the automated cycle runs.) The option of holding the centrifuge dirty for more than 12 h presented challenges to validating cleaning of the system. Consequently, the manual HPW rinse immediately after centrifuge use was implemented to allow for validation of centrifuge cleaning with a 12-plus hour hold time after rinsing.

Table 6.4: CIP cycle details for the Alfa Laval LAPX 404 unit shown in Figure 6.7.

Phase	Cleaning fluid	Details	Purpose
Initial rinse	HPW	– Single pass[a] – 24 min – 7.5–8.0 L/min – 30 °C	Remove free-rinsing soils from equipment and piping surfaces. Cold or ambient water is desirable to avoid denaturing any proteinaceous soils.
Alkaline wash	CIP 100®, KOH-based detergent	– ~3% CIP 100® – Recirculated – 120 min – 7.5–8.0 L/min – 75 °C	Remove organic (e.g., proteinaceous) soils that are not easily rinsed.
Post-wash Rinse[b]	HPW	– Single pass – 25 min – 7.5–8.0 L/min – 30 °C	Remove alkaline wash chemicals.
Acid wash[b]	CIP 200®, phosphoric acid-based detergent	– ~3% CIP 200® – Recirculated – 120 min – 7.5–8.0 L/min – 75 °C	Neutralize traces of alkali residual on equipment surfaces; remove inorganic soils.
Final rinse	HPW	– Single Pass – 60 min, >0.9 µS/cm – 7.5–8.0 L/min – 30 °C	Remove all cleaning agents.

[a]Single pass means cleaning fluid only flows through the centrifuge once and is then sent to drain. Recirculated means that cleaning fluid flows from CIP skid to the centrifuge and back to the CIP skid continuously throughout the phase.
[b]CIP was validated without a post-wash rinse step and acid wash step, which are both optional.

6.6 Disc-stack centrifuge process and performance parameters

To begin the discussion of parameters that impact performance of a disc-stack centrifuge, let's write an equation similar to 6.6 for a centrifuge; that is, an equation that relates the maximum volumetric flow rate that allows adequate removal of solids, Q_{max}, to properties of the particle, liquid, and centrifuge. It is important to be able to estimate Q, because process time depends on it. For centrifuges, these equations were first presented by Hebb and Smith [166], then Ambler [167, 168]. We do not derive these equations here, as the derivations are available in the previously cited references and in textbooks [169]. For a disc-stack centrifuge, the applicable equation is:

$$Q_{max} = \left[\frac{d_p^2\left(\rho_p - \rho_l\right)}{18\mu}\right]\left[\frac{2N_{discs}\pi\omega^2\left(R_0^3 - R_1^3\right)\cot\theta}{3}\right] \tag{6.9}$$

where the variables in the first bracket have been defined previously in equation (6.3), and the remaining variables are defined as follows: N_{discs} is the number of discs, ω is the angular velocity, R_0 and R_1 are the radial distances from the center of rotation to the outer and inner radii of the discs, and θ is the angle the disc makes with the bowl axis.

Equation (6.9) suggests the following about properties of the feed to the centrifuge and their impact on disc-stack centrifuge performance:

- Increasing the particle diameter to be separated increases the maximum volumetric flow rate that can be used.
- Increasing the density difference between the fluid and particle and/or decreasing the fluid viscosity allows for larger volumetric flow rates through the centrifuge; alternatively, for a fixed flow rate, a larger density difference allows separation of smaller particles.

When recovering or removing cells from fermentation or cell culture broth, making adjustments to the particle size, particle density, liquid density, and liquid viscosity is unlikely. Those parameters are dictated by the fermentation or cell culture step.

We can also use equation (6.9) to understand the effect of input parameters that are readily controllable (i.e., process parameters). Ranges or set points for these parameters should be determined in development studies and provided in a batch record.

- Feed flow rate/residence time in the bowl. Generally speaking, increasing residence time of the slurry in the bowl by decreasing the feed flow rate results in better separation. Consider equation (6.9) from a different perspective. If that equation is solved for d_p, it becomes evident that d_p is proportional to $Q_{max}^{1/2}$. This relationship means that as Q_{max} is decreased, thereby increasing residence time in the centrifuge bowl, even smaller particles may be separated (i.e., the clarity of the clarified liquid, a performance parameter, may be improved). It is worth noting that disc-stack centrifuges typically cannot effectively remove particulates of <1 μm in diameter [157]. To achieve complete removal of solids, clarified liquid from the centrifuge typically undergoes further processing using filtration – the topic of the next chapter – to remove residual particulate before subsequent downstream processing.
- Rotational speed. From equation (6.9), the maximum volumetric flow rate that allows for separation of particles of diameter d_p is proportional to the angular velocity (ω) squared. Therefore, for a given feed, higher rotational speeds enable larger values of Q and shorter processing times; alternatively, for a fixed value of Q, a higher rotational speed enables smaller particles to be separated. Note that high

rotational speeds create shear. Shear forces created by the liquid act along the surface of the cell and may cause cell damage. Shear is particularly a problem at the centrifuge inlet. Cell damage is undesirable in intracellular products because it means that product ends up in the clarified liquid waste. In the case of extracellular products, cell damage results in undesirable soluble impurities being released into the clarified liquid product stream. Disc-stack centrifuges for use in biopharmaceutical processes are designed to minimize the effects of shear [159].

– Particle concentration in the feed. The settling velocity equations (6.3) and (6.8) presented previously were derived under the assumption that particles settle freely through a fluid. As the particle concentration increases, however, particles becomes obstructed by other particles, which lowers the settling velocity and results in difficult separation of solids from liquid (i.e., more solids end up in the clarified liquid and step yield is reduced for an intracellular product). As a result lower-than-expected feed flow rates would have to be used to achieve the desired separation. To avoid the effects of so-called hindered settling, feed to the centrifuge can be diluted to reduce particle concentration. Of course, diluting the centrifuge feed leads to greater feed volume, which leads to a longer processing time. Additional information on hindered settling can be found elsewhere [170].

– Feed interval/discharge frequency. Time between discharges in a disc-stack centrifuge with intermittent discharge is referred to as the feed interval or discharge interval and designated t_{feed} here. This value may have to be entered into the centrifugation system prior to a run and can be estimated based on the following calculation:

$$t_{feed} = \frac{\text{bowl capacity for solids}}{Q_{solids}} = \frac{V_{bowl} F_{bowl}}{C_{solids, f}/100 \times Q} \tag{6.10}$$

where Q_{solids} is the volumetric flow rate of solids to the centrifuge, V_{bowl} is the volume of the centrifuge bowl, F_{bowl} is the fraction of bowl volume filled with solids before discharging, $C_{solids,f}$ is the solids content of the feed in volume percent, and Q is the total volumetric feed rate (solids + liquid) to the centrifuge. It is common to allow the bowl to fill to 50–70% solids before discharging; that is F_{bowl} is 0.50–0.70 [171]. If the feed interval is too long, solids will accumulate to the point that they are carried into the clarified liquid and will eventually block the space between discs, which must remain open for fluid flow. Solids in the clarified liquid translate to a reduced step yield for intracellular products and reduced clarification efficiency for extracellular products. Further, if the feed interval is too short (i.e., the frequency of discharge too high), then for extracellular products, more liquid (i.e., product) will be lost with the discharged solids, which results in a decreased step yield.

Example: Calculating the feed interval for recovery of cells from an *E. coli* fermentation using a disc-stack centrifuge

Cells are recovered from a 500 L *E. coli* fermentation with a disc-stack centrifuge. The fermentation contains 6% cells by volume. The flow rate to the centrifuge is 120 L/h, and the centrifuge bowl has a volume of 2.2 L. The centrifuge requires that a feed interval be entered in units of seconds. What value do you enter?

Solution

Calculate the feed interval by substituting the given values into equation (6.10) as follows:

$$t_{feed} = \frac{2.2\ L \times 0.5}{6/100 \times 120\ L/h} \times 3600\ s/h$$

$$= 550\ s\ (\text{or } 9.2\ \text{min})$$

– Discharge ratio. Bowl discharge in a disc-stack centrifuge can be full or partial. A full discharge empties most or all bowl contents – both solids and liquid. A partial discharge, in which the bowl is open for less time, allows for only a fraction of the bowl contents to be discharged, thereby minimizing loss of liquid to the solids stream. When product is dissolved in liquid, partial discharges are particularly useful for minimizing the amount of product lost with the solids waste (i.e., maximizing step yield). The discharge ratio is the ratio of partial to full discharges. By setting a discharge ratio, buildup of solids in the bowl is not as excessive as if partial discharges alone are used, and loss of liquid is less than if only full discharges were utilized. The discharge strategy should be optimized during process development and the discharge ratio (if used) specified in the batch record.

6.7 Sigma factor, centrifugation step design, and scale-up

Central to centrifuge scale-up is sigma analysis, which uses a Σ value to characterize the centrifuge. The Σ value is arrived at as follows. If we multiply the first part of the equation (6.9) in brackets by g, the gravitational acceleration, and divide the second part in brackets by g, the following equation results:

$$Q_{max} = \left[\frac{d_p^2 \left(\rho_p - \rho_l \right) g}{18\mu} \right] \left[\frac{2N_{discs}\pi\omega^2 \left(R_0^3 - R_1^3 \right) \cot\theta}{3g} \right] \tag{6.11}$$

The first bracketed expression is simply the settling velocity due to gravity. The second bracketed expression is referred to as sigma, Σ, and the equation can be written more simply as:

$$Q_{max} = v_g \Sigma \tag{6.12}$$

where $\Sigma = 2N_{discs}\pi\omega^2(R_0^3 - R_1^3)\cot\Theta/3\,g$, for a disc-stack centrifuge. As you can see, v_g depends on particle and fluid properties, while Σ depends on the geometry and speed of the centrifuge. Also, despite the fact that equation (6.12) applies to a centrifuge, v_g instead of v_c shows up, due to the fact that we intentionally multiplied and divided the right side of equation (6.9) by g. Importantly, note the similarity between equations (6.6) and (6.12), which is written for a gravity settling tank. The similarity between the two equations results in the following interpretation for Σ: the area in a gravity settling device, such as that shown in Figure 6.3, that would be required to give the same performance as a centrifuge at equal volumetric flow rates.

To put Σ in perspective, consider the disc-stack centrifuge in Figure 6.7, which has a Σ value of 5,230 m². To match the performance of this centrifuge, a gravity settling tank would need 5,230 m² of surface area. For comparison, the area of an American football field is approximately 5,350 m². Disc-stack centrifuges are available in sizes much larger than the one shown in Figure 6.7. Machines with Σ values in excess of 100,000 m² that handle flow rates in excess of 100,000 L/h are in use [159]. Also, it is important to point out that the mathematical expression for Σ depends on the centrifuge type (i.e., the expression for Σ for a disc-stack centrifuge is different than that for a tubular centrifuge).

Equation (6.12) can be used for scale-up as follows. Because v_g is only a function of particle properties, fluid properties, and the constant g, which are all scale independent, the following is true for centrifuges at any two scales, designated 1 and 2 in the following equations:

$$v_{g,1} = v_{g,2} \tag{6.13}$$

Therefore

$$\frac{Q_{max,1}}{\Sigma_1} = \frac{Q_{max,2}}{\Sigma_2}. \tag{6.14}$$

Equation (6.14) can be used as the basis for centrifuge scale-up as illustrated in Figure 6.9, which shows determination of Q_1/Σ_1 as a first step. This can be done experimentally by using a bench-scale centrifuge of the same type to be used at production scale. Once that value is determined, there are two scenarios for scale-up depending on whether an existing production centrifuge is available. If there is, the flow rate to the production centrifuge is calculated by solving equation (6.14) for Q_2 and plugging in known values for Q_1, Σ_1, and Σ_2. Note that the centrifuge manufacturer should be able to provide a value for Σ_2. Alternatively, if a new centrifuge is to be purchased, Σ_2 is unknown. A value for Q_2 will have to be determined based on an acceptable processing time, and from that value, Σ_2 is estimated using equation (6.14). An example calculation is shown below.

Example: Scaling up a disc-stack centrifugation step from bench-scale data
You are purchasing a disc-stack centrifuge for recovery of cells from an *E. coli* fermentation used to produce an intracellular therapeutic protein. A study using a bench-top disc-stack centrifuge ($\Sigma = 1,900$ m^2) suggests a flow rate of 1,200 mL/min is suitable to meet your centrifugation objectives. This value was determined by varying the flow to the bench-top unit and selecting a value that resulted in no more than 2% loss of cell mass. You don't currently have a disc-stack centrifuge to use at manufacturing scale. Answer the following:
a. What flow rate will you operate at to process a 5,000 L fermentation?
b. What size centrifuge (Σ value) will you purchase?

Solutions
a. To determine a feed flow rate, assume an acceptable processing time and then use equation (6.7) to solve for Q, the volumetric feed rate to the centrifuge. Let's assume a processing time of 3 h, which does not include centrifuge setup and cleaning; these steps will lengthen the total time to complete the centrifugation step to approximately 8 h:

$$Q = \frac{V_{feed}}{t_{process}} = \frac{5,000 \text{ L}}{3 \text{ h}}$$

$$= 1,667 \text{ L/h}$$

b. Now that Q is known, equation (6.14), solved for Σ_2, can be used to determine the size of production centrifuge to carry out the step:

$$\Sigma_2 = \frac{Q_2 \, \Sigma_1}{Q_1} = \frac{1,667 \text{ L/h}}{72 \text{ L/h}} \times 1,900 \text{ m}^2$$

$$= 43,990 \text{ m}^2$$

It is worth noting that the ratio of settling velocity in a centrifugal field to that in a gravitational field has been found to fit the following equation for a disc-stack centrifuge [159]:

$$\frac{v_c}{v_g} = \left[\frac{\omega^2 R}{g}\right]^{0.75} \tag{6.15}$$

Based on equations (6.3) and (6.8), we would expect the exponent to be 1 rather than 0.75. However, the relationship in equation (6.15) results from increased shear forces in a disc-stack centrifuge that act to drag particles to the clarified liquid outlet, ultimately resulting in a reduced particle settling velocity in the centrifuge [159]. As a result of the reduced settling velocity, which is not accounted for in the derivation of Σ for a disc-stack centrifuge, the KQ value was developed as an empirical alternative to characterize disc-stack centrifuge size [172] and is given by:

$$KQ = 280 \left(\frac{\text{rpm}}{1,000}\right)^{1.5} N_{discs} \cot \theta \left(R_0^{2.75} - R_1^{2.75}\right) \tag{6.16}$$

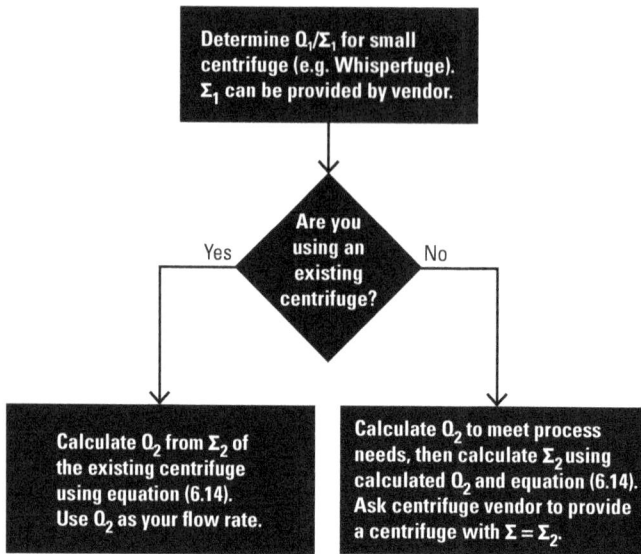

Figure 6.9: Using equation (6.14) for centrifuge scale-up. Image © NC State University; reprinted with permission.

6.8 Summary

There are numerous solid-liquid separation applications in biopharmaceutical processes. These include:

- removal of cells and cell debris from the cell culture broth, leaving a product that is dissolved in the spent clarified liquid media;
- recovery of cells from the bioreactor for further processing;
- clarification of lysate streams;
- recovery of inclusion bodies in processes in which the host system (e.g., *E. coli*) produces the product as an inclusion body;
- recovery of precipitated product or removal of precipitated impurities in processes that use precipitation for purification;
- removal of microorganisms to produce a bioburden-free or sterile process stream.

The three most common methods for solid-liquid separation in biopharmaceutical processes are centrifugation, covered in this chapter, depth filtration, and tangential-flow microfiltration, both of which are covered in the next chapter. Centrifugation relies on centrifugal force and a density difference between solids and the surrounding liquid to cause suspended solids to settle rapidly in their surrounding liquid. Once settled, the clarified liquid and solids can be recovered separately. The centrifugal force used in centrifuges for biopharmaceutical applications is typically around 5,000–10,000 × g; thus, particles settle much faster than by gravity, which allows for higher feed rates to be used, thereby resulting in relatively short process times for the separation.

There are numerous types of centrifuges in use industrially. For biopharmaceutical applications, the disc-stack centrifuge is the most common. Discs stacked in a bowl provide surface area for settling, while rotation of the bowl provides the necessary centrifugal force. The feed slurry flows continuously to the centrifuge and into the rotating bowl, while a clarified liquid continuously flows out. Accumulated solids are discharged from the bowl either intermittently at predetermined intervals or continuously, depending on the centrifuge design.

There are a number of process parameters for which ranges must be determined and which are typically written into a batch production record and/or built into the process automation. These include feed flow rate, rotational speed of the bowl, particle concentration in the feed, and the feed interval in the case of a centrifuge with intermittent discharge. The procedure for a disc-stack centrifugation run includes setup steps that precede the actual separation followed by multiple feed-discharge cycles (for centrifuges with intermittent discharge). Once the entire feed slurry has been processed, the system can be flushed with a buffer for product recovery (if the product is in the clarified liquid phase), then cleaned and sanitized. For CGMP production of biopharmaceuticals, cleaning/sanitization is performed using a validated method, often with an automated CIP system.

To properly scale up results from bench-scale development studies to larger scale, sigma analysis is used. The relevant equation is $Q_{max,1}/\Sigma_1 = Q_{max,2}/\Sigma_2$, where Q is the volumetric flow rate at two different scales designated 1 and 2. While Σ is represented by a relatively complex equation (see the second bracketed expression in equation (6.11)), its interpretation is straightforward: it is the area in a gravity settling device that would be required to give the same performance as the centrifuge at equal volumetric flow rates. Centrifuge manufacturers can provide Σ values for their models.

6.9 Review questions

1. A swing-out test tube centrifuge can be used to estimate v_g for a particle. To show how this is done, do the following. (a) First derive an expression for v_g as follows. Starting with equation (6.8) and recognizing $v_c = dR/dt$ for the particle settling, solve the resulting differential equation to show that $v_g = g \times \ln (R/R_o)/(\omega^2 t)$, where R is the radius (from center of rotation) at which a particle is located at time t and R_o is the radius (from center of rotation) at which a particle starts. (b) Then explain how you would experimentally determine v_g based on your derived equation.

2. You are trying to recover *E. coli* cells from fermentation broth using an Eppen-dorf 5418 centrifuge, similar to the one shown in Figure 6.1(a). The cells are sus-pended in a liquid with a density of 1.00 g/cm^3 and a viscosity of 1.0 centipoise (1 cp = 0.01 g/cm s).
   (a) How much time in *hours* is required for all the *E. coli* cells to settle by gravity to the bottom of an Eppendorf tube, which has a height of 4 cm? Assume that settling is not hindered by neighboring cells.
   (b) How much time in *seconds* is required for all the *E. coli* cells to settle to the bottom of an Eppendorf tube while rotating in the Eppendorf 5418 centrifuge at 14,000 rpm? Assume that the Eppendorf tube is positioned horizontally in the centrifuge. Also, the top of the tube is 4.5 cm from the center of rotation. Again, assume that settling is not hindered by neighboring cells.

3. A lab exercise is being designed to determine the effect of feed flow rate on the concentration of solids in the clarified liquid from a bench-top disc-stack centri-fuge. The unit does not have automated discharge. You will use flow rates of 200 mL/min, 700 mL/min, and 1,000 mL/min. You anticipate operating at 200 mL/min for 10 min followed by 700 mL/min for 7 min followed by 1,000 mL/min. How long could broth flow at 1,000 mL/min before the bowl is half full of solids? The con-centration of solids in the feed is 6% by volume, and the bowl volume for the unit is 1 L.

4. You are working with a suspension of cells that are approximately 2 µm in diame-ter. The cells are lysed, and the resulting cell debris has a diameter of approxi-mately 1/6 of the intact cells. The density of the intact cells and cell debris is about the same. In a swing-out test tube centrifuge in which the tubes are perfectly hor-izontal during rotation, how many times faster do the whole cells settle than the cell debris? Hint: begin by recognizing that $v_c = dR/dt$ for a particle, assume no hindered settling, and solve the resulting differential equation for time.

5. *Saccharomyces cerevisiae* cell broth is fed to a bench-top disc-stack centrifuge with $\Sigma = 1,900$ m^2. You conduct a study on the impact of flow rate on separation. At what flow rate would you expect to see yeast cells in the clarified liquid, as-suming that all cells are exactly the same size? The bowl is not allowed to fill sig-nificantly with solids.

6. You are a process engineer tasked with designing a disc-stack centrifugation step for recovery of cells from a 2,000 L *E. coli* fermentation. You will be using an ex-isting production-scale disc-stack unit with $\Sigma = 5,230$ m^2. In development studies using a bench-top disc-stack centrifuge ($\Sigma = 1,900$ m^2), you determined $Q = 1,200$ mL/min to be a suitable flow rate. Calculate the time required in hours to process the 2,000 L *E. coli* fermentation using the production centrifuge.

7. You are working on a centrifugation step for the clarification of CHO culture in a monoclonal antibody (mAb) production process. Recall that mAbs are typically secreted by CHO cells. The mAb process you are working on requires that you centrifuge 5,000 L of CHO cell culture broth. Estimate the minimum size for a centrifuge to clarify the CHO cell culture.

8. In problem 6, a Q_{max} value = 1,200 mL/min was observed for separation of *E. coli* using a bench-top disc-stack centrifuge (Σ = 1,900 m^2). Using equation (6.12) and the settling velocity for *E. coli* presented in this chapter, calculate the theoretical Q_{max}. Give three possible reasons for the difference in the observed and calculated values.

9. A 1,000 L *Pichia pastoris* fermentation at a cell concentration of 50% by volume is clarified (product is secreted) using a disc-stack centrifuge. In order to achieve the required liquid clarification, fermentation broth is diluted to five times its original volume. The bowl volume is 3 L and you will allow the bowl to fill to 1/2 of its solids capacity before discharging.
   (a) If the flow rate to the centrifuge is 850 L/h, what is the feed interval in seconds?
   (b) Estimate the time required for the run, in hours.

10. Show that the expression for Σ in a tubular centrifuge is given by $[\pi \times L \times (R_o{}^2 - R_1{}^2) \times \omega^2]/[g \times \ln(R_o/R_1)]$. (Note that R_o is the distance from the center of rotation to the liquid surface within the bowl, and R_1 is the distance from the center of rotation to the bowl wall.) To do this requires developing a differential equation for the rate that a particle moves in the radial direction (dR/dt) and the rate at which it moves in the axial direction, pushed by the flow of fluid (dz/dt). The equations can then be combined in the form $dR/dz = \ldots$ to give a particle trajectory equation (i.e., an equation describing the radial and axial position of a particle) from which a relationship in the form of $Q = v_g \times \Sigma$ can be written.

Chapter 7
Harvest operations, part 3: solid-liquid separation by filtration

In addition to centrifugation, filtration is a commonly used operation for solid-liquid separations in biopharmaceutical processes. In this chapter, we cover two distinctly different but common types of filtration steps in biopharmaceutical processing: depth filtration and tangential-flow microfiltration (MF). We begin by discussing fundamental aspects of filtration and then turn our attention to details related to depth filtration and tangential-flow MF. Specific questions addressed are:

- What is filtration?
- What is the difference between tangential-flow and normal-flow filtration?
- What is the difference in how solids are retained by a depth filter and an MF membrane?
- What materials are depth filters made from and how are they "packaged" for large-scale processing?
- What is the basic equipment setup and procedure for a depth filtration step? A tangential-flow MF step?
- What are the important process and performance parameters for depth filtration?
- What bench-scale studies are used to select a depth filter and how are depth filtration steps scaled up?
- In addition to centrifugation (covered in the last chapter), depth filtration, and tangential-flow MF, what other operations can be considered for solid-liquid separations in biopharmaceutical processes?

You may notice that this chapter emphasizes depth filtration over MF. This is intentional. While we introduce the basic concepts of microfiltration (MF) here, we reserve a more detailed discussion for Chapter 9, which focuses on ultrafiltration (UF). UF is a close "cousin" to MF, and certain concepts – such as process parameters, development studies, and scale up – are similar for both and, therefore, are covered together.

7.1 Filtration definition

Let's start our discussion with a definition of filtration. A good one is given by Harrison et al. [162] as follows: any technique used to separate *particulate* in a fluid suspension or *components in solution* according to their size by flowing under a *pressure differential* through a *porous medium*. The first part of the definition refers to the separation of solid particles from a fluid suspension and covers applications such as

https://doi.org/10.1515/9783111112459-007

removing Chinese hamster ovary (CHO) cells from a cell culture broth to produce an-tibody-containing clarified liquid ready for the purification stage. This type of applica-tion is what most of us think about when we use the term filtration: removal of solids from a liquid. But the definition also includes the separation of components *in solu-tion*. An example is the use of UF to concentrate (i.e., remove water from) a protein solution or to exchange one buffering solution for another. UF is commonly used as a formulation step for drug substance and, as previously mentioned, is discussed in de-tail in Chapter 9. This chapter focuses on the first part of the definition: removal of solids from a fluid suspension.

7.2 Filtration applications in biopharmaceutical processing

There are numerous examples in which filtration is used for solid-liquid separation in biopharmaceutical processes, as summarized in Table 6.1 in the previous chapter. Let's take a closer look at one example in Figure 2.3, which shows a typical process for production of a monoclonal antibody in CHO cells. In the process represented by that figure, CHO cells secrete the monoclonal antibody into the culture medium, and a combination of centrifugation and depth filtration is used to clarify (i.e., remove solids from) the cell culture broth in preparation for loading onto a Protein A chromatogra-phy column. This is a common use of depth filtration – removal of residual particles from the clarified liquid from a centrifuge, with the dissolved product passing through the filter in the filtrate stream. In addition, depth filtration alone – that is, without centrifugation – can be used to produce clarified cell culture broth. Without centrifugation, it is common for depth filtration to be performed in two stages: a first stage with a larger pore size filter medium followed by a second stage with a smaller pore size. Alternatively, tangential-flow MF could be used in place of both the centri-fugation and depth filtration steps.

As described in Table 6.1, depth filtration and tangential-flow MF can also be used for other applications in biopharmaceutical processes. For example, they can be used to clarify lysate streams generated from lysis steps used to release intracellular prod-uct. Examples of processes requiring lysate clarification include *E. coli*-based pro-cesses for protein therapeutic production and production of certain viral vectors for gene therapy using human embryonic kidney 293 (HEK293) cells. Both types of filtra-tion can also be used for removal of precipitated impurities from process streams. And normal-flow MF is commonly used for bioburden removal, but we focus on tan-gential-flow MF in this chapter.

7.3 Normal-flow versus tangential-flow filtration

There are two common modes of operation for filtration steps: normal flow, also referred to as dead end, and tangential flow, also referred to as cross flow. As the names suggest, in normal-flow filtration, feed flows perpendicular to the filter medium, in the same direction as filtrate flow. Filtrate refers to the fluid that passes through the filter and is therefore relatively free of solids. In tangential-flow filtration, feed is pumped parallel to the surface of the filter, perpendicular to the filtrate flow, with only a portion of the feed permeating the filter. The two scenarios are illustrated in Figure 7.1. It is important to note that even though the images depict what appears to be solid particles being removed from the filtrate, the same illustrations apply to the case in which dissolved solutes, such as proteins, are retained by a filter. In this chapter we discuss solid particles and discuss dissolved solutes in Chapter 9. As you can see, in normal-flow filtration solid particles accumulate at the surface of the filter,

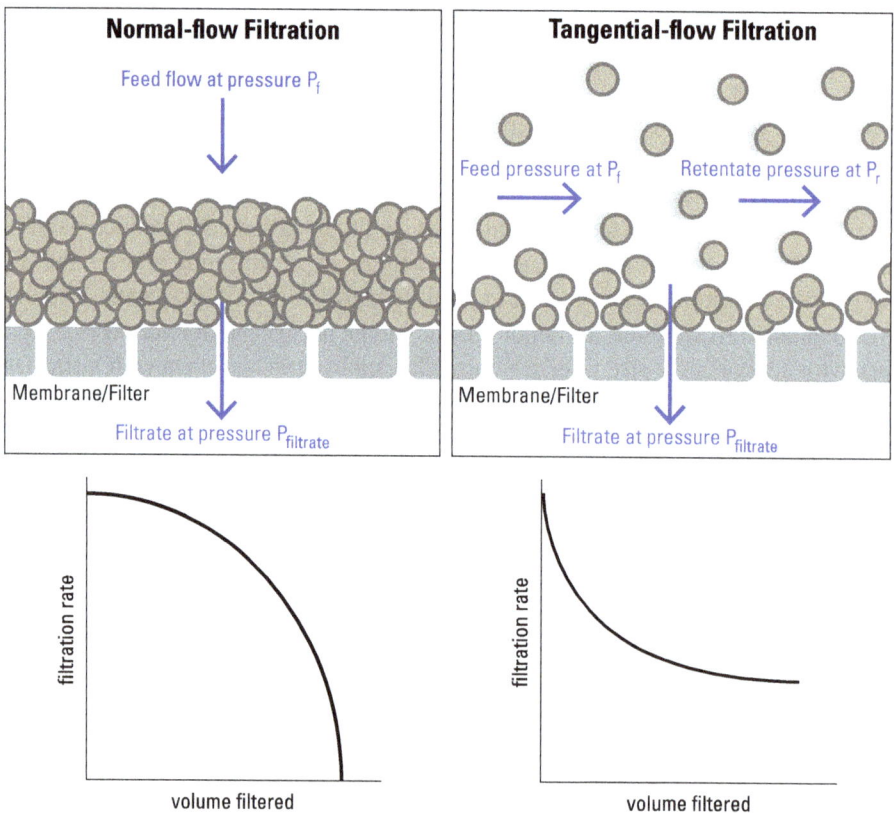

Figure 7.1: Normal-flow versus tangential-flow filtration. Note that the graphs apply to the operating scenario in which P_f, the feed pressure to the filter, is constant. Image © NC State University; reprinted with permission.

and the filtrate rate quickly decreases with time when the feed pressure is held constant. In tangential-flow filtration, a portion of the feed passes through the filter as
filtrate (also referred to as permeate), carrying particles with it to the filter surface.
However, the buildup of particles is minimized by the sweeping action of the tangential feed flow. As a result, the filtrate rate stabilizes over time instead of dropping to
zero. Depth filtration is operated in normal-flow mode, while MF is run in tangential-
flow mode when the solids load is high.

As described in the definition previously presented, filtration is a pressure-driven
process. In the normal-flow filtration diagram in Figure 7.1, the pressure differential
that drives filtrate flow is the difference between the feed and filtrate pressures,
given by

$$\Delta P = P_f - P_{filtrate} \tag{7.1}$$

where ΔP is the pressure difference, and P_f and $P_{filtrate}$ are the pressures of the feed
and filtrate streams, respectively. Note that the filtrate is typically at atmospheric
pressure. As long as $\Delta P > 0$ and the filter is not completely plugged, flow through the
filter will occur. In tangential-flow mode, the relevant pressure differential is defined
differently. Because feed flows tangentially along the filter surface, there is a pressure
drop that occurs from the feed inlet to the feed outlet. The effluent from the feed side
of the membrane is referred to as the retentate – think slurry that is *retained* by the
filter. Assuming a linear decrease in pressure along the length of the filter feed channel, an average pressure on the feed side of the filter is calculated as: $(P_f + P_r)/2$, where
P_r is the pressure of the retentate. The relevant pressure differential that drives permeate across the filter is known as the transmembrane pressure, TMP, defined as:

$$\text{TMP} = \frac{P_f + P_r}{2} - P_{filtrate} \tag{7.2}$$

7.4 Solids retention by filters

Two ways in which particulates are retained by a filter are illustrated in Figure 7.2.
With surface filtration, particles are sieved out at the surface of the filter medium. An
MF membrane is an example of a surface filter, with pore sizes as small as 0.1 μm or
as large as 10 μm [173]. Depth filters, on the other hand, trap particles not only at the
surface, but within the filter medium. Depth filters are typically operated in normal-
flow mode, while surface filters are operated in either normal-flow or tangential-flow
mode. Depth filters have a high solids capacity relative to surface filters because they
are able to retain a large number of particles before becoming plugged; consequently,
depth filters are often the normal-flow filtration method of choice to clarify feeds containing a large solids load, such as cell culture broth. Using a surface filter in normal-

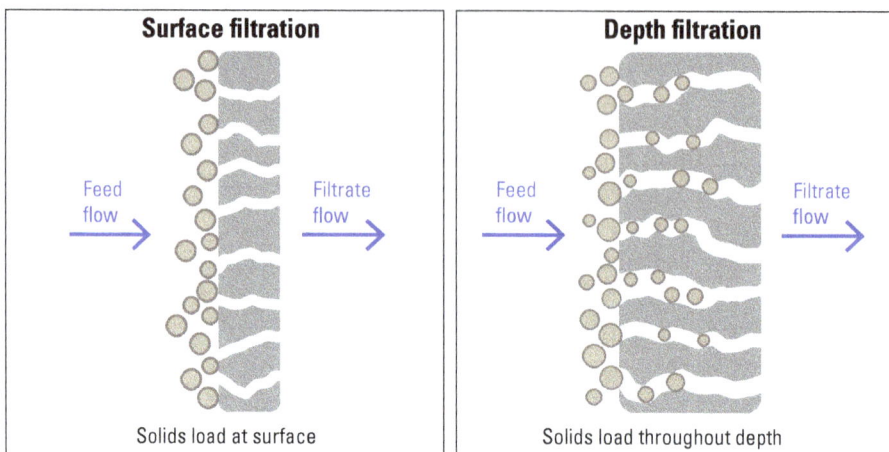

Figure 7.2: Depth versus surface filtration in normal-flow mode. Image © NC State University; reprinted with permission.

flow mode for this type of application is rare beyond the bench scale, as the surface area required would be large and therefore prohibitively expensive. Using a surface filter in tangential-flow mode, on the other hand, minimizes solids build-up at the surface and allows filtration to proceed for a longer time, which means less filter area is required relative to normal-flow operation.

In biopharmaceutical manufacturing, *membranes*, such as previously mentioned MF membranes, are commonly used as surface filters. These thin, semi-permeable materials are designed to retain particles or molecules, primarily based on size. In contrast, depth filters are much thicker and are typically referred to simply as *filters* rather than membranes. It is also worth noting that the filtrate from a membrane is often referred to as permeate. The term permeate is typically used with MF membrane- and UF membrane-based steps.

7.5 Depth filtration

We now turn our attention to details on depth filtration. As discussed previously, depth filtration is used to remove particles from feeds such as cell culture broth; thus, removing the solids and achieving the necessary clarity in the resulting filtrate is a primary objective. In addition, the design and implementation of a depth filtration step focuses on maximizing step yield (i.e., ensuring that product easily passes through the filter and is present in the filtrate) and filter capacity for solids. Let's begin by focusing on the depth filters themselves.

7.5.1 Depth filters and filter modules

Depth filters used in biopharmaceutical processes are most commonly made of cellulose fibers bound by a polymeric resin that adds a positive charge to the filter medium [174]. Depth filters may also contain a filter aid, such as perlite or diatomaceous earth, to provide increased surface area for improved solids capacity. There are also depth filters made from polypropylene fibers, polyacrylic fibers, and activated carbon that find use in biopharmaceutical applications [175].

Figure 7.3(a) shows an image of cellulosic depth filtration media. Note the wide distribution of pore sizes, a common characteristic of depth filters. Filters for solid-liquid separations are typically rated by a pore size, which equates to the size of the largest particle that can pass through the filter. For depth filters, the pore size rating is described as "nominal" rather than "absolute." A filter with a nominal rating can retain some percentage of solid particles of greater than the stated pore size. For example, a nominal rating of 0.8 μm may mean that the filter retains only 60% of particles with a diameter larger than 0.8 μm. Filters with a relatively wide distribution of pore sizes, such as depth filters, are typically given a nominal rating. An absolute rating is applied to a filter that will reliably remove nearly all particles of a size greater than or equal to the pore size rating. It is usually used for a filter that has a fairly narrow pore size distribution. MF membranes are an example.

Figure 7.3(b) shows an example of how the depth filtration media is packaged to create a process-scale single-use depth filter, in this case a Millistak+® HC Pro D0SP Pod device from MilliporeSigma. Flow through this type of device is illustrated in Figure 7.3(c), which shows that feed through the filter is split equally among multiple identical filtration media "packets" within the Pod (in the case of Figure 7.3(c), two packets); that is, the Pod is configured so that each packet is in parallel. These packets include one or more layers of filter media and filter aid. Filtrate from the packets combine within the Pod to create a single filtrate stream. This type of filter construction, in which feed is equally distributed to each packet in a module, is common among depth filters currently on the market. However, not all filters have a rectangular geometry like the MilliporeSigma Pod devices shown.

To provide the filtration surface area required for a given application, modules are typically placed in a parallel configuration into a holder, as shown in Figure 7.3(d) for Pod devices. Filtration area may be scaled by increasing the number of Pod devices with the Pod rack. For the filter module shown in Figure 7.3(b), the process-scale holder can accommodate up to 16 m^2 (= 0.77 m^2/Pod × 7 Pods/rack × 3 racks) of filtration area. Depth filters from other vendors are commonly used in biopharmaceutical applications as well, including Stax™ capsules from Pall (part of Cytiva), Sartoclear® depth filters from Sartorius Stedim Biotech, and 3M™ Zeta Plus™ depth filters from Solventum (formerly part of 3M).

In modern biopharmaceutical applications, depth filtration is typically a single-use operation. All the depth filters mentioned in this chapter are sold as single-use

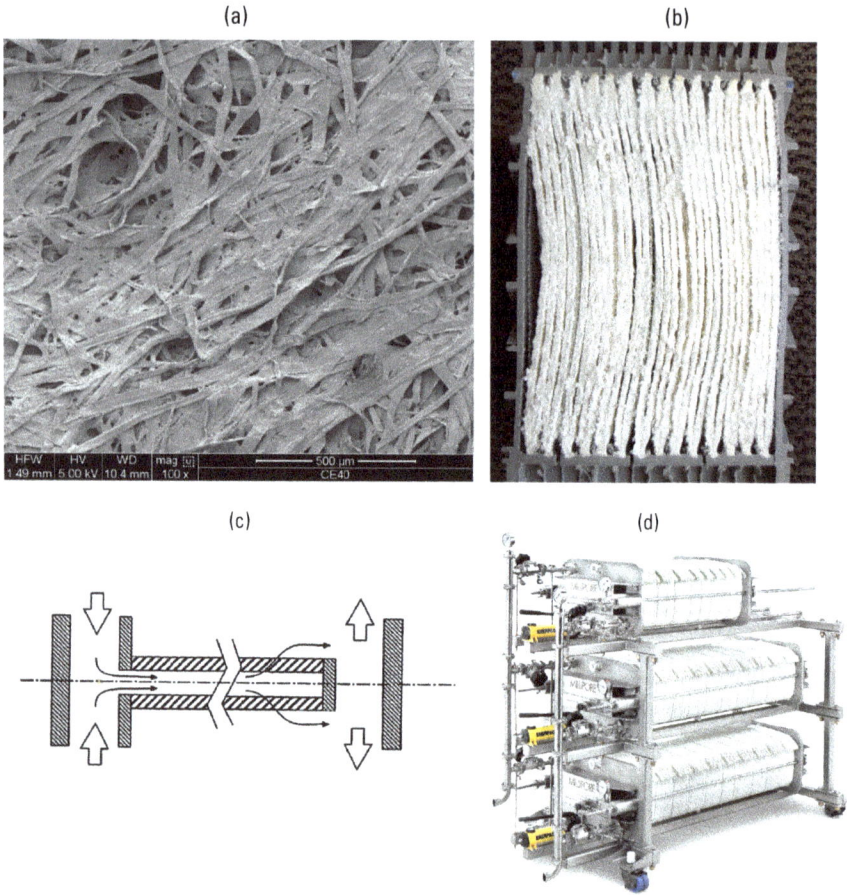

Figure 7.3: Depth filter images. (a) An electron micrograph of cellulose fibers that make up Millistak+®
CE40 depth filtration media. (b) MilliporeSigma's Millistak+® HC Pro D0SP Pod device with 0.77 m² of filter
area. The media is made from silica filter aid with polyacrylic fiber [176]. (c) A schematic showing how the
feed stream is equally distributed to layers (only two are shown) of depth filtration media within a Pod
device. (d) A photograph of a Millistak+® Pod filter holder showing how multiple filtration modules are
placed in a parallel configuration to provide the required surface area [177]. Images reproduced with
permission from Merck KGaA, Darmstadt, Germany and/or its affiliates.

encapsulated modules. The equipment required to execute the step is also available in
a single-use format, with details provided in the next section.

Depth filters come in a range of pore sizes, and, as already mentioned, for a given
filter type, there is a wide distribution of pore sizes. This range of pore sizes is shown
in Figure 7.4 for depth filtration products offered by MilliporeSigma. The chart shows
depth filtration media with nominal pore size ratings from slightly less than 0.1 µm
up to greater than 10 µm. It is also worth pointing out that multi-layer depth filtration
media is available. Typically, multi-layer media is made of a more open first layer fol-

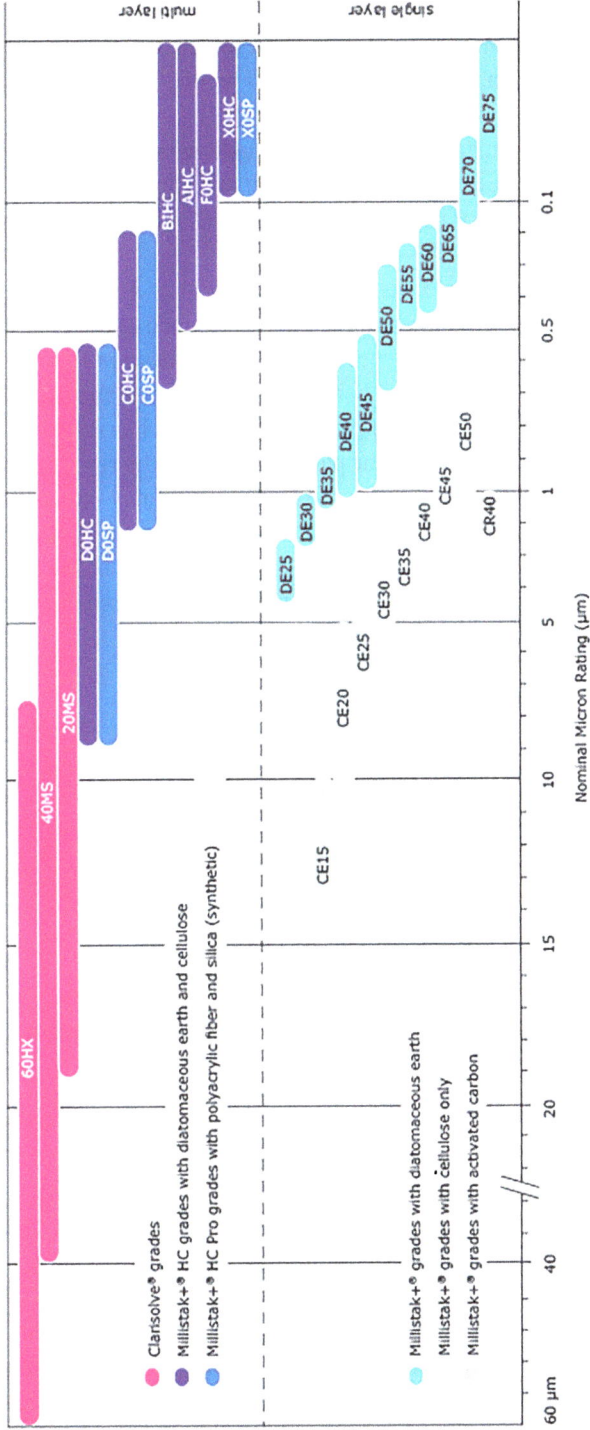

Figure 7.4: Various depth filtration media grades for Millistak+® family and Clarisolve® depth filters (from MilliporeSigma) and their nominal pore size rating in microns [178]. Image reprinted from "Millistak+® family and Clarisolve® depth filters at a glance" 2017, p. 1 © by MilliporeSigma Corporation and reproduced with permission from Merck KGaA, Darmstadt, Germany and/or its affiliates.

lowed by tighter second layers. For example, from Figure 7.4, MilliporeSigma's Millistak+® HC D0HC media combines a layer of CE25 media, which has a nominal pore size of approximately 6 µm, with a layer of DE40 media, which has a nominal pore size between 0.5 and 1 µm.

While solids retention in a surface filter, such as an MF membrane, is due mainly to particle sieving at the surface, a number of retention mechanisms in a depth filter enable it to trap particles. Primary mechanisms for retention are described below [175] and illustrated in Figure 7.5:

– Sieving. Particles cannot pass into the filtrate because they are larger than the filter pores. Sieving may occur at the entrance to a pore, such as at the surface, or in a restriction within the pore. Both types of sieving are shown in Figure 7.5.
– Adsorption. Electrostatic interactions between the particles and filter occur, which results in adhesion of particles to the filter surface and enables removal of particles smaller than the size of the filter pores. Recall that the resin binder used in depth filters imparts a positive charge to the filter that may attract negative charge on the surface of the particles. It is worth noting that if the depth filter media has a positive charge and the active ingredient dissolved in the feed has a negative charge, product loss (i.e., reduction in step yield) may occur during filtration due to binding of product to the filter media.

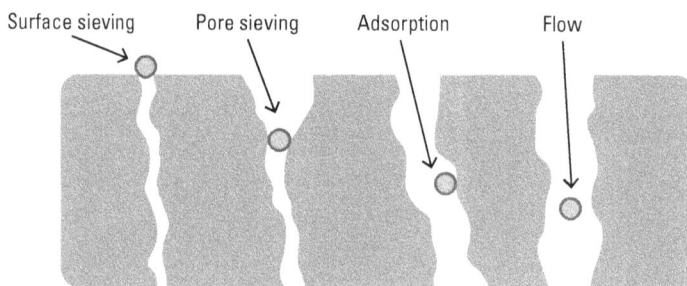

Figure 7.5: Particle retention mechanisms in a depth filter. Note that an actual depth filter is more porous than the image shown. Image © NC State University; reprinted with permission.

7.5.2 Depth filtration equipment and procedures for production

A schematic of a depth filtration setup is shown in Figure 7.6. Feed from a bioreactor or process vessel is pumped to the depth filter. Gauges on the feed and filtrate streams monitor pressure, in particular the pressure difference across the depth filter. For many applications, including clarification of cell culture broth to prevent plugging of a subsequent chromatography column, a 0.2 µm MF membrane filter (i.e., a filter with 0.2 µm pores) is placed in series with the depth filter to ensure the filtrate is largely free of solids.

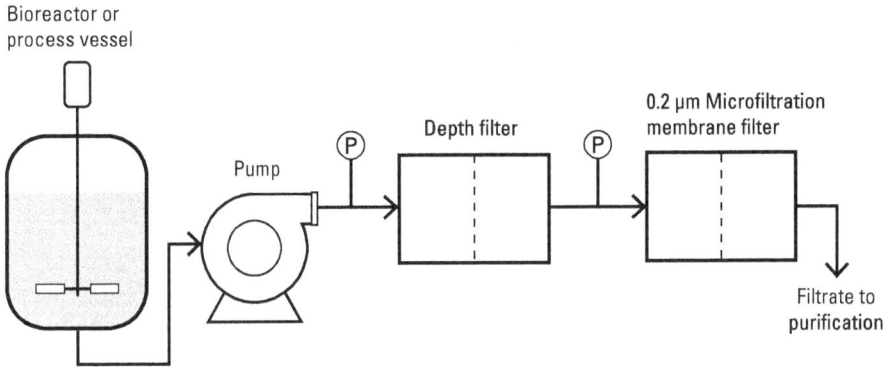

Figure 7.6: Schematic of a depth filtration setup for biopharmaceutical processing. Image © NC State University; reprinted with permission.

As mentioned previously, in addition to the filters themselves, the equipment required to execute a depth filtration step is available in single-use format. Specifically, single-use skids that contain all the components shown in Figure 7.6 are available. Examples include the Allegro™ MVP Single-use System from Cytiva and the Mobius® FlexReady Solution for Large Scale Clarification from MilliporeSigma. These systems rely on disposable flow paths that are replaced after each run. The flow paths can be equipped with disposable sensors as well. For example, the Allegro™ MVP system includes both single-use pressure and flow sensors [179]. It is also worth noting that depth filtration steps can be carried out without a skid by bringing together the individual components, including tubing, valves, pressure gauges, and a pump, and making proper connections to a holder that contains the encapsulated filters.

A typical depth filtration procedure is summarized by the diagram in Figure 7.7. Initially, the filters are flushed with high purity water (HPW) to remove entrapped air and leachable components, followed by a buffer to condition the filter prior to use. Then filtration of the feed stream begins, with product (i.e., filtrate) collection beginning after all HPW or buffer held up in the system from the previous step is purged to drain. While purging the flushing liquid is not mandatory, skipping this step can result in product dilution if the flushing liquid mixes with the filtrate. Upon completion of the product filtration, buffer is pumped through the system to recover product that remains held up in the system. A blowdown step in which pressurized air is used to displace the buffer held up in the system may be performed afterwards to reduce the weight of the filter modules and to facilitate a cleaner disassembly. Finally, the filters and other single-use components are disassembled, decontaminated (e.g., in a waste autoclave), and discarded as biohazardous waste. Any reusable components that may have been used such as a multi-use filter housing, must be cleaned and sanitized with a validated method (when executing a process for a commercial product) because there is a potential that some of their surfaces have contacted product.

HPW/Buffer flush	Reduces leachables from depth filters, removes trapped air that would otherwise reduce adsorptive capacity, and conditions the filter prior to use
Filtration	Removes particulate (e.g., cells, cell debris, precipitates) from feed stream, filtrate (product) collected
Buffer chase	Recovers product in the system hold up
Post-filtration blowdown	Reduces weight of the filter modules and facilitates a clean disassembly
Decontamination	Prepares filters and other single-use components for disposal, cleaning/sanitization of any reusable equipment used in the step also done to prepare for next run

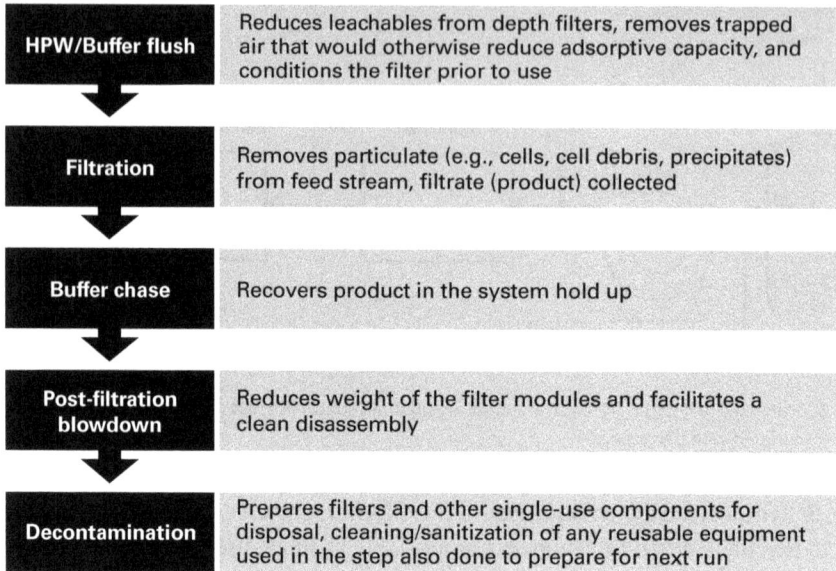

Figure 7.7: Steps involved in a depth filtration procedure after the filter(s) has been installed in the system. Image © NC State University; reprinted with permission.

7.5.3 Depth filtration parameters, performance, and design

Before beginning the discussion of depth filtration performance and development, it's necessary to define a few terms:

– *Feed flux*: the volumetric feed rate, Q_f, across the filter per unit area, A_{filter}. It is controlled by the system pump and typically expressed in units of LMH (L of suspension fed per m^2 of filter area per hour). If, for example, the feed flux across the Pod filter in Figure 7.3(b) is 150 LMH, the total flow rate being fed during filtration is 150 L/m^2/h × 0.77 m^2 or 116 L/h. The feed flux, J_f, across the depth filter can be generally represented by the following expression [180]:

$$J_f = \frac{Q_f}{A_{filter}} = \frac{\Delta P}{\mu R_{total}} \tag{7.3}$$

where ΔP – the driving force for feed flux – is defined by equation (7.1), μ is the viscosity of the liquid making up the slurry, and R_{total} is the resistance to flow through the filter. Contributions to resistance include the filter medium, particles blocking the entrance to pores, and particles restricting pores internally. The resistance increases as the depth filtration run proceeds because particles accumulate on or in the filter and increasingly impede flow. Depth filtration is typically operated with a constant feed flux; therefore, based on equation (7.3), the driving force for flow, ΔP, will increase to keep the flux constant because the resistance, R, increases as the depth filtration run proceeds.

- *Capacity* (also referred to as throughput or loading): the maximum volume of suspension that can be filtered (because a maximum pressure difference has been reached, for example) per unit area of filter. It is typically expressed in units of L/m^2. Note that we use the term *throughput* to more generally mean the volume of material filtered per unit area of filter – but not necessarily the maximum volume that can be filtered. Thus, capacity represents the maximum throughput for the filter system.

Example: Calculating depth filtration area requirements and process time

Consider the removal of cells from a CHO culture broth. 2,000 L of broth is to be clarified using a depth filter with a capacity of 250 L/m^2 at a feed flux of 150 LMH. Answer the following questions:

(a) What is the depth filtration area, $A_{depth,min}$, required?

(b) The depth filtration medium in use is packaged in modules with a filtration area of 0.77 m^2. How many modules are required?

(c) How much time would be required to process this broth?

Solutions

(a) The area required is the ratio of the volume of broth fed to the depth filters to the capacity of the filter:

$$A_{depth,\,min} = \frac{V_{feed}}{capacity} \qquad (7.4)$$

$$= \frac{2,000\ L}{250\ L/m^2}$$

$$= 8\ m^2$$

(b) The number of modules is calculated by dividing the depth filtration area calculated previously by the area per module:

$$\#\,modules = \frac{8\ m^2}{0.77\ m^2/module}$$

$$= 11\ modules\ (rounding\ up\ from\ 10.4)$$

(c) 11 depth filtration modules at 0.77 m^2 per module leads to 8.47 m^2 of filtration area. The feed rate, Q_f, is the product of the feed flux (150 LMH) and the depth filtration area:

$$Q_f = J_f A_{depth} \qquad (7.5)$$

$$= 150\ L/m^2h \times 8.47\ m^2$$

$$= 1271\ L/h$$

The process time, $t_{process}$, is calculated from equation (6.7) as

$$t_{process} = \frac{V_{feed}}{Q} = \frac{2,000\ L}{1,271\ L/h}$$

$$= 1.57\ h\ (or\ 94\ min)$$

Note that often a safety factor is used in the design of filtration steps. For example, a safety factor of 1.5 means 50% more filter area is used, $A_{depth,actual}$, than is calculated using data from bench-scale studies, $A_{depth,min}$; that is

$$A_{depth,actual} = \text{safety factor} \times A_{depth,min} \qquad (7.6)$$

Safety factors account for potential variability in feed streams between product batches, variability in filter media across lots, and filter performance differences on scale-up [181]. So, if a 1.5 safety design factor were used here, $A_{depth,actual}$ would be 12 m² (= 1.5 x 8 m²), and the number of modules and processing time would be calculated based on this value.

- *Turbidity*: cloudiness of solution due to the presence of particles. It is typically measured using a nephelometer, which measures the scattered light in an illuminated sample by placing a detector at a specific angle (often at a right angle) to the incident light. The more scattered the light, the higher the turbidity. It is expressed in nephelometric turbidity units (NTU). See Figure 7.8 for a photo of suspensions with different turbidity values.
- *Turbidity breakthrough*: the point where particles have made it all the way through the filter and are observed in the filtrate.
- *Fouling*: plugging of the filter due to accumulation of solids on and in the depth filtration media.

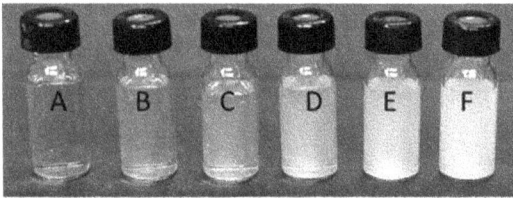

Figure 7.8: Standards with turbidity values in vials A-F of 0, 50, 100, 250, 500, and 1,000 NTUs, respectively. Reprinted from "Evaluation of Microflow Digital Imaging Particle Analysis for Sub-Visible Particles Formulated with an Opaque Vaccine Adjuvant" by G. Frahm, A. Pochopsky, T. Clarke, M. Johnston, 2016, PLoS ONE, under license by Creative Commons Attribution 4.0 International [182].

Let's begin by considering input parameters and output (performance) parameters in a depth filtration step. Input parameters include the following: (1) type of depth filter, (2) feed flux, (3) solids concentration of the feed, and (4) particle size distribution of feed. Note that the first three are controllable and therefore fit our definition of process parameter presented in Chapter 2. The last is a property of the process intermediate fed to the depth filter and may not be controllable. Performance parameters include (1) filter capacity, (2) filtrate turbidity or turbidity reduction (= turbidity$_{feed}$ − turbidity$_{filtrate}$), (3) pressure differential across the filter (ΔP defined by equation (7.1)) at constant feed flux, and (4) percent recovery of product (i.e., step yield).

A number of studies have been conducted that shed light on the impact of process parameters and attributes of the feed material on depth filter performance. For example, Figure 7.9 presents data on the impact of feed flux on filter capacity and filtrate turbidity [183]. Filter capacity is presumably measured as the volume fed until the recommended maximum pressure difference across the filter is reached. These data were obtained by varying the feed flux of a CHO culture, at a density of 3.6×10^6 cells/ mL and viability of 70%, to a depth filter with a pore size rating of 8 μm. The data show that as feed flux increases, filter capacity decreases and turbidity increases. The increase in turbidity is attributed to adsorptive (electrostatic) capture of particles becoming less efficient as the residence time within the filter medium decreases (due to increased flow rate). This increase in filtrate turbidity with increasing feed flux would also result in a decrease in the throughput of a 0.2 μm surface filter placed after the depth filter, due to increased solids loading to the 0.2 μm filter. Thus, a larger filtration area would be required of a 0.2 μm filter as the feed flux to the preceding depth filter increases.

Attributes of the feed may have an impact on depth filtration performance as well, as illustrated in Figure 7.10. These data were obtained from a process in which a yeast (*Pichia pastoris*) fermentation broth was clarified using a centrifuge. The resulting clarified liquid was processed for additional solids removal using MilliporeSigma's Millistak+® A1HC depth filtration media at a constant feed flux of 250 LMH. Two different batches of clarified liquid were used: one with a solids content of 0% (that is, very low solids concentration) and another at 0.7%. Figure 7.10 shows pressure differential, ΔP, and turbidity as a function of throughput for each feed. We might expect that the 0.7% feed would show greater ΔP and filtrate turbidity values, relative to the 0% feed, given the higher particle concentration. While ΔP for the 0.7% feed is always higher than that for the 0% feed, the rise in turbidity for the 0% feed is more rapid than that of the 0.7% feed. So what's going on? The feed with the low solids concentration likely contains smaller particles than the feed with the higher solids concentration. These small particles bind mainly by an adsorption mechanism rather than a sieving mechanism. The adsorption binding sites quickly become saturated, resulting in solids breakthrough for the 0% solids feed. On the other hand, for the feed with the higher solids content, sieving is the main retention mechanism, and as pores become blocked, ΔP increases much more rapidly than for the low-solids feed. This interesting result clearly demonstrates that properties of the feed to the depth filter, in this case solids concentration and particle size, can impact filter performance.

Table 7.1 presents a summary of the impact of process parameters and material attributes on depth filtration performance, evaluated by considering their impact on capacity, filtrate turbidity, and pressure drop across the filter. Note that in addition to the parameters shown, ionic strength of the feed may impact particle retention for cases in which adsorption is a primary retention mechanism. At high concentrations, ions may occupy binding sites that would otherwise be used for adsorption of particles. Lot-to-lot variability of filters may also impact depth filtration performance.

8µm Depth Filter Capacity as a Function of Different Flow Rates
3.6 x 10^6 CHO cells/ml at 70% viability level

Capacity (liter/m^2)

| | Constant flow mode at 100l/m^2/h flow | Constant flow mode at 200l/m^2/h flow | Constant flow mode at 300l/m^2/h flow | Constant flow mode at 500l/m^2/h flow |

Effluent Turbidity as a Function of Flow Rate
3.6 x 10^6 CHO cells/ml at 70% viability level into 8µm depth filter

Filter effluent turbidity
[NTU]

| | Constant flow mode at 100l/m^2/h flow | Constant flow mode at 200l/m^2/h flow | Constant flow mode at 300l/m^2/h flow | Constant flow mode at 500l/m^2/h flow |

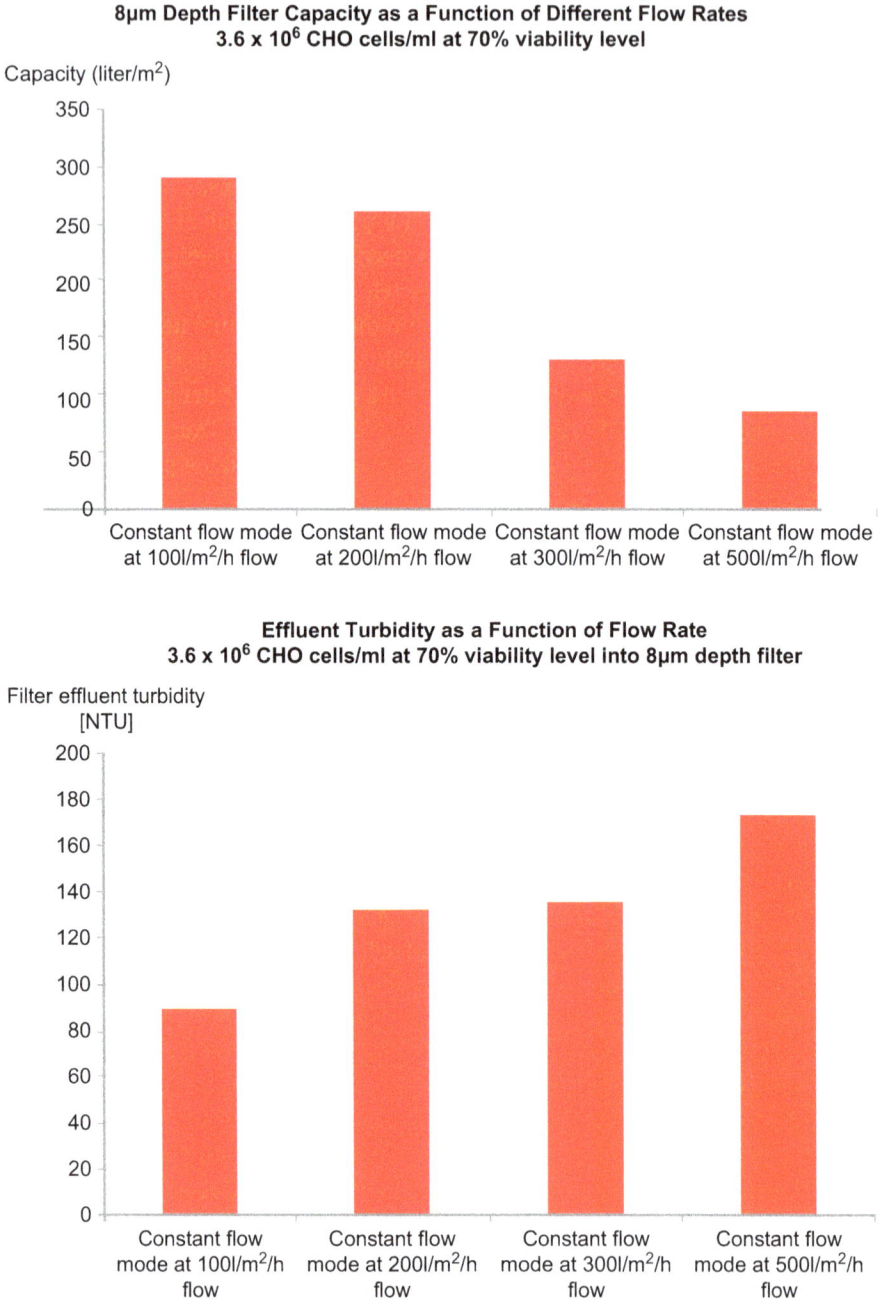

Figure 7.9: Depth filter capacity and filtrate turbidity as a function of feed flux for filtration of a CHO culture at a density of 3.6 × 10^6 cells/mL and viability of 70%. Note that a depth filter with a pore size rating of 8 µm was used. Adapted from "Depth filtration: Cell clarification of bioreactor offloads," by M. Prashad and K. Tarrach, 2006, *Filtration and Separation, 43*(7), 29. © (2006) Elsevier B.V [183].

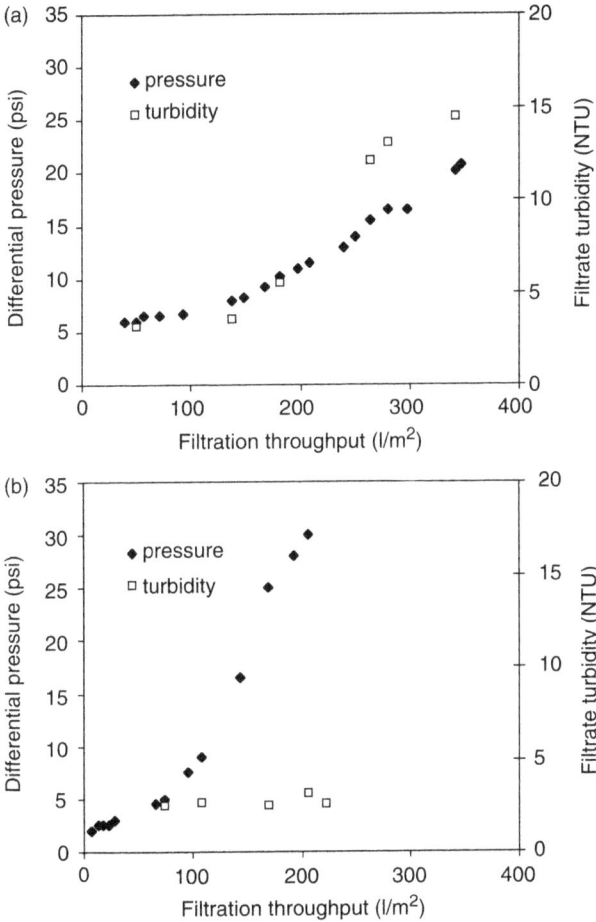

Figure 7.10: Pressure difference and filtrate turbidity versus throughput for depth filtration of centrifuge clarified liquid from a *Pichia pastoris* fermentation broth. Millistak+® A1HC depth filtration media was used at a constant feed flux of 250 LMH. (a) Results for a clarified liquid feed with very low solids concentration (0% solids). (b) Results for a clarified liquid with 0.7% solids. Image copyright 2007 from *Process Scale Bioseparations for the Biopharmaceutical Industry* by Shukla, A; Etzel, M; Gadam, S., eds. Reproduced by permission of Taylor and Francis Group, LLC, a division of Informa plc [184].

How do we select a depth filter for a given application and know how much filter area will be required for a production process? If information on the size distribution of particles to be filtered is known, initial selection of depth filtration media may be done based on pore size using charts like the one shown previously in Figure 7.4. However, ultimately it is prudent to conduct a bench-scale study given that the amount of filter area required is difficult to predict theoretically. Most filter manufacturers offer small-area depth filtration devices for these bench-scale studies.

Table 7.1: Depth filtration inputs and their impact on performance.

Process parameter or material attribute	Impact on depth filter performance
Depth filter type	For a fixed feed flux and a feed stream with fixed properties (solids concentration, particle size distribution), as pore size decreases – Filter capacity (based on pressure drop) decreases – Filtrate turbidity decreases – ΔP rises more rapidly as filtration proceeds
Feed flux	For a fixed filter and feed stream, as feed flux increases – Filter capacity decreases – Filtrate turbidity increases – ΔP rises more rapidly (either plotted against time or throughput) as filtration proceeds
Solids concentration in feed	For a fixed feed flux and filter, as solids concentration increases (at constant particle size) – Filter capacity (based on pressure drop) decreases – Filtrate turbidity may increase – ΔP rises more rapidly (either plotted against time or throughput) as filtration proceeds Note that solids concentration in the feed can be impacted by bioreactor operating conditions
Particle size distribution in feed	For a fixed feed flux and filter, as particle size decreases – Filter capacity (based on filtrate turbidity) decreases – Filtrate turbidity may increase – ΔP rises less rapidly as filtration proceeds Note that particle size distribution in feed can be impacted by bioreactor operating conditions

The best approach to conducting performance studies on depth filters at bench scale is a constant-flow method, which MilliporeSigma refers to as a Pmax™ study. Pmax™ stands for maximum pressure [185, 186]. Recall that under constant flow conditions, the pressure drop across the filter increases as the pores becomes blocked and plugged. The constant-flow method measures the filter capacity by loading feed until a maximum pressure drop, ΔP_{max}, for the filter is reached. The procedure is straightforward. A representative feed is pumped at a constant flow rate through a small-scale version of the depth filter. Feed flux values of 100–200 LMH are common. The filtrate volume, feed pressure, filtrate pressure, and filtrate turbidity are measured with time and recorded. The test is stopped when the feed pressure reaches the chosen ΔP_{max}. Typically, the product concentration in the feed, product concentration in the filtrate, turbidity in the feed and the turbidity of the final filtrate pool are also measured to determine product step yield and overall clarification efficiency. The filter capacity is calculated simply as the liters fed per square meter of filter area at ΔP_{max}.

The maximum pressure drop can be based on the allowable pressure for the filter, accounting for the fact that at large-scale, the depth filter is likely to be followed by a 0.2 µm surface filter that has its own pressure limitations. For example, MilliporeSigma's Millistak+® process-scale Pod filters are rated for a maximum operating pressure of 50 psig and a maximum pressure differential of 30 psi [187]. You might choose to limit the pressure drop across the depth filter to 25 psi, estimating that the 0.2 µm surface filter in the setup would contribute an additional 15 psi of pressure drop. This additional pressure drop would require a feed pressure of 40 psig for operation, just below the allowable maximum pressure to the depth filter. (Note that the depth filter filtrate pressure would be 15 psig with a 25 psi pressure drop, and the 0.2 µm filter filtrate would be at 0 psig with a 15 psi pressure drop.)

Used for screening, the constant-flow method provides data (ΔP vs throughput, filter capacity, product recovery, and turbidity) that enables comparison of the performance of different depth filtration media. Used for sizing, the data provide a capacity value from which a production system can be sized. To size a production system based on a specific batch volume and production time, the constant-flow testing should be conducted over a range of feed flux values that includes the lowest practical feed flow rate for production-scale processing [185].

There is one last point to make regarding depth filtration before moving to tangential-flow MF. Depth filtration is not only capable of clarification, but it can remove soluble impurities from process streams as well. The charge on depth filters allows them to be effective at removing soluble process-related impurities such as host cell proteins, host cell DNA, and virus, presumably because many of these impurities have a negative charge at process conditions, which promotes binding to the positively charged depth filtration media. A number of studies have demonstrated clearance of these impurities [188, 189]. Note that these impurities are discussed in greater detail in Chapter 8.

At this point, we turn our attention to another common filtration method used for separation of solids from liquid: tangential-flow MF. As mentioned at the beginning of the chapter, tangential-flow MF is not covered in as much detail as depth filtration in this chapter. Instead, we wait to present some details, particularly those related to membrane materials, parameters, process development, and scale up, until we cover the related topic of tangential-flow UF in Chapter 9.

7.6 Tangential-flow MF overview

As the name suggests, tangential-flow MF involves the use of MF membranes operated in tangential-flow mode for solid-liquid separations. In contrast to depth filtration, in which only a clarified filtrate can be collected as product (i.e., solids cannot be recovered from a depth filter), either the clarified permeate or the retained solids, in the form of a concentrated cell slurry, can be recovered as product with a tangential-flow

MF system. The primary mechanism of particle retention for MF membranes is sieving at the membrane surface. MF membranes have pore sizes as small as 0.1 μm or as large as 10 μm [173]. They are one of a number of different membrane types used for separations and classified by their pore size. Others include UF membranes, the topic of Chapter 9, which have a smaller pore size; they retain solutes with a diameter of 0.005–0.15 μm [190]. Reverse osmosis membranes, another example, have pores small enough to separate water from ions such as Na^+ and Cl^-. Reverse osmosis is commonly used in the biopharmaceutical industry not as a production step but as one of several steps used to purify water. A graphic comparing different types of membrane separations is shown in Figure 7.11.

Tangential-flow MF steps are designed and implemented to deliver the desired clarity in the resulting permeate if the product is dissolved in the liquid phase or to achieve a target concentration of solids if solids recovery is the primary objective. Maximizing step yield, short and reproducible processing times, acceptable membrane area, and, of course, acceptable product quality are additional goals.

A typical equipment setup for tangential-flow MF is shown in Figure 7.12. To produce tangential flow through the MF membrane module, a different setup is used than that shown for depth filtration in Figure 7.6. A pump moves feed slurry from the feed tank to the MF membrane module. The retentate valve is used to increase pressure on the feed side of the module, which creates the TMP, defined previously in equation (7.2), that is required to force liquid to permeate the membrane. Two streams are created as a result of the process: the permeate, which passes through the MF membrane, and the retentate, the portion of the feed that does not permeate the membrane and is recirculated to the feed vessel. Note that in MF, a pump may also be present in the permeate line to lower the permeate flow rate thereby reducing the buildup of solids on the membrane. The retentate stream passes through the system multiple times and becomes concentrated in solids, while the permeate stream is essentially void of solids but contains the product if it is present in the liquid phase (e.g., CHO culture broth in which the monoclonal antibody is secreted to the surrounding liquid). To maximize recovery of product in the permeate, the retentate remaining in the vessel may be washed (also referred to as diafiltration) with a buffer. More specifically, as liquid containing the product permeates the MF membrane, wash buffer is added to the feed, often at the same rate that liquid permeates, thereby maintaining constant volume in the feed tank. This procedure washes product from the retentate and into the permeate and results in enhanced recovery of the product from the initial feed (i.e., enhances step yield). In addition, a tangential-flow MF setup will have a variety of sensors in place. For example, the pressure of the feed, retentate, and permeate are typically monitored to measure the TMP. We continue discussion of TFF equipment in Chapter 9, including more information about MF membranes and membrane modules.

In tangential-flow MF, the permeate rate is not equal to the feed rate (also referred to as the recirculation rate) as it is in depth filtration because only a portion of the liq-

Reverse Osmosis

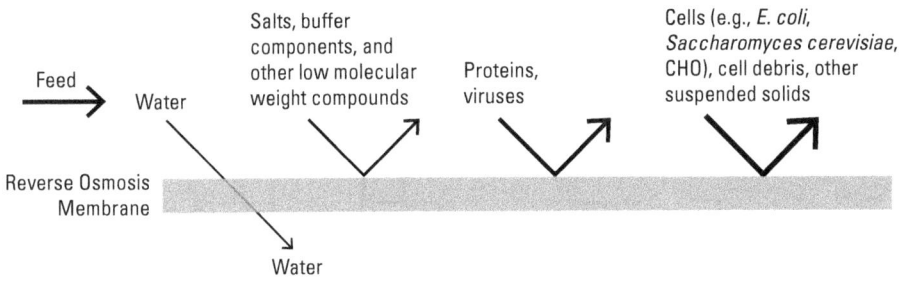

Ultrafiltration

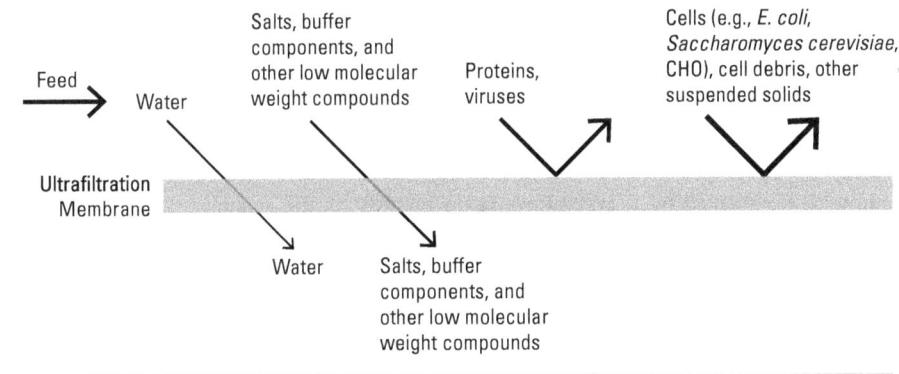

Microfiltration

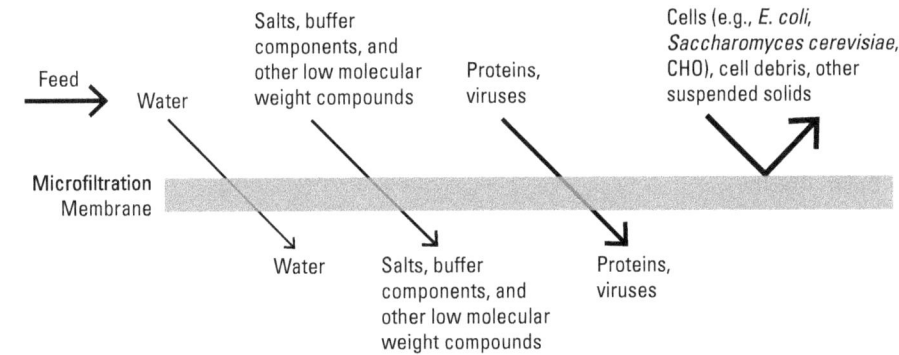

Figure 7.11: A comparison of different types of membrane separations. Image © NC State University; reprinted with permission.

uid in the feed permeates the MF membrane. The permeate flow rate is usually expressed as a permeate flux (i.e., the volumetric permeate rate across the membrane per unit area, often in units of LMH). It can be represented by the same general equation written for the feed flux across a depth filter (equation (7.3)) but in a slightly modified

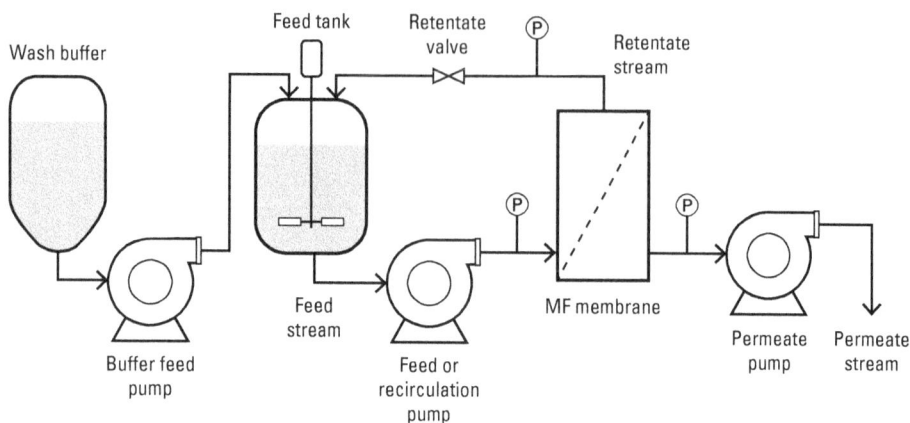

Figure 7.12: Equipment setup for a tangential-flow MF step. Note that the wash buffer vessel is present to recover product in the liquid phase from the retentate by diafiltration. Image © NC State University; reprinted with permission.

form, with J_f replaced by J_p and ΔP replaced by TMP as the relevant driving force for permeate through the membrane (refer to Figure 7.1)

$$J_p = \frac{Q_p}{A_m} = \frac{\text{TMP}}{\mu R_{\text{total}}} \tag{7.7}$$

Note that Q_p and A_m represent the volumetric flow rate of the permeate and the membrane area, respectively. In the case of MF, the total resistance has three contributions: the intrinsic membrane resistance (i.e., the resistance of a clean, unfouled membrane), R_m, resistance due to formation of a filter cake on the membrane surface, R_{fc}, and resistance from internal membrane fouling, R_{if} [191]. Both R_{fc} and R_{if} initially increase with filtration time; therefore, permeate flux decreases with time at constant TMP, although in tangential-flow mode, the permeate flux levels off with time after an initial decline.

As we have mentioned, there is much more to be covered on the topic of MF, including procedures for current good manufacturing practice (CGMP) production, parameters that impact and quantify performance, process development, and scale up. These topics are addressed in Chapter 9.

7.7 Unit operations for solid-liquid separation methods: a comparison

Having considered in detail the three most common methods for solid-liquid separations in biopharmaceutical processes, each of which is well suited to the harvest stage, the logical question is, how do we choose one? To help in that decision, a comparison among the three main operations considered in the last two chapters is shown in Table 7.2. Note that normal-flow MF has not been included in this table because its solids capacity is too small to be useful for feeds with a relatively high concentration of solids. However as was mentioned previously, normal-flow MF is by far the most common method for a very important solid-liquid separation application in biopharmaceutical processing: removal of bioburden.

Table 7.2 shows that each of the operations discussed is effective at separating solids from liquids and, if designed properly, can produce high product step yields. If recovery of solids is desired, depth filtration is not an option; centrifugation and tangential-flow MF are. Note that because cells can be recovered by microfiltration, MF membranes are often used as cell-retaining devices in perfusion bioreactor setups. Tangential-flow MF typically produces a filtrate with low solids concentration relative to centrifugation or depth filtration. Generally, the solids capacity for centrifuges is higher than either depth or tangential-flow MF. Expect the capital cost for a centrifuge to be higher than equipment for the other two operations. Consumables costs will be higher for depth filtration, given the need to purchase new depth filters for every batch of product. Because MF membranes may be reused for the same product, their costs may be distributed across many product batches, which lowers the consumables cost associated with membrane module purchases.

Table 7.2: Comparison among the solid-liquid separation operations covered in Chapters 6 and 7.

Unit operation	Effectiveness at separating solids from liquids	Product yield	Collection of solids possible	Particle-free stream	Solids capacity	Capital cost	Consumables cost
Centrifugation	High	High	Yes	No	High	High	Low
Depth filtration	High	High	No	No	Medium	Low	High
Tangential-flow MF	High	High	Yes	Yes	Medium	Medium	Medium

7.8 Other operations for solid-liquid separation

Centrifugation, depth filtration, and tangential-flow MF will continue to be used for solid-liquid separation applications in biopharmaceutical processes into the foreseeable future. Advances in these well-used technologies will continue, such as development of enhanced filter/membrane materials that improve performance and less expensive single-use options. However, there are also a number of other solid-liquid separation technologies with potential for greater adoption over the coming years. These include acoustic separation, flocculation followed by filtration or centrifugation, and magnetic cake filtration. We provide a synopsis of each in the following sections.

7.8.1 Acoustic wave separation

Acoustic wave separation is a technology that continuously removes solids, such as CHO cells, from a feed. The feed slurry flows to a flow channel within an acoustic chamber. A three-dimensional standing acoustic wave is created within the flow channel, and the resulting forces cause solids flowing through the chamber to migrate to the nodes of the standing wave. As the solid cell clusters grow, they quickly settle by gravity, resulting in a clarified liquid exiting the acoustic chamber [192]. The cells can then be pumped out of the chamber.

A few products on the market utilize acoustic wave separation technology. MilliporeSigma offers the Ekko™ Acoustic Cell Processing System, designed for cell therapy manufacturing, which enables cell washing and concentration [193]. Another example is the Applikon BioSep, a cell retention device used in perfusion cell culture [194].

7.8.2 Flocculation

Flocculation is a method that has been implemented for solid-liquid separation in other industries for years and is gaining increasing interest for biopharmaceutical manufacturing. It is a pre-treatment method in which a flocculating agent is added to a slurry and causes the dispersed individual particles to adhere to one another, thereby creating larger particles. This shift to larger particles can enhance separation by methods such as centrifugation and filtration. A number of different flocculating agents have been studied over the years. Because animal cell surfaces are known to be negatively charged [195], cationic polymers have been found to serve as a good flocculating agents by neutralizing the cell surface, thus allowing agglomeration of cells and/or bridging between cell surfaces. Care must be taken in choosing a flocculating agent as strongly cationic polymers can be toxic to cells [195]. One cationic polymer that has been marketed for CGMP use and has been shown as effective for flocculating CHO cells is poly (diallyldimethylammonium chloride) (pDADMAC) from MilliporeSigma. A study comparing depth filtration of a CHO culture broth treated with pDADMAC against depth filtration of untreated CHO culture broth showed that filter capacity of the broth treated with pDADMAC was significantly higher than could be achieved using untreated broth [196]. Of course, a process that incorporates a polymer flocculation step must be designed to provide sufficient clearance of the flocculating agent, particularly if the agent is cytotoxic at the levels used in the process.

7.8.3 Magnetic cake filtration

Magnetic separations use polymeric beads that are magnetic and to which a ligand has been attached with a high affinity for a component (e.g., a product). The bead selectively binds the product from a crude feed and is removed by some method that takes advantage of its magnetic properties. The product is then recovered from the beads. To date, magnetic separations have been limited to analytical and diagnostic applications in which only a small number of beads are required [197].

For biomanufacturing applications, a relatively large number of beads is needed. To make this separation work at larger scale, magnetic cake filtration has been proposed [197]. The magnetized beads are mixed with a feed and allowed ample contact time to adsorb the target compound. The beads are then trapped in the filter with the help of a magnet and washed for removal of cells, cell debris, and unwanted soluble impurities. Finally the product is recovered by eluting it from the beads. While studies have only been performed at bench scale [197], the method holds promise for larger scale biopharmaceutical process applications as it streamlines operations by combining product harvest and initial purification step.

7.9 Summary

Filtration is any technique used to separate particulate in a fluid suspension or components in solution according to their size by flowing under a pressure differential through a porous medium. This chapter continued the discussion from Chapter 6 on solid-liquid separation in biopharmaceutical processes by considering methods for separation of solid particles from a fluid suspension (i.e., the first half of the definition) by filtration. There are numerous solid-liquid separation applications in biopharmaceutical processes for which filters are well suited, including clarification of cell culture broths, clarification of lysate streams, removal of precipitated impurities from process streams, and removal of bioburden.

Depth filters operated in normal-flow mode are commonly used and offer the advantage of high capacity for solids, particularly when compared to surface filters used in normal-flow mode. They come in a range of pore sizes, from approximately 0.1 μm to greater than 10 μm. Depth filter media is typically packaged into single-use modules that are available from numerous vendors. The modules are placed in a parallel array using holders, and the number of modules in the holder can be increased or decreased to achieve the appropriate surface area for a given application. In addition to the filter modules and holder, equipment required for depth filtration includes a feed vessel, pump, and pressure gauges to measure pressure drop, as shown in Figure 7.6. A typical depth filtration procedure for CGMP processing is shown in Figure 7.7. Depth filtration steps are typically operated at a constant feed flux (volumetric flow rate through the filter per unit area, measured in $L/m^2/h$). To select an appropriate filter for a given application, bench-scale screening studies are conducted. These studies are also designed to determine the capacity of the filter – the maximum volume that can be filtered per unit area, typically measured in L/m^2 – for a given feed for scale-up purposes. A constant-flow method is commonly used that measures the volume of feed that can be loaded until a maximum allowable pressure drop is reached, from which the capacity is determined. The resulting capacity value can then be used to estimate the filter area (i.e., number of filter modules) required for a particular application.

In addition to normal-flow depth filtration, tangential-flow MF is also a common filtration method for solid-liquid separation in biopharmaceutical processes. MF membranes come in pore sizes similar to depth filters, although the pore size distribution is narrower. They are surface filters that are run in tangential-flow mode when processing feeds with relatively high solids concentration. A typical process setup is shown in Figure 7.12. More information on tangential-flow MF, including a discussion of process and performance parameters, procedures in a CGMP environment, development studies, and scale up, is provided in Chapter 9.

There are a number of advantages and disadvantages to the three solid-liquid separation methods described in Chapters 6 and 7: centrifugation, depth filtration, and tangential-flow MF. For example, if recovery of solids is required, depth filtration is not an option. For applications in which product is dissolved in the liquid phase of a suspension, tangential-flow MF is capable of producing a product stream with low solids concentration relative to centrifugation and depth filtration; however, the clarified liquids from centrifugation and depth filtrations can be further processed using normal flow MF if needed. Capital costs are likely to be significantly higher for centrifugation than the other methods, while consumables costs are typically higher for depth filtration and tangential-flow MF. In addition, several other methods for solid-liquid separation in biopharmaceutical manufacturing are available and may see wider adoption in coming years. These include acoustic wave separation, polymeric flocculation, and magnetic cake filtration.

7.10 Review questions

1. Qualitatively plot ΔP versus throughput and the feed flux, J_f, versus throughput for the following two scenarios: a depth filtration step operated at constant feed flux and a depth filtration step operated at constant inlet pressure.

2. *Pichia pastoris* is used to produce an extracellular therapeutic protein. Centrifugation is used as a primary clarification step for the fermentation broth and depth filtration as a secondary step (i.e., clarified liquid from the centrifuge is fed to the depth filter). The depth filter that provides the best performance has a capacity of 160 L/m^2 at a feed flux of 250 LMH [198].

(a) How much depth filtration area is required to process a 2,000 L fermentation? Note that the centrifuge removes most of the yeast, which is at a concentration of 50% by volume in the fermenter. Use a 50% safety factor in your calculation.

(b) How many depth filtration modules are required if the largest available has a surface area of 1.1 m²?

(c) How much time is required to perform the filtration?

3. The capacity of a depth filter used to clarify a CHO culture is significantly less at production scale than what was measured in bench-scale studies. High pressure drop across the production setup has resulted in a premature end to filtration, and the feed has to be directed to a backup filtration setup. Give three possible reasons for the less-than-expected capacity.

4. Bench-scale studies are being planned to mimic a depth filtration step used to clarify a production CHO culture, and an estimate of the volume of feed material that will be required is needed. In the production process, 1,000 L of CHO culture is fed to Zeta Plus™ 60SP02A depth filter capsules from 3M. Seven capsules with a surface area of 1.6 m² per capsule are used. The small-scale setup will use a capsule of the same filter media with an area of 170 cm². What minimum volume of culture is required at bench scale to equal the throughput of the large-scale operation? (Data from Thomas P. O'Brien, "Large-Scale, Single-Use Depth Filtration Systems for Mammalian Cell Culture Clarification," Bioprocess International, 10(5), May 2012 [174].)

5. Data from a study to determine the filter area required for a production-scale depth filtration step for clarification of an insect cell (Sf9) lysate to produce a gene therapy vector is shown below.

Table 7.3: Data required to determine depth filtration area for clarification of an insect cell lysate.

Time (min)	Volume filtered (mL)	Feed pressure (psi)	Filtrate turbidity (NTU)
3.5	40	1.4	
8	49.7	1.6	
10	71.1	1.8	
14	114.7	2.1	39.2
21	189.6	2.5	
25	233	2.7	38

Table 7.3 (continued)

Time (min)	Volume filtered (mL)	Feed pressure (psi)	Filtrate turbidity (NTU)
32	308.3	3	
38	372.1	3.2	39
46	459.8	3.5	
50	502.9	3.7	39.4
58	588.8	4.2	
61	621.8	4.6	39.9
69	707.8	5.4	
74	762	5.9	41
83	863	7	
91	948.8	8.1	40.5
103	1,081	10.7	
117	1,223.6	14.3	
127	1,329.4	17.7	40.3

This study used a Millistak+® HC depth filter in μPod® format, D0HC media series. The experimental setup is shown in Figure 7.13.

(a) What is the nominal pore size rating Millistak+® HC D0HC filtration media?

(b) What is the surface area of the depth filter used? An Internet search is required.

(c) What average feed flux, in LMH, was used?

(d) Plot the pressure (psi) versus throughput in L lysate fed/m² filter area.

(e) The maximum pressure difference for the filtration step is 15 psi. From the plot, estimate the filter capacity at this pressure.

(f) How much filter area is required to process 2,000 L of Sf9 lysate assuming the filters are loaded to the capacity estimated previously?

(g) You purchase filter modules with a 1.1 m² area for production. How many modules are required if 50% more area (safety factor) is purchased than calculated above?

(h) What will the process flow rate be in L/h?

(i) What time will be required to process 2,000 L of Sf9 lysate?

Process Vessel

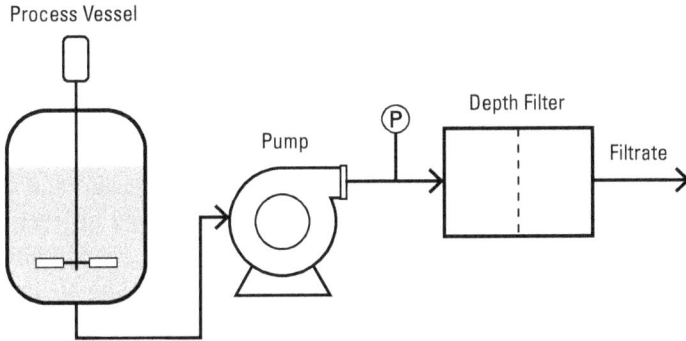

Figure 7.13: Experimental setup used to generate the data shown in Table 7.3. Image © NC State University; reprinted with permission.

6. To determine the intrinsic resistance of a 0.1 μm hollow fiber MF membrane, R_m, a water permeability test is conducted using HPW. HPW permeates a new membrane at a rate of 200 L/h, measured using a graduated cylinder and stopwatch, at a pressure of 5 psi using a hollow fiber module with 0.16 m² of area. What is the value of R_m, in m^{-1}? (Data from [184].)

7. To recover an extracellular protein therapeutic produced in *Pichia pastoris*, a tangential-flow MF wash (diafiltration) step was performed, using the same 0.1 μm hollow fiber MF membrane from the previous problem. If a constant permeate flux of 55 LMH results in an average TMP of 2.5 psi, what percentage of the resistance to permeate flow is due to the cake on the surface and internal membrane fouling (i.e., $R_{fc} + R_{if}$)? (Data from [184].)

8. The tangential-flow diafiltration step referred to in the previous problem was executed using two diafiltration volumes (DVs); DV is defined as the ratio of the volume of buffer added to the volume of product processed. The diafiltration was performed by maintaining constant volume in the feed vessel. What % step yield would be achieved with two DVs, assuming that the product, a protein therapeutic, passes freely across the MF membrane? Note that the fermentation broth being diafiltered contains 50% by volume *P. pastoris*; therefore two DVs based on total volume correspond to four DVs based on liquid volume. Hint: to complete this problem, a product balance in the form of input – output = accumulation is required on the MF system. First derive the equation for the simple case in which no wash is done.

Chapter 8
Purification operations: chromatography

At the end of the harvest stage, the active product is typically dissolved in a clarified – that is, solids-free – aqueous solution. A sample would likely look "clean" and clear, yet numerous soluble impurities are present that must be removed to levels low enough to ensure no risk to a patient. Liquid chromatography is the unit operation in biopharmaceutical processing most commonly used to remove these soluble impurities from the product and is therefore essential for meeting critical quality attributes related to purity.

The topic of chromatography in biopharmaceutical processes is a big one. The processes occurring at the molecular level to create separation are complex. There are numerous types of chromatography media to choose from, numerous process parameters that can be varied in the design of chromatography steps to optimize performance, and a number of validation considerations required for current good manufacturing practice (CGMP) processing. This chapter attempts to succinctly explore all of these topics, by addressing the following questions:

- What are the impurities that must be removed in a process to produce a drug product safe for patients?
- How does separation of the product from impurities occur in chromatography?
- What different chemistries are utilized on chromatographic stationary phases to achieve separation of product from impurities?
- In addition to resin beads used to pack chromatography beds, what other stationary phase formats are in use and how does their performance compare to a bed packed with resin?
- What factors describe the performance of a chromatography step?
- What is a typical sequence of phases (i.e., steps) used to carry out a chromatography run in a biopharmaceutical production process?
- What are key components of a chromatography system for CGMP biopharmaceutical manufacturing and how is the quality of a packed bed assessed prior to its use?
- What are the important process parameters that potentially impact chromatographic performance?
- How are chromatography steps developed at bench scale and scaled up to manufacturing?
- What is required to validate chromatography steps for a CGMP biopharmaceutical process?

We also note that in biopharmaceutical manufacturing, liquid chromatography is commonly used not only for production but also for testing samples of raw materials, product intermediates, and final product, in the form of high performance liquid

https://doi.org/10.1515/9783111112459-008

chromatography (HPLC) and ultra-high-performance liquid chromatography (UHPLC). While the basic modes of interaction between chromatography media and solutes used in analytical and process chromatography are the same, there are a number of differences between the two techniques. Analytical chromatography is a bench-scale activity, unlike most process chromatography, and is used for quantitative and/or qualitative analysis of a sample, which often requires resolving individual components as separate chromatographic peaks. To provide necessary resolution, analytical chromatography uses relatively small resin beads of 5 μm diameter or smaller. In contrast, liquid chromatography steps for biopharmaceutical production are designed to produce a product intermediate with the necessary purity and potency (rather than trying to resolve each individual component in the intermediate), while also providing a high step yield, high productivity, and a low cost.

8.1 Soluble impurities

Because the purpose of chromatography is purification, we begin the chapter with a discussion of some of the impurities of concern in biopharmaceutical processes. As mentioned previously, biopharmaceutical process streams contain a number of soluble impurities, which are usually removed for the sake of patient safety and for efficacy of drug product. Referring back to the release specification for a monoclonal antibody (mAb) presented in Table 1.5, you see that a number of tests are performed to quantify impurity levels in drug product. Generally speaking, an *impurity* is "any component present in the drug substance or drug product which is not the desired product, a product-related substance, or excipient including buffer components" [199]. Further, impurities may be classified as either process- or product-related [199]. *Process-related impurities* are impurities originating from the manufacturing process [199]. Common process-related impurities include cell-derived impurities such as host cell proteins (HCPs) and host cell DNA, cell culture/fermentation-derived impurities such as antibiotics and induction agents, and downstream impurities such as leached protein A from protein A chromatography resin. Common process-related impurities in biopharmaceutical processes are described more fully in Table 8.1.

    Product-related impurities refer to molecular variants of the product that may form during manufacture and storage. They are often physically and chemically similar to the product but not as safe or effective [199]. Separation of these impurities from product can be challenging due to the similarity between each. There are numerous examples of product-related impurities. Modification of amino acid residues during upstream production steps, downstream processing, and storage may take place, leading to oxidized, reduced, and deamidated forms of protein-based products. Product may become fragmented or aggregated as a result of processing, both of which may result in reduced product potency and undesirable immune response, particularly in the case of protein therapeutics [200, 201]. Further, as new biopharmaceutical modalities are approved, different product-related impurities must be considered. For example, viral vectors used in gene therapies have as a quality attribute the concentration (or percentage) of viral capsids that contain the gene of interest. Those capsids without the gene of interest or capsids that contain the incorrect DNA are product-related impurities and are undesirable for a number of reasons, including that they are a source of potentially antigenic material [202]. It is important to note that the U.S. FDA has no specific guidance on acceptable levels of product-related impurities in biopharmaceutical drug products. We will not go into detail on the formation mechanisms of these product-related impurities, as that is beyond the scope of this book and discussion on those mechanisms can be found elsewhere [203].

A term related to the discussion on impurities is *contaminant,* defined as an external component introduced into the process that is not intended to be part of the process [199]. For example, bioburden that enters a process stream from the environment is termed a contaminant rather than an impurity based on this definition.

Table 8.1: Common process-related impurities in biopharmaceutical processes.

Impurity	Description	Acceptable level
Cellular deoxyribonucleic acid (DNA)	Nucleic acids, including DNA, reside in any host cell and can be released. Host cell DNA is a risk due to the possibility that it might cause tumors or transmit viral infections [124].	<10 ng/parenteral dose [204] or decrease biological activity by reducing size (<200 base pairs) [124]
Endotoxin	Endotoxins (literally, toxins from within) are lipopolysaccharides that reside in the outer membrane of gram-negative bacteria like *E. coli*, as discussed in Chapter 5. Endotoxins produce a variety of responses in humans, including fever and a lowering of blood pressure. Even in processes using hosts other than gram-negative bacteria, endotoxins can be introduced as a contaminant and should therefore be cleared.	<5.0 endotoxin units/kg body weight/h [205]
Fermentation/ cell culture components	A variety of components, including serum, and other additives (selection agents, antibiotics, induction agents, transfection agents) that may reside in spent media. Some of these may be removed from the process prior to the purification stage, but some may not.	None reported

Table 8.1 (continued)

Impurity	Description	Acceptable level
Host cell proteins (HCPs)	Refers to the native proteins from the host cell that make their way into product intermediates. They are a heterogeneous group made up of hundreds and possibly even thousands of different proteins, with widely different properties [206]. They may cause an unwanted immune response, and, for those with biological activity, there may be a risk from the direct activity of the impurity. They are also a risk to the process due to the ability of some to degrade product.	Levels are typically expected to be <100 ng HCP/mg product [207], although the USP suggests achieving levels as low as reasonably possible [206].
Leachables	As defined in Chapter 2, leachables are compounds that migrate (i.e., leach) from any product-contact material under normal process conditions. As the use of single-use technologies increases, leachables have become a bigger concern due to the prevalence of plastics used to make single-use components.	Evaluation required as part of process validation
Virus	Production materials, especially those that are animal derived, are a source of unwanted viruses that could infect a patient. These unwanted viruses may fit the definition of contaminant as well.	Level of clearance depends on the potential viral load [208]

Note that salts and buffer components introduced during the process and not intended to be a component in the final product may also be classified as process-related impurities. However, these small molecular weight components are all typically removed during the final diafiltration step, if not earlier in the process, and are not discussed in this chapter as impurities.

8.2 Chromatographic principles

Chromatography's dominant use as a purification method is due largely to its ability to separate a variety of impurities from product while ensuring that product remains potent and safe. Chromatography systems can readily and effectively be cleaned and sanitized, ensuring that both batch-to-batch product contamination and microbiological contamination are minimized.

To begin the discussion on how chromatography works, consider Figure 8.1(a). A solution containing three colored solutes (fictitious blue, green, and red solutes) is introduced (i.e., loaded) to a cylindrical column that is packed with spherical particles. In chromatography, these particles are referred to as resins or beads, and they are added to an empty column to create a packed bed. The packed column contains a support on each end that keeps the resin in place but allows liquid to flow through. (Note that in addition to resin beads, the column always contains liquid in the space between the beads.) Once packed into a column, the beads do not move and therefore are referred to as the *stationary phase.* Liquid flowing through the column is referred to as *mobile phase.* The solution containing the blue, green, and red solutes is loaded under conditions in which each solute has an affinity for the resin and therefore binds to it. After the feed solution is loaded to the column, a change to the liquid flowing through the packed column is made. The mobile phase used at this step of the procedure is chosen to weaken the interaction between the solutes and the resin so that the column flow can push the solutes downward through the column. The greater the affinity of a solute for the stationary phase, the longer the time it spends bound to the stationary phase and the slower its movement through the bed. Each solute has a different affinity for the resin and therefore moves through the column at a different speed, which leads to separation. In the example shown in Figure 8.1(a), the blue solute has the greatest affinity for the resin and the red solute the least. The individual components originally mixed together in solution have been separated because they each move through the column with a different speed. If the green component is product and the red and blue components impurities, it is easy to see how product that is free from impurities can be collected in the column effluent.

It is important to note that not all types of chromatography involve binding of a mobile phase solute to the stationary phase. In particular, size exclusion chromatography is based on differences in the size of different components, as is discussed shortly. Regardless, the idea that different solutes travel through the packed column at different speeds still applies.

Figure 8.1(b) shows a plot of the concentration of each component in the column effluent (i.e., flow from the column outlet) versus time or versus the volume. The concentration trace results in a chromatographic peak for each solute. This type of plot is referred to as a *chromatogram,* which is a graphical presentation of any precolumn or postcolumn measurement – for example, UV absorbance, pH, or flow rate – plotted versus cumulative volume of mobile phase that has flowed through the column. Note also that throughout this chapter, we will continue to use the term *solute* or *component* to refer to molecules – products or impurities – in the mobile phase that may or may not bind or interact with the stationary phase. We use the term *target solute* specifically to refer to the molecule to be recovered – usually the product – in the mobile phase. The target solute may or may not bind to the resin, as will become clear shortly.

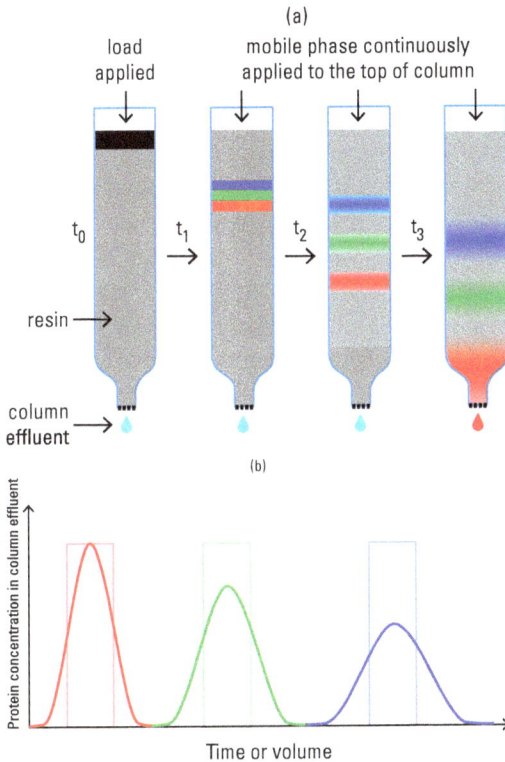

Figure 8.1: Illustration of a chromatographic separation. (a) The movement of three solutes through a packed bed. (b) The solute concentration in the column effluent versus cumulative volume of mobile phase fed (or time), referred to as a chromatogram. The bell-shaped curves show the concentration trace that results for each solute, while the rectangular curves are the traces that would result in the absence of band-broadening effects. Images © NC State University; reprinted with permission.

8.3 Creating separation: stationary phases, mobile phases, and properties of the solutes to be separated

In this section, we explore the nature of the interaction between a mobile phase solute and the stationary phase. The degree of interaction depends on properties of the solutes to be separated, properties of the stationary phase, and properties of the mobile phase. Let's start by considering properties of proteins (i.e., the solutes) that impact separation, given that the majority of biopharmaceuticals are protein therapeutics, such as insulin or Humira®, or are protein based, such as inactivated influenza virus used in the flu vaccine or adeno associated virus used for gene therapies.

8.3.1 Protein properties

Recall from Chapter 1 that proteins are macromolecules made up of amino acids and have a three-dimensional structure that is tied to their function. Three protein properties particularly relevant to chromatography are charge, hydrophobicity, and size.

Proteins are charged mainly because some amino acids have side chains, designated as "R" in Figure 1.1, that carry a charge, as shown in Figure 1.2. The net charge on a protein is the sum of charges of individual amino acids. Note that the amine and carboxyl groups on amino acids are also charged when in a solution at neutral pH; however, with the exception of the terminal amine and carboxyl groups of the protein, these groups form the peptide bonds with other amino acids and therefore don't contribute to the protein charge. The pH at which the net charge on a protein is zero is referred to as the isoelectric point, written pI. The pH of the solution in which a protein is dissolved determines its charge. At pH values below the pI, the net protein charge is positive due to the relative abundance of hydronium ions (H_3O^+) in solution. If the solution is at a pH value above the pI, the net protein charge is negative. The dependence of charge on pH is shown for three common proteins – bovine serum albumin (BSA), lysozyme, and ovalbumin – in Figure 8.2. Note that the net charge vs. pH curve is different for each of the proteins; therefore, at a given pH, different proteins have different net charges. For example, if BSA and lysozyme are both dissolved in a buffer at pH 7.0, BSA carries a negative charge, while lysozyme carries a positive charge. This difference in charge can be exploited to separate proteins by chromatography as becomes clear shortly.

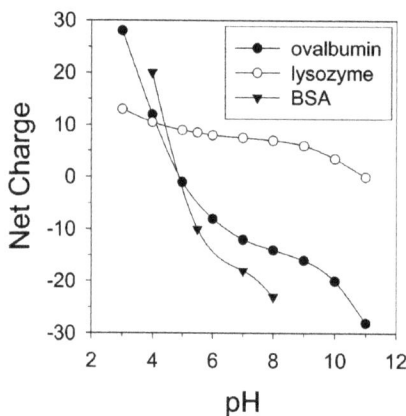

Figure 8.2: Net protein charge versus pH for ovalbumin, lysozyme, and BSA. The charge on the y-axis is in e^0. Reprinted with permission from "Specific Ion Effects in Solutions of Globular Proteins: Comparison between Analytical Models and Simulation," by M. Boström, F.W. Tavares, D. Bratko, B.W. Ninham, 2005, *The journal of physical chemistry.B, 109*(51), 24491. Copyright (2005) American Chemical Society [209].

Another property that differentiates proteins is their degree of hydrophobicity, which literally means fear of water. The term is used to describe molecules that have portions repelled by water. Figure 1.2 shows that amino acids such as phenylalanine and isoleucine are hydrophobic, given the nonpolar nature of their side chains. The presence of hydrophobic amino acids like these in proteins results in hydrophobic patches on proteins. Again, because different proteins are made up of different amino acids, hydrophobicity varies from protein to protein. Like charge, differences in hydrophobicity can be exploited to separate proteins in solution by chromatography.

Finally, in addition to charge and hydrophobicity, size is a property with significant variation among proteins, as clearly shown in Table 1.1. The molecular weight of insulin, a small protein with only 51 amino acids, is about 5,808 Da, while the molecular weight of a mAb, like the active ingredient in Humira®, is approximately 150,00 Da. Further, viruses, such as adeno-associated virus, are made up of protein shells and are significantly larger than even mAbs.

8.3.2 The most common stationary phase: resins used in packed beds

We begin this discussion by focusing on the interaction of solutes with the most common chromatographic stationary phase: resin beads. Resins are typically spherical and porous, as shown in Figure 8.3, to provide ample surface area for binding. Most resin beads, with the exception of those used in size exclusion chromatography, are made up of a support that is chemically modified by attaching specific ligands. This chemical modification, which is often referred to as the bead being functionalized, enables the interaction between soluble components and stationary phase illustrated in Figure 8.1. Resin supports are made from a variety of materials, including natural polymers such as agarose, synthetic polymers such as acrylate polymers, and inorganic materials such as silica. Diameters are typically 30–100 μm [290]. In process chromatography for biopharmaceutical applications, the area on the outer surface of the particle does not contribute significantly to the total area available for binding; that is, most of the area for binding is inside the resin beads, associated with the pores [210]. Therefore, for a solute to bind in significant amounts to the resin, the pore must be accessible to the solute, which means the pore diameter must be larger than the size of the solute. Pore size depends on the type of resin and can range from 30–400 nm [210]. Desirable properties of a resin include resistance to the many chemicals to which the resin is exposed, mechanical rigidity to withstand pressure drop, low non-specific binding to ensure adsorption occurs mainly by the mechanism for which the resin is designed (a large amount of non-specific binding affects the degree of purification achievable), high capacity for protein binding, and minimal leachable components.

As mentioned previously, the type of ligand attached to the support defines the interaction mode. Common chromatographic interaction modes are:

- **Ion exchange (IEC)**. Adsorption of a component from the mobile phase is based on attraction of opposite charges: a negatively charged solute binds to a positively charged resin bead and vice versa. Separation between components in the mobile phase is based on differences in charge between components. For example, a solute with a strong positive charge binds more strongly to a negatively charged resin than a solute with less positive charge or one with a negative charge.

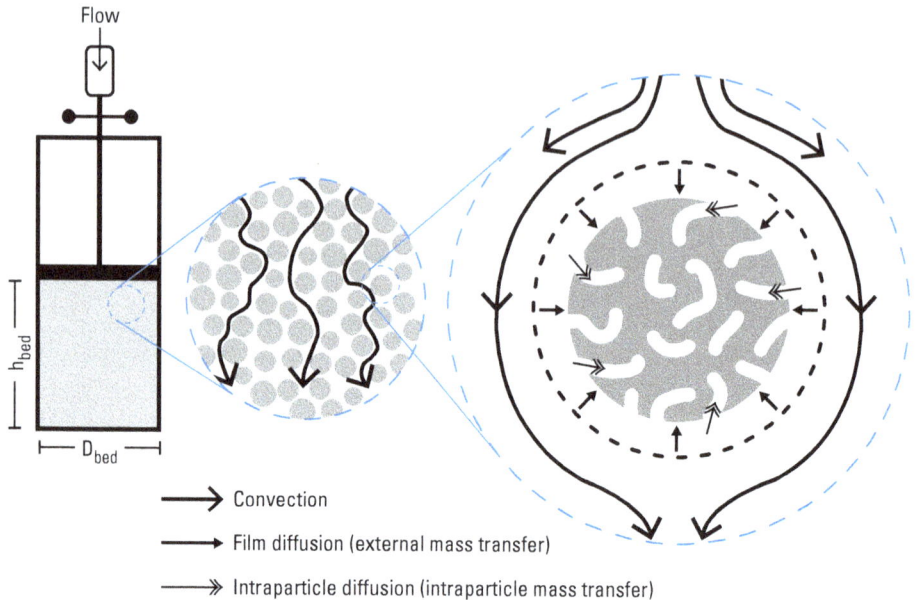

Figure 8.3: Steps (convection, film diffusion, and intraparticle diffusion) involved in the transport of solutes from the mobile phase to the surface of the resin pores (in a packed column) where most adsorption occurs. Note that the dotted line around the resin particle (closest to the particle) on the right represents the stagnant film layer through which solutes diffuse. Image © NC State University; reprinted with permission.

– **Affinity (AC)**. Adsorption relies on a highly specific interaction between a target solute and the resin, due to specific chemical groups on the solute. Think of the interaction like a key fitting in a lock. Solutes without the specific chemical group do not bind to the resin.
– **Hydrophobic interaction (HIC)**. Adsorption of a solute is based on hydrophobicity: a hydrophobic solute binds to a hydrophobic resin. The greater the hydrophobicity, the stronger the interaction between the solute and resin. Addition of certain salts is required to adsorb hydrophobic solutes to the resin. Solutes are eluted by decreasing the concentration of these salts.
– **Reversed phase (RPC).** Binding of a component to a resin is based on hydrophobic interaction, similar to HIC, but the binding is stronger and elution of bound solutes typically requires an organic solvent.
– **Size exclusion (SEC).** Separation is not driven by solute binding to the resin; instead, separation is based on size differences between solute molecules. Larger molecules in the mobile phase cannot diffuse into pores within the resin particles and elute after a volume of buffer equal to the column void volume (i.e., the volume between resin particles) has passed through the column. Smaller molecules can access the resin pores and are therefore held up longer in the column. SEC is

different from the other modes in that interaction of the solute with the resin is not based on affinity of a component to the resin bead.

- **Multimodal or mixed mode (MMC).** Adsorption of a solute is based on more than one type of interaction. A common type of MMC resin includes a ligand that is both charged and hydrophobic, which results in both ion exchange and hydrophobic interaction.

A summary of these different interaction modes, including examples of common ligands for each, appears in Table 8.2. There are numerous suppliers of chromatography resins, including Cytiva, Bio-Rad, Tosoh Bioscience, MilliporeSigma, Thermo Scientific, and Repligen. Given the number of companies producing resins and the fact that some companies offer multiple different resins for a given interaction mode, there are many resins to choose from when designing a purification step.

Table 8.2: Major chromatography techniques by interaction mode and examples of each.

Technique	Description	Ligand examples	Mobile phase conditions that promote solute adsorption and desorption
Ion exchange (IEC)	Adsorption of a component (solute) to a resin is based on opposite charge. Separation is based on differences in charge between components.	Anion exchange (AEC) [211, 212] quaternary amine (Q): $-CH_2N^+(CH_3)_3$ also: diethylaminoethyl (DEAE), diethylaminopropyl Cation exchange (CEC) [211, 212] Sulfopropyl (SP): $-CH_2CH_2CH_2SO_3^-$ also: carboxymethyl (CM), sulfonate (S)	Adsorption: pH that produces a charge on the solute that is opposite that of resin; low ionic strength. Desorption: pH that produces a solute charge that is the same as the resin; high ionic strength.
Affinity (AC)	Adsorption of a component to the resin based on a specific interaction ("lock and key") due to the presence of a specific chemical group on the component adsorbed. Other components do not interact with the resin and flow through.	Protein A, Protein G, Protein L for human IgG containing Fc region [74] Single-domain [V_HH] antibody fragment for adeno-associated virus [213]	Mobile phase conditions leading to adsorption or desorption of solute vary according to specific affinity resin.

Table 8.2 (continued)

Technique	Description	Ligand examples	Mobile phase conditions that promote solute adsorption and desorption
Hydrophobic Interaction (HIC)	Adsorption of a component (solute) to a resin is based on hydrophobic interaction between component and resin. Separation is based on differences in hydrophobicity between components.	Butyl: $-O-(CH_2)_3 -CH_3$ [214] Octyl: $-O-(CH_2)_7-CH_3$ [214]	Adsorption: high ionic strength resulting from including salts like $(NH_4)_2SO_4$ or Na_2SO_4. Desorption: low ionic strength resulting from lowering salt concentration.
Reversed Phase (RPC)	Adsorption of a component (solute) to a resin is based on hydrophobic interaction between component and resin (stronger than HIC). Separation based on differences in hydrophobicity between components	Matrix surface made up of nonpolar polystyrene/divinyl benzene [214]	Adsorption: aqueous buffer that includes an ion pairing agent and organic modifier. Desorption: nonpolar, organic solvent required.
Size Exclusion (SEC)	There is no adsorption in SEC. Solutes smaller than the resin pore diffuse in; solutes that are larger do not. Separation is based on size difference between solutes.	NA	No binding involved. Separation by SEC should be independent of mobile phase composition.
Mixed Mode (MMC)	Adsorption based on more than one type of interaction, typically charge and hydrophobicity	Often a ligand with ion exchange and hydrophobic chemistry	See IEC and HIC.

We have already mentioned that chromatography resins are packed into a cylindrical column to create a packed bed. The volume of the bed is simply the volume of the cylinder formed by the bed (refer to Figure 8.3) and is calculated as:

$$V_{bed} = A_{bed}h_{bed} = \pi R_{bed}^2 h_{bed} = \pi \frac{D_{bed}^2}{4} h_{bed} \tag{8.1}$$

A_{bed} is the cross-sectional area of the bed, h_{bed} is the packed bed height, R_{bed} is the packed bed radius, and D_{bed} is the packed bed diameter or, equivalently, the inner diameter of the column. The total bed volume, V_{bed}, is the sum of multiple components: the volume of void space between resin beads (interparticle pores, V_{inter}) and the volume taken up by the resin particles themselves. The resin particle volume can be further divided into the open volume within the pores (intraparticle pores, V_{intra})

and the volume taken up by the polymer that forms the resin particle (V_p). Therefore, V_{bed} can also be written as:

$$V_{bed} = V_{inter} + V_{intra} + V_p \qquad (8.2)$$

The different volume contributions are more commonly expressed as porosity values, as explained and illustrated in the example that follows.

Example: Calculating the total porosity in a packed bed

The total porosity in a packed bed, ε_t is defined as the ratio of the total void volume within the bed, (= $V_{intra} + V_{inter}$), to the total bed volume ($\varepsilon_t = (V_{intra} + V_{inter})/V_{bed}$). Calculate the total bed porosity in a packed bed with an interparticle porosity, ε_{inter}, of 0.4 and an intraparticle porosity, ε_{intra}, of 0.8.

Note that ε_{intra} is defined as the ratio of the pore volume within a resin bead to the total volume of the resin bead, and ε_{inter} is the ratio of the interparticle void volume to the total bed volume (= V_{inter}/V_{bed}).

Solution

First, derive an expression for ε_t as a function of ε_{inter} and ε_{intra}, then substitute in the values to calculate ε_t. Note that the volume of resin, V_{resin}, is the volume of resin particles – including both porous and solid parts – in the bed:

$$\varepsilon_t = \frac{V_{intra} + V_{inter}}{V_{bed}} = \frac{(\varepsilon_{intra} V_{resin} + \varepsilon_{inter} V_{bed})}{V_{bed}}$$

$$= \frac{\varepsilon_{intra}(1 - \varepsilon_{inter})V_{bed} + \varepsilon_{inter} V_{bed}}{V_{bed}}$$

$$= \varepsilon_{intra}(1 - \varepsilon_{inter}) + \varepsilon_{inter} \qquad (8.3)$$

$$= 0.8(1 - 0.4) + 0.4 = 0.88$$

So, based on the porosity values given, 88% of the packed column is voids. Note that typical values for ε_{inter} range from 0.3–0.4, while values for ε_{intra} are as high as 0.9 [210].

Because most of a resin's surface area is associated with the resin pores, the majority of ligands for binding are attached at the pore surface rather than the outer surface of the resin bead. Therefore, for a solute to bind, it must be transported from the mobile phase into the pore where it can be adsorbed by the resin. Likewise, the solute must also be desorbed and make its way back to the mobile phase flow to be recovered. The transport steps involved are illustrated in Figure 8.3. A solute is transported through the porous packed bed by the mobile phase flow, usually created by a pump and therefore referred to as convection. As the mobile phase moves through the packed bed, a binding solute diffuses from the bulk mobile phase through a stagnant film surrounding the particle surface in a process commonly referred to as external mass transfer. (The term *diffusion* refers to the random movement of a component from an area of high concentration to low concentration.) From the particle outer surface, the solute then diffuses into the liquid-filled pores of the resin in a process referred to as intraparticle mass transfer (or intraparticle diffusion). It then adsorbs on the pore surface as a result of the interaction with a ligand attached to the resin. In chromatography steps for biopharmaceutical processing, intraparticle diffusion is typically the transport step that requires the most time [210], and we say that intraparticle diffusion provides the greatest resistance to mass transfer.

8.3.3 Other stationary phase formats: membrane adsorbers and monoliths

In addition to resin beads packed in a column, membrane adsorbers and monoliths are also common stationary phase options for biopharmaceutical manufacturing. Membrane adsorbers are porous membranes, like the microfiltration membranes discussed in Chapter 7, whose surfaces have been functionalized by attaching ligands to provide interaction chemistry, much like resins. In fact, many of the same chemistries used for resin bead ligands are also used in the membrane adsorber and monolith formats. Many membrane adsorbers are housed in a cylindrical (annular) module geometry, in which multiple layers of membranes are wrapped around a central core. Monoliths are similar, but instead of multiple layers of membrane, a monolith consists of a continuous porous stationary phase made up of a network of pores through which fluid flows [210]. Both formats are most often made from polymers such as reinforced cellulose or modified hydrophilic polyethersulfone (for membrane adsorbers) or polymethacrylate (for monoliths).

Membranes and monoliths are used in much the same way as packed chromatography columns. The process intermediate to be purified is fed to the membrane or monolith module, and product and/or impurities bind. In the case of a cylindrical geometry, it is common that feed enters the module, flows through an outer channel, then moves radially inward through the pores of the membrane layers or monolith to an inner channel and out of the module at the end opposite the feed. Membrane adsorbers in cassette format are also available [215].

There is an important difference between solute transport through a bed packed with resin beads, shown in Figure 8.3, and a membrane adsorber or monolith, illustrated in Figure 8.4. In membrane/monolith systems, ligands are directly exposed to the flowing mobile phase within the pores. Consequently, solute access to the ligands primarily occurs through convective transport, unlike resin beads in which solute exposure to the ligand typically requires intraparticle diffusion. Convective flow through pores of a membrane or monolith results from their relatively large size and their interconnectedness relative to pores in resin beads. Pore sizes from 0.4 μm to greater than 3 μm are common [215–217]. These differences in pore size and geometry lead to the following important differences in performance between the resin and membrane adsorber/monolith formats [218].

- Binding capacity (the amount of a component that binds per unit volume of stationary phase, as will be discussed shortly) and resolution of the chromatographic peaks are independent of flow rate, allowing higher flow rates to be used. In packed beds, both binding capacity and resolution typically decrease with increasing flow rate. We will come back to this point shortly.
- Membrane and monolith adsorbers have lower pressure drops than packed beds for a given volumetric flow rate.
- Membrane and monolith adsorbers are particularly good for binding large biomolecules, including larger proteins and viruses, because diffusion into small pores is not required; however, binding capacity for "normal sized" (and smaller) biomolecules is typically higher in packed columns due to the higher surface area for binding per unit volume of column.

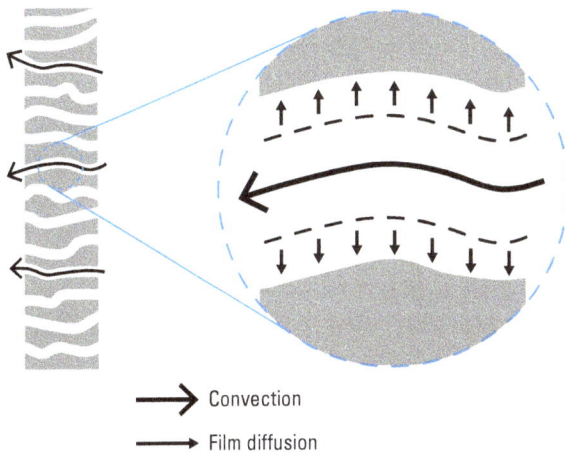

Convection

Film diffusion

Figure 8.4: Transport of solutes from the mobile phase to the surface of a membrane adsorber or monolith pore where adsorption occurs. Note that intraparticle diffusion, which limits the rate at which solutes bind to resin beads in packed beds, is not present in membrane adsorbers and monolith devices. Image © NC State University; reprinted with permission.

Membrane adsorbers are offered by a number of vendors, including Sartorius, MilliporeSigma, and Cytiva. Sartorius BIA Separations is a major supplier of monoliths.

8.3.4 Mobile phase properties and their impact on separation

To complete our discussion of how and why mobile phase components interact with chromatographic stationary phases, let's turn our attention to the impact of the mobile phase. In Table 8.2, we have summarized key points related to mobile phase properties required for solute binding and desorption with each interaction mode and briefly elaborate here.

A common feature of chromatography steps in biopharmaceutical processes is that products are typically applied to chromatography stationary phases in a buffered solution, regardless of interaction mode. pH fluctuations can occur for a number of reasons, and surrounding the product with a buffer during chromatography (and other processing steps) minimizes these and helps ensure product stability.

In ion exchange, because binding is based on opposite charge, the solute to be bound must be at a pH that creates this opposite charge. So, for AEC, the solute to be bound must be in a buffered solution with a pH > pI of the binding solute during loading. For CEC, the buffered solution must be at a pH < pI of the binding solute during loading. Therefore, design of an ion exchange step requires that product be in a solution with buffering capacity at the specific pH necessary for binding. Further, the concentration of the buffering agent is usually kept low (10–50 mM) to minimize conductivity, so buffer ions are not competing for binding with the target solute. Because most buffering agents have a limited useful pH range, careful selection of a buffering agent to match the needed load pH is required. To help make this selection, tables are available that provide information on the useful pH range for a variety of buffering agents [211].

For AC, there are a variety of different ligands in use, as shown in Table 8.2, and the mobile phase requirements depend on the nature of binding between the affinity ligand and mobile phase solute. Commonly used protein A resins (i.e., resins that have protein A attached as a ligand) have a high affinity for most immunoglobulin G antibodies, which include mAbs used as biopharmaceuticals. Protein A binds at neutral pH to a specific region of the antibody known as the Fc region. The antibody is eluted by lowering the pH [219].

From Table 8.2, the mobile phase for binding a solute to a HIC resin should have a high conductivity, and in particular should include a type of salt referred to as a kosmotropic salt such as ammonium sulfate ($(NH_4)_2SO_4$) or sodium sulfate (Na_2SO_4). Removal of those salts – and therefore a lowering of conductivity – favors desorbing conditions. Why are these salts required for binding? When hydrophobic solutes and ligands are in aqueous solution, the water molecules form an energetically unfavorable ordered structure at the surface of the hydrophobic substance, hindering ligand binding. Kosmotropic salts make it less favorable for water molecules to surround hydrophobic solutes. This increases the tendency of hydrophobic solutes and ligands to associate with each other, which allows for binding to occur. More in-depth discussion on the nature of protein binding on HIC resins is offered elsewhere [220].

Example: Separation by ion exchange chromatography

Consider the separation of the following proteins from a mixture: ribonuclease (pI = 9.5), cytochrome C (pI = 10.3), and lysozyme (pI = 11). The proteins are dissolved in a solution of 50 mM MES buffer at pH 6.0. (Note that MES is the common name for 2-(N-morpholino)ethanesulfonic acid, a good buffering agent at pH 6.0). Answer the following:

(a) What type of ion exchange chromatography should be used for the separation – anion or cation?

(b) The concentration of NaCl in the MES buffer is slowly increased to elute the proteins from the ion exchange resin. In what order will the proteins elute from the resin?

Solutions

(a) At pH 6.0, all three proteins have a net positive charge. None will bind to an anion exchange resin, therefore separation by AEC is not possible. All will bind to a cation exchange resin for separation, and because each protein has a different pI, each binds with a different strength, which can be exploited for separation. So use CEC for the separation.

(b) Proteins with a lesser positive charge elute before proteins with a greater positive charge. Further, the farther away the pI for a given protein is from the surrounding pH, the more positively charged the protein. Therefore, the order of elution is ribonuclease, cytochrome C, then lysozyme. Referring back to Figure 8.1, ribonuclease would be the red component, cytochrome C the green, and lysozyme the blue (even though these proteins are not really colored).

8.4 Chromatography performance

In this section, we turn our attention to chromatography performance. Note that many concepts covered in this section are described in terms of packed beds but can be applied to membrane adsorbers and monoliths as well. Because column performance is often related to mobile phase flow rate, we begin by discussing velocity and residence time – both related to flow rate – in a column. In chromatography at process scale, flow rate typically refers to the volumetric flow rate, Q, through a packed bed in units of L/h. However, we also refer to the superficial velocity, v_s, of the mobile phase in units of cm/h. The superficial velocity represents the velocity of a mobile phase fluid element if no resin is present in the column and is calculated as the ratio of the volumetric flow rate to cross sectional area of the bed, A_{bed}:

$$v_s = \frac{Q}{A_{bed}} \tag{8.4}$$

For a cylindrical column, A_{bed} is calculated as $\pi \times R_{bed}^2$ or equivalently $\pi \times D_{bed}^2/4$, as shown in equation (8.1). The actual velocity of a fluid element in the mobile phase is higher than that calculated by equation (8.4) because a significant amount of the cross sectional area in a column is occupied by resin particles and therefore not available for liquid flow. If the interparticle porosity is known, then the actual velocity, v_{actual}, can be calculated as:

$$v_{actual} = \frac{Q}{\varepsilon_{inter} A_{bed}} \qquad (8.5)$$

Related to fluid velocity is the residence time of a mobile phase fluid element in the packed column. The residence time is calculated by taking the distance that a fluid element must flow – the height of the packed bed – and dividing by its velocity through the column:

$$t_{residence} = \frac{h_{bed}}{v_{actual}} \qquad (8.6)$$

In practice, the superficial velocity is often used when calculating $t_{residence}$ because the value of ε_{inter} is not always known:

$$t_{residence} = \frac{h_{bed}}{v_s} = \frac{V_{bed}}{Q} \qquad (8.7)$$

Calculating $t_{residence}$ by equation (8.7) is common, but it should be recognized that the residence time calculated is that required for a fluid element to travel through an empty column. It is also worth noting the residence time, defined by either of the two previous equations, is different from the time required for a binding solute to travel through a column. That time is more commonly referred to as the solute's retention time, t_r. For example, the blue, green, and red solutes in Figure 8.1 are moving more slowly than the mobile phase because they interact with the stationary phase; therefore, their retention time is longer than the mobile phase residence time.

Example: Calculating the mobile phase residence time in a packed column
A chromatography step is optimized at bench scale using a 1.5 cm diameter column packed to a bed height of 25 cm and operated at a flow rate of 15 mL/min. What is the residence time of the liquid phase in minutes?

Solution
To use equation (8.7), first calculate the superficial velocity, v_s, using equation (8.4) and recognize that 15 mL/min is the same as 15 cm³/min:

$$v_s = \frac{Q}{A_{bed}} = \frac{15\ cm^3/min}{\pi \times (1.5\ cm/2)^2}$$

$$= 8.4883\ cm/min$$

Now calculate the residence time using equation (8.7):

$$t_{residence} = \frac{h_{bed}}{v_s} = \frac{25\ cm}{8.4883\ cm/min}$$

$$= 2.9\ min$$

Thus, a fluid element in the mobile phase moving at a volumetric flow rate of 15 mL/min through the packed column would spend 2.9 min in the column if the column is not packed. As discussed previously, the true residence time is less, given that the volume available for mobile phase flow is less than then total bed volume.

You may have asked yourself previously why the bands in Figure 8.1 that represent each solute in the chromatography column broaden instead of maintaining their original rectangular shape (the shape that represents the load). Solutes moving through a packed column get dispersed. It is desirable to minimize this dispersion as it can lead to loss of resolution between solutes; that is, the concentration traces shown in Figure 8.1(b) begin to overlap as the dispersion occurs, which eventually makes it difficult to collect the product in a pure form, thus compromising purity. The height equivalent of a theoretical plate (HETP) is a theoretical concept used to describe this broadening of solute bands in a column. It is based on the idea that a packed column contains fictitious plates that allow for equilibration between the mobile and stationary phases. Statistically, the number of plates in the column, N_{plates}, can be defined from the concentration trace for a single band eluting from the column as [221]:

$$N_{plates} = \frac{V_r^2}{\sigma_v^2} \tag{8.8}$$

where V_r is the retention volume of the peak (refer to Figure 8.8) and σ_v^2 is the peak variance in volume units. The greater the peak variance, the broader the chromatographic peak, and the more likely it is that the solute band represented by the peak will overlap with other bands and compromise separation. Thus, N_{plates}, serves as a measure of the efficiency of the packed bed.

HETP is simply the height of a single plate and is given by the equation

$$HETP = \frac{h_{bed}}{N_{plates}} = h_{bed}\frac{\sigma_v^2}{V_r^2} \tag{8.9}$$

HETP is often expressed in units of cm. Given that peak variance is a measure of the width of a chromatographic peak, HETP is a direct measure of band broadening; that is, the wider the solute band in a column, the larger the HETP. However, equation (8.9) provides no insight as to the processes occurring within a column that lead to the band broadening. To gain insight into those processes, consider the well-known van Deemter equation [222] given by the following equation:

$$HETP = A + \frac{B}{v_s} + Cv_s \tag{8.10}$$

A, B, and C are constants related to hydrodynamic dispersion, axial diffusion of solute, and transport of solute from the mobile phase to the stationary phase, respectively. Each of these processes is described in greater detail below.

- **Hydrodynamic (or eddy) dispersion (A).** As illustrated in Figure 8.3, multiple paths are available for mobile phase to travel through a packed column. Some liquid elements take a longer, more winding path, while other liquid elements take a shorter path. Therefore, solute molecules of a specific type (e.g., product molecules) in the mobile phase may each take a different path through the inter-particle voids in the column, which results in broadening of the solute band.
- **Axial molecular diffusion (B/v_{actual}).** Solutes diffuse axially in the mobile phase – that is, up and down the length of the packed bed, away from the center of solute concentration in the band. The result of this movement is broadening of the solute band as the chromatography step proceeds. Axial diffusion is typically not significant in liquid chromatography applications because diffusion of solutes in liquids is relatively slow. Generally, the longer the solute band resides in the column, the greater the impact of axial diffusion on band broadening. Thus, at low fluid velocities, the B/v_{actual} term becomes larger and the HETP increases (i.e., band broadening increases).
- **Mass transfer of solute to the stationary phase (Cv_{actual}).** Among the three contributions to HETP, the impact of solute transfer from the mobile phase to the resin surface may be the most challenging to understand, but in process-scale liquid chromatography, it is the most important. As a solute band moves through the column, equilibration with the stationary phase takes place gradually due to the relatively slow process of solute transport into the resin pores, which is required for binding. At any axial position within the column, as the solute in the mobile phase moves toward equilibrium with the stationary phase, some unbound solute is not adsorbed to the stationary phase until it moves further down the column. As the solute band passes the stationary phase, time is also required for solute to be desorbed from the resin and transported back to the mobile phase. These lags in solute transport into and out of the resin pore result in broadening of the solute band. Band broadening due to mass transfer resistance increases linearly with mobile phase velocity, as faster flow rates reduce the time available for solute-stationary phase interaction. Importantly, the time required for a solute to adsorb once it has diffused to the resin surface is usually fast relative to the time required for mass transport to the resin surface [210] and is not a significant source of band broadening.

From equation (8.10), it is clear that at low mobile phase velocities, the second term in the equation becomes large, and the third term becomes negligible, resulting in a high HETP. Likewise, at high velocities, the second term in the equation goes to zero, while the third term becomes large, resulting in a high HETP. Thus, there is an optimal mobile phase velocity that results in a minimum (i.e., desirable) HETP value. Minimizing HETP leads to sharper chromatographic peaks, which result in good resolution between solutes in the feed and enhanced purity of product. We'll come back to the concept of HETP shortly when we discuss measurements to test that a column has been packed correctly.

Related to HETP is resolution, a term that we have used several times already, but have yet to specifically define. In chromatography, the term *resolution* is a measure of the separation between two components and is specifically defined as the distance between the center of two neighboring peaks (in time or volume) divided by the average of the peak widths (in time or volume). Poor resolution between a biopharmaceutical product and impurities results in impure product eluting from a chromatography column. Both the number of plates in a column and the selectivity of the stationary phase impact the resolution between two components to be separated. Selectivity is a measure of how much time one solute (e.g., product) spends on the stationary phase relative to another (e.g., impurities). The greater the selectivity, the better the resolution. We mention it here because selectivity is often used to screen chromatography resins for use in biopharmaceutical processes, as is discussed shortly.

Another important performance parameter for chromatography steps is binding capacity. Generally, the term refers to the amount of a molecule that can be adsorbed by a stationary phase under a specific set of conditions and is typically expressed as the mg of solute (e.g., protein) bound per mL of resin. It is desirable to maximize the binding capacity of the target component(s) to reduce the amount of stationary phase required to perform the separation. There are two types of binding capacities that are measured: the equilibrium binding capacity (EBC) (also referred to as total binding capacity or static binding capacity) and the dynamic binding capacity (DBC). The EBC is the total amount of a molecule that the resin is capable of binding under conditions of no flow. Imagine that resin is added to a beaker filled with a solution that contains a solute that binds to the resin. The resin is mixed with the solution long enough to ensure that equilibrium between the target solute in solution and resin is achieved. The mass of solute adsorbed (e.g., mg solute) per unit volume of resin (e.g., mL resin) is the EBC at the concentration of unbound solute in the liquid phase. A plot of EBC versus the concentration of solute (i.e., unbound solute) with which the stationary phase is in equilibrium at a given set of conditions (e.g., at fixed temperature, pH, or ionic strength) is referred to as an adsorption isotherm.

The DBC for a specific solute is the amount of that solute that is adsorbed onto the stationary phase while operating at a specific flowrate. To measure the DBC, a solution that contains the solute of interest (e.g., product) at a known concentration is fed to a chromatography column at the desired flow rate. When the concentration of product in the column effluent reaches some specified percentage of the product concentration in the column feed, we say that breakthrough has occurred. Therefore, this type of test is often referred to as a breakthrough study. It is common to use 10% of the target solute concentration in the feed as the acceptable concentration in the column effluent, which leads to the notation $DBC_{10\%}$. Note that the concept of DBC can be applied to membrane adsorbers and monoliths as well, even though the previous description referred to "columns."

Figure 8.5 illustrates the DBC concept more fully. The data in that figure were generated by feeding BSA in 50 mM tris, pH 8.0 to a 3 mL anion exchange membrane capsule (Figure 8.5(a)) and 3 mL packed column packed with anion exchange resin (Figure 8.5(b)) until the concentration of BSA in the column effluent became constant – that is, until the stationary phase became saturated with BSA. To calculate the $DBC_{10\%}$ for the target solute, in this case BSA, the mass adsorbed, m_{ads}, is approximated by the total mass of target solute fed to the point of 10% breakthrough and divided by the bed volume to give:

$$DBC_{10\%} = \frac{m_{ads,10\%}}{V_{bed}} = \frac{C_f V_{loaded,10\%}}{V_{bed}} \tag{8.11}$$

where C_f is the concentration of the target solute (BSA) in the column feed and $V_{loaded,10\%}$ is the volume of solution fed to the column up to the point of 10% breakthrough. The equation is only an approximation of the mass adsorbed because it includes not only solute that binds to the stationary phase but also solute in the column effluent (i.e., solute that did not bind because binding sites in the stationary phase are not available), up to the point of 10% breakthrough. Looking at Figure 8.5(b), we estimate the $DBC_{10\%}$ for BSA in the packed bed at a flow rate of 1.28 mL/min to be 67 mg BSA/mL Q Sepharose FF resin (= 1 mg/mL × 200 mL)/3 mL). Likewise, at 3.83 mL/min, the $DBC_{10\%}$ from Figure 8.5(b) is estimated as 37 mg/mL.

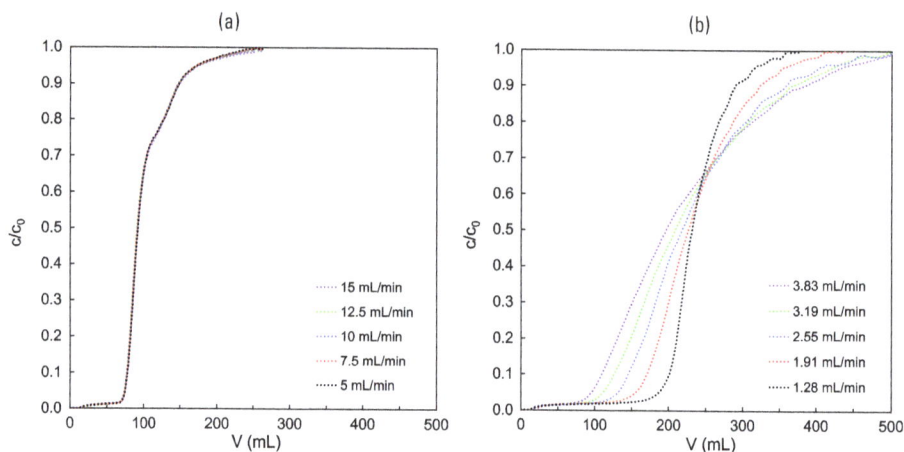

Figure 8.5: Breakthrough curves for adsorption of BSA at 1 mg/mL in 50 mM tris, pH 8.0 on (a) a Sartobind® Q nanomembrane capsule with a 3 mL volume and (b) a Q Sepharose® FF packed column with 3 mL volume. Note that measurements for both the membrane capsule and packed bed were performed at equal superficial velocity values. Note also that for these graphs, the notation is slightly different that defined previously in this chapter. c is the concentration of BSA in the column effluent, c_o is the concentration of BSA in the column feed, and V is the volume of BSA solution fed in mL. Reprinted from "Direct comparison between membrane adsorber and packed column chromatography performance," by C. Boi, A. Malavasi, R. Carbonell, G. Gilleskie, 2020, *Journal of Chromatography A, 1612, 6.* Copyright 2019 Elsevier B.V [223].

Comparison of Figure 8.5(a) and 8.5(b) also leads to the following observations regarding performance of packed beds compared to membrane adsorbers:

- Breakthrough curves for each flow rate tested with the packed bed are not identical; curves at low feed flow rates are much steeper than the curves at the higher flow rates. Therefore, $V_{loaded,10\%}$ increases with decreasing flow rate, and, consequently $DBC_{10\%}$ for BSA in the packed bed increases with decreasing flow rate. This relationship between $DBC_{10\%}$ and feed flow rate is a direct consequence of slow intraparticle diffusion: slower feed rates allow more time for intraparticle diffusion to take place, and therefore greater binding of BSA to binding sites on the resin occurs before BSA shows up in the column effluent.
- Breakthrough curves for each flow rate tested with the membrane capsule are nearly identical and result in a $DBC_{10\%}$ of approximately 21 mg BSA/mL membrane capsule at each flow rate tested. The fact that $DBC_{10\%}$ does not depend on flow rate – in stark contrast to the $DBC_{10\%}$ values measured for the packed bed – is a result of the rapid transport of BSA to binding sites on the membrane (i.e., intraparticle diffusion is not needed for a solute to bind to the membrane). The independence of the slope of the breakthrough curve with flow rate is also predicted by the van Deemter equation (equation (8.10)), as the Cv term for the membrane adsorber goes to zero due to rapid mass transfer of solute to the stationary phase, and the B/v term is negligible due to the fact that diffusion in liquids is slow. This leaves the HETP to be a fixed value, independent of flow rate.
- Binding capacity values are higher for the packed bed than the membrane capsule, likely a result of the greater surface area per unit volume in the packed bed.

DBC has practical implications for sizing of chromatography columns, as the capacity of stationary phase for a solute dictates the amount of stationary phase required to bind a solute (e.g., product) without product ending up in waste. The example below illustrates the relationships between DBC and column sizing.

Example: Estimating column size (diameter) from the DBC

10 kg of a protein therapeutic is produced by a 2,000 L working volume cell culture step. Assuming no loss of product in the harvest steps, calculate the diameter of the column used for the first chromatography step if:

- the column is loaded to 70% of its $DBC_{10\%}$ (note that this amount of loading is significantly greater than what is illustrated in Figure 8.1).
- the $DBC_{10\%}$ at the flow rate used for this step has been measured as 50 g product/L resin.
- the bed is packed to a height of 25 cm.

Solution

The packed bed volume required to capture all product is calculated as the ratio of the mass of product loaded, m_{load}, to the product load, defined as the amount of product loaded per unit volume of bed:

$$V_{bed} = \frac{m_{load}}{product\ load}\tag{8.12}$$

$$= \frac{10,000\ g\ product}{0.7 \times 50\ g\ product/L\ resin}$$

$$= 286\ L\ of\ resin\ or\ 285,714\ cm^3\ of\ resin$$

From equation (8.1):

$$V_{bed} = \pi \frac{D_{bed}^2}{4} h_{bed}$$

$$D_{bed} = \left(\frac{4V_{bed}}{\pi h_{bed}}\right)^{1/2} = \left(\frac{4 \times 285,714\ cm^3}{\pi \times 25\ cm}\right)^{1/2}$$

$$\approx 120\ cm$$

Note that DBC and EBC are related, but DBC is impacted by the band broadening factors resulting from flow through a packed bed (or membrane adsorber or monolith). As flow rate becomes smaller in a breakthrough study, the DBC should approach the EBC, all other conditions (e.g., temperature, pH, and conductivity) being the same. The amount of solute that binds to a stationary phase measured either under static or flow conditions depends on solute properties (e.g., solute charge in ion exchange), mobile phase properties (concentration of binding solute, pH, and ionic strength), and stationary phase properties (pore size in resins, porosity of stationary phase, ligand, etc.). For example, in IEC, if the mobile phase contains a high concentration of ions then the DBC for the target solute on ion exchange media is reduced. A low concentration of certain salts, such as NH_4SO_4, in the mobile phase decreases the binding capacity of target solute on HIC media. Note that we have only scratched the surface of the topic of solute binding to resins. This topic, and in particular the relationship between the concentration of adsorbed solute on the stationary phase in equilibrium with the solute in the mobile phase as it applies to liquid chromatography is covered in greater detail in other references [210].

In addition to HETP, resolution, selectivity and binding capacity, other measures of chromatography performance include step yield and productivity. Building on the definition of percent yield provided in equation (2.1), the percentage step yield in a chromatography step can be calculated as the amount of product at the end of the step divided by the amount of product loaded to the column converted to a percentage:

$$\%\ step\ yield = \frac{C_{eluate}V_{eluate}}{C_{load}V_{load}} \times 100\tag{8.13}$$

where C_{eluate} and C_{load} are the concentration of product in the eluate from the column and load to the column, respectively, and V_{eluate} and V_{load} are the volumes of the eluate from and load to the chromatography step. Productivity, P, of a chromatography step is defined as the amount of product collected per bed volume per time, and is given by the following equation:

$$P = \frac{C_{\text{eluate}} V_{\text{eluate}}}{V_{\text{bed}} t_{\text{process}}} \tag{8.14}$$

where t_{process} is the time for the chromatography step. Clearly, maximizing productivity is desirable.

8.5 Operational modes

The most common operational modes for chromatography in biopharmaceutical purification are bind-and-elute and flow-through. Bind-and-elute mode is designed and operated so that product binds to the stationary phase during the load step. Impurities may or may not bind. The product is recovered during the elution step by making a change in mobile-phase (i.e., elution buffer) composition that disrupts interaction between the stationary and mobile phases. The mobile phase composition is either changed in a step or in a stepwise fashion and referred to as a *step elution*, or changed linearly with time and referred to as a *linear gradient*. These different elution methods will be discussed in greater detail shortly. Note that the mobile phase requirements that allow for release of components from the resin have been previously discussed (refer to Table 8.2).

Flow-through chromatography is designed and operated so that product is not bound to the stationary phase, but impurities are. Loading of the product solution continues until impurities in the effluent from the column (or membrane adsorber or monolith) exceed an acceptable limit. A common flow-through step is the use of AEC as the final chromatography step in a process for mAb production, as shown in Figure 2.3, often performed with a membrane adsorber. mAbs have pI values ranging from approximately 6 to 9 [224]. If the pH of the buffer in which the mAb product is dissolved is less than the pI, then the mAb is positively charged and flows through the anion exchange bed or membrane, while impurities that are negatively charged at even these lower pH values, such as endotoxins, some HCPs, and DNA, bind to the chromatography media [225]. Membrane adsorbers are attractive for flow-through applications meant for polishing – that is, steps that remove low levels of remaining impurities – because their relatively low binding capacities are suitable for binding the residual impurities that remain, and the membrane systems can be operated at higher flowrates than packed beds.

8.6 Typical chromatography procedures for CGMP manufacturing

Once a chromatography column is properly packed, cleaned and sanitized, it is ready for use. Every chromatography step, including steps using membrane adsorbers and monoliths, in a downstream bioprocess is actually made up of a series of steps – that we refer to as *phases* for clarity – necessary to ensure that purification objectives are met and that the packed column can be reused. A typical sequence of phases for a batch bind-and-elute chromatography step, regardless of interaction mode, is: equilibration, product load, wash, elution, cleaning/regeneration, sanitization, buffer rinse, HETP/asymmetry measurement, and storage. Each of these phases is described in Table 8.3. Note that the sequence of phases in Table 8.3 can be adapted to SEC by removing the wash step. The specific design of each phase varies, of course, according to the product and specific resin used. And there are a number of variations on the typical sequence of phases that are used. For example, cleaning and sanitization may be done with the same solution; therefore the cleaning and sanitization phases become one. Further, multiple wash steps may be incorporated to maximize removal of all weakly bound impurities from a column prior to elution.

Table 8.3: Phases in a typical bind-and-elute chromatography step.

Phase	Description
(1) Equilibration ↓	– A buffer fed to the column removes storage solution from the previous column run and prepares the column for binding product. For example, in IEC, equilibration ensures that resin charge groups are in equilibrium with counter ions of binding buffer and that column pH and ionic strength are appropriate for product binding. The equilibration buffer is often the same buffer that the protein loaded to the column (in the next step) is dissolved in. – Buffer flows through the column until the pH and/or conductivity of the effluent leaving the column equals that of the solution fed to the column. Typically, 3–5 CVs of buffer is sufficient.
(2) Load ↓	– Product is applied to the column so that it binds. Other components (i.e., impurities) may or may not bind to the stationary phase. – Overloading beyond the binding capacity of the column is avoided to ensure no loss of product to drain.
(3) Wash ↓	– Buffer, often the same as equilibration buffer, is fed to the column to wash away any unbound impurities, i.e., mobile phase impurities in the voids between resin particles, or weakly bound impurities. The product remains bound. – Typically 3–5 CVs of wash buffer is sufficient.

Table 8.3 (continued)

Phase	Description
(4) Elution ↓	– A change in mobile phase composition is made that weakens interaction between the stationary phase and target solute (e.g., product). Elution is designed and executed in such a way that the target component (product) is collected separately from bound impurities. Elution conditions for each chromatographic interaction mode are shown in Table 8.2. – Elution can be executed making a step change in mobile phase composition (most common) or changing the mobile phase composition linearly.
(5) Cleaning/ regeneration ↓	– Product is no longer bound to the column, but other strongly bound impurities may be. A solution is fed to the column that is capable of removing these impurities. For example, in ion exchange, a relatively high concentration of NaCl solution is fed so that Na^+ ions (in cation exchange) or Cl^- ions (in anion exchange) displace bound components. The solution used varies according to type of chromatography used. – Typically 3–5 CVs of cleaning solution is adequate.
(6) Sanitization ↓	– A solution is fed to the column to remove any microbiological contamination without degrading the chromatography resin. A common sanitizing agent is 0.5 M NaOH solution; however, not all resins – most notably some protein A affinity resins – are compatible with these high concentrations of NaOH. Sanitization and cleaning phases may be combined into a single phase. – Typically 3–5 CVs of sanitization solution is adequate. In addition, a static hold of the column in sanitizing solution may be performed.
(7) Buffer rinse ↓	A buffer – often that used for equilibration – is used to quickly bring the pH to more neutral conditions if an acid or caustic cleaning/sanitization agent is used.
(8) HETP/ asymmetry ↓	Some biopharmaceutical manufacturers may choose to measure HETP/asymmetry after every use of a column to assess bed stability over the lifetime of the column. More information on this measurement is given in the next section.
(9) Storage	– A storage solution that inhibits growth of bioburden is fed to the column. Solutions of ethanol in water are often used. – Typically 3–5 CVs of storage solution is adequate.

The volume of solution fed to the column at each phase is typically described using the scale-independent parameter "column volumes," often referred to as "CVs." As the name implies, the number of CVs of a solution corresponds to the volume of solution fed relative to the volume of the stationary phase (e.g., packed bed), calculated as:

$$CVs = \frac{V_{\text{mobile phase fed}}}{V_{\text{bed}}} \tag{8.15}$$

where $V_{\text{mobile phase fed}}$ is the volume of a solution fed at any phase in the chromatography step and V_{bed} is the volume of packed column, membrane adsorber, or monolith.

For example, 600 L of equilibration buffer fed to a column packed to a volume of 200 L corresponds to 3 CVs of equilibration buffer.

Also, as described in Table 8.3, elution can be carried out in a stepwise fashion or using a linear gradient, each referring to a different method to make the change in mobile phase composition required to elute bound solutes. For example, in HIC, the concentration of $(NH_4)_2SO_4$, which is required for binding, must be reduced to elute product from the column. If the concentration of $(NH_4)_2SO_4$ is decreased slowly, with a constant slope over time, we refer to this type of elution as a linear gradient. This gradient is typically described as going from a starting concentration of $(NH_4)_2SO_4$ to an ending concentration of $(NH_4)_2SO_4$ over the applicable number of CVs, which is the total volume of elution buffer fed during the gradient. In a step elution, a single step change in the buffer composition is made (e.g., reducing the $(NH_4)_2SO_4$ concentration from 2 M to 0 M at the beginning of the elution step). Generally, linear gradients provide better resolving capability between product and impurities but may be more difficult to control in a production environment than a step elution. An example of a linear NaCl gradient for an AEC step is shown in Figure 8.7 and can be identified by the increase in conductivity during the elution phase.

Chromatography resins are typically reused and dedicated to a single product. Membrane adsorbers and monoliths may also be reused. Consequently, after storage, the sequence of phases can be repeated as many times for each new batch of a product as the column lifetime allows. The column lifetime must be determined as part of process validation. It is also worth mentioning that instead of the full amount of product from the previous process step being loaded to the chromatography column, the batch from the previous step may be split into sub-batches that are fed individually. In so-called intrabatch cycling, the equilibration phase would begin the cycle for each sub-batch. But the sequence of phases would likely proceed only to the sanitization step and then return to equilibration for the next cycle. This pattern continues until all sub-batches have been purified. Upon completion of the final sub-batch, the buffer rinse, HETP/asymmetry, and storage phases are completed, and the column is held until the next batch is ready for processing. Because the amount of product loaded in a sub-batch is less than in a full batch, less resin is needed per batch and a smaller, less expensive column is used. However, a smaller column that sees more cycles for a given batch of product may use up its lifetime more quickly than a larger column that processes the entire batch in a single cycle. In addition, multiple cycles per batch increase process time relative to single-cycle operation. However, in some scenarios, intrabatch cycling may lead to better utilization of resin lifetime. For example, if producing product for clinical trials, the total number of runs required to produce the necessary amount of product is likely small and resin is likely to be disposed after the production campaign given that it may be years before the clinical trial is complete and more product is needed. In this case, cycling a column may lead to better utilization of resin lifetime than would be achieved otherwise.

8.7 Chromatography equipment for biopharmaceutical production

Let's turn our focus to equipment required to conduct chromatographic separations. Consider the schematic of a typical chromatography system for biopharmaceutical manufacturing in Figure 8.6(a).

Moving from left to right in Figure 8.6(a), there are vessels, connected to the chromatography skid, that contain the different solutions (including the product) required to execute the multiple phases of a chromatography run. Referring back to Figure 2.5, you see an example of one type of vessel – buffer totes – lined up for use in a chromatography run. Stainless steel portable vessels may also be used. The skid is equipped with at least one pump to move fluid to the chromatography column, and often there are two. The second pump is necessary to create step and linear gradients. The pumps on the AKTAprocess system in Figure 8.6(b) are diaphragm pumps, which move fluid by the back-and-forth motion of a diaphragm. The skid is also equipped with an air trap, sometimes referred to as a bubble trap, to remove air that inadvertently makes its way into the chromatography system with the various feed solutions. Air trapped in a packed bed is undesirable because the mobile phase flows around the air pockets and prevents uniform flow distribution throughout the entire column cross section, thus compromising separation. A filter housing is also in place to house a surface filter – typically 0.2 µm – to protect the column from particulate. There are sensors both precolumn and postcolumn that are used to measure a number of different parameters, which we will discuss shortly. And, of course, there is a vessel to collect product during the load step (for flow-through chromatography) or elution step (for bind-and-elute chromatography). The column is connected to the skid through a number of valves that enable mobile phase to flow downward, upward, or around the packed column. There are also numerous other valves on the system, including multiple inlet valves to select feed from one or more of the feed vessels and outlet valves to properly direct flow out of the system (e.g., to the product collection vessel or to waste).

Sensors in place upstream of the column are used to measure flow rate, pressure, conductivity, temperature (usually with conductivity), pH, and the presence of air. Postcolumn, there are likely to be sensors for measurement of conductivity, temperature (usually with conductivity), pH, and UV absorbance. The chromatography system may also be equipped with a postcolumn pressure sensor. Two of the most important to chromatography operations – conductivity and UV absorbance – are discussed in more detail.

Conductivity is a measure of a fluid's ability to conduct electric current and is related to the fluid's ionic strength. Conductivity is readily measured using a bench-top conductivity meter or an inline conductivity element, as is common on chromatography systems. The SI unit for conductivity is Siemens (S)/m; in biopharmaceutical applications, however, mS/cm or µS/cm is more commonly used. Solutions that contain sol-

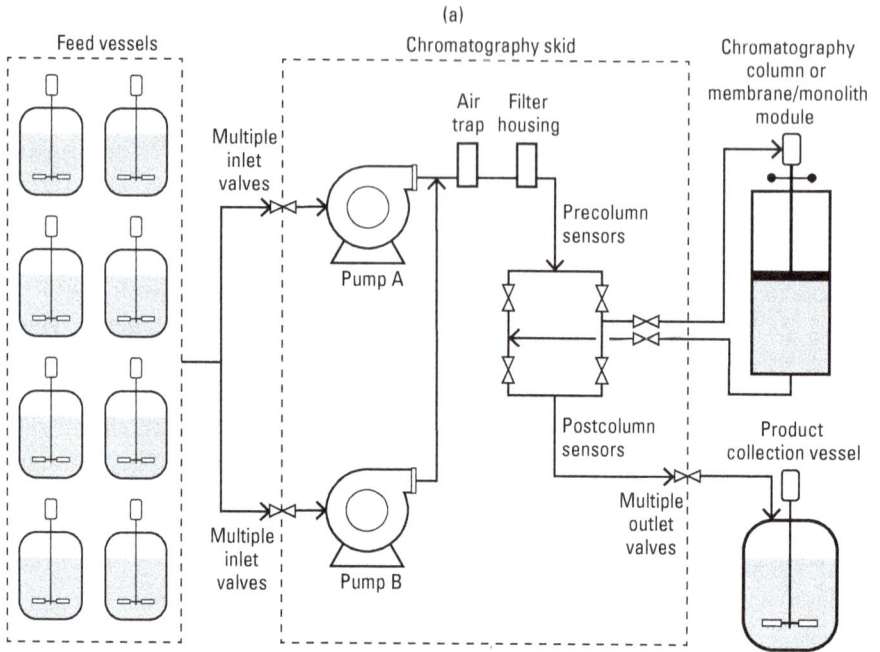

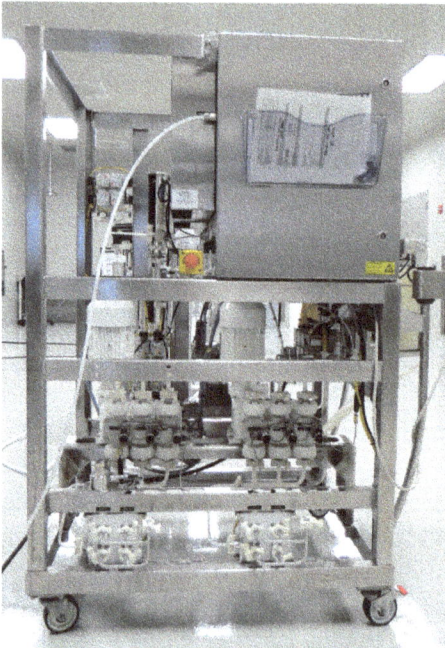

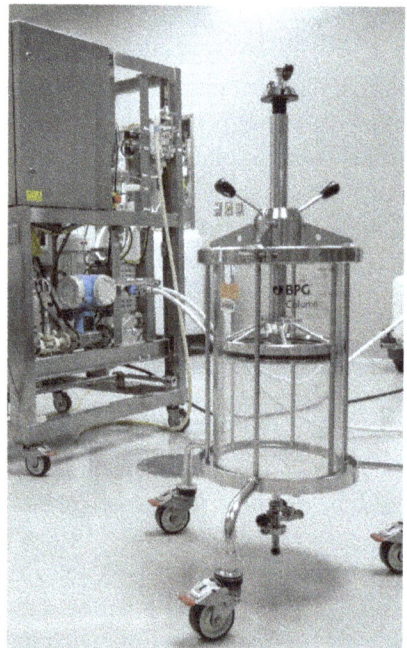

Figure 8.6: A typical production chromatography system: (a) a basic flow diagram for a production-scale chromatography system and (b) the front view and rear view of a Cytiva

utes that readily ionize in water have relatively high conductivity values because of their ability to conduct current; for example, 0.5 M NaOH has a conductivity of approximately 100 mS/cm. Solutions that contain solutes that do not dissociate in aqueous solutions (e.g., ethanol) have relatively low conductivity values; for example, a solution of 20% (by volume) ethanol in water has a conductivity of approximately 1 µS/cm or less.

The measurement of conductivity in chromatography operations is extremely important as it provides a measure of a fluid's ionic strength, which plays a key role in protein binding for several of the chromatographic interaction modes. Conductivity measurement is also used to monitor the progress of a chromatography run. Because different solutions have different conductivity values, a chromatogram showing either precolumn or postcolumn conductivity plotted versus time or cumulative volume of fluid fed can serve as confirmation that the correct solutions were introduced at each phase of the chromatography step and in the proper amount.

Ultraviolet (UV) absorbance is the common method for detecting components in the effluent from a chromatography device. Proteins absorb UV light – electromagnetic radiation with a wavelength range of 10–400 nm – and do so strongly at a wavelength of 280 nm due to the presence of amino acids with aromatic rings (tyrosine, phenylalanine, and especially tryptophan). Monitoring UV absorbance therefore serves as a method for detecting proteins and protein-based components in the column effluent. Most chromatography systems are equipped with inline UV sensors, many of which have the capability of monitoring multiple wavelengths simultaneously. At low enough concentrations of protein, the relationship between UV absorbance, A, and concentration of an absorbing solute such as a protein, C_{solute}, is given by the Beer-Lambert law:

$$A = \epsilon l C_{solute} \tag{8.16}$$

where l is the optical path length and ϵ the extinction coefficient. This equation suggests that a plot of A versus C_{solute} results in a straight line, making absorbance measurements useful for quantifying protein concentration. However, the A versus C_{solute} curve can deviate from this linear relationship for a number of reasons, including that at higher concentrations, molecular interactions increase, which results in proteins not contributing to absorbance independently. Proteins and protein-based biopharmaceuticals may be at high enough concentrations in chromatography steps to be outside the linear relationship defined by the Beer-Lambert law.

————————

Figure 8.6 (continued)
AKTAprocess chromatography skid that operates at flow rates of up to 180 L/h. In (b) an unpacked BPG 300/500 column is also shown, which has a diameter of 30 cm (300 mm) and allows a maximum bed height of 50 cm (500 mm). Images © NC State University; reprinted with permission.

Figure 8.7 shows a chromatogram for the purification of green fluorescent protein (GFP) in a clarified lysate generated from *E. coli* cells. Precolumn conductivity, post-column conductivity, and postcolumn UV absorbance at 280 nm and 395 nm are all plotted against cumulative volume (in L) of mobile phase fed to the column. The equilibration and wash buffer is 50 mM tris, pH 8.0. As expected for an AEC step, conductivity during the equilibration, load, and wash phases is kept low so that few ions are present to disrupt product binding. Tris was chosen due to its high buffering capacity at pH 8, a pH that results in a net negative charge for GFP ($pI_{GFP} \approx 5.8$, an average of values determined by various online isoelectric point calculators) so that it binds to the resin. The method is programmed with a linear NaCl gradient – 0.1 M NaCl in tris buffer, pH 8.0 to 0.2 M NaCl in tris buffer, pH 8.0 – which is reflected by the linear increase in precolumn conductivity during elution. Conductivity increases significantly during the regeneration phase, which uses a 2 M NaCl solution. GFP absorbs UV light strongly at a wavelength of 395 nm (A395) in addition to 280 nm (A280). The elevated A280 during the load and wash phases, with no absorbance at 395 nm, indicates that many impurities are not bound to the column and are present in the column effluent, which is directed to drain during these phases. The A280 is also high during the initial portion of the elution phase, during which effluent is directed to drain. GFP elutes near the end of the linear gradient as indicated by the A395 peak. Product collection is initiated when the A395 value rises to 0.1 absorbance units (AU) and ends as the A395 decreases to 0.1 AU. Based on the A280 and A395 traces, it is clear that significant separation of impurities from product has taken place. From the chromatogram, it is also clear that about 7 L of eluate (product) was collected. (Note that a pilot-scale column with a 14 cm diameter and bed height of 25 cm was used in this run).

We've already discussed packed beds in some detail, but let's briefly discuss column hardware. A chromatography column is basically a tube that holds packed chromatography resin in place. Columns are designed to provide even flow distribution across the column's cross section. For example, Cytiva BPG columns use both nets, which also keep resin from being displaced, and distributor plates at each end of the column to ensure even flow distribution [291]. Most columns allow for adjustable bed heights. Some characteristics of columns used for biopharmaceutical production, pulled together from a search of the websites (December 2024) of two leading suppliers of columns for the biopharmaceutical industry (Cytiva and MilliporeSigma) are:

- Diameter: up to 1.6 m (larger custom sizes available)
- Bed height: up to 100 cm
- Column materials of construction: plastic, glass, or stainless steel
- Pressure ratings: up to 5 bar

As you can see, the pressure rating for most columns used for biopharmaceutical applications is relatively low pressure. Columns for medium (20–40 bar) and high pressure (>40 bar) are available but less commonly used.

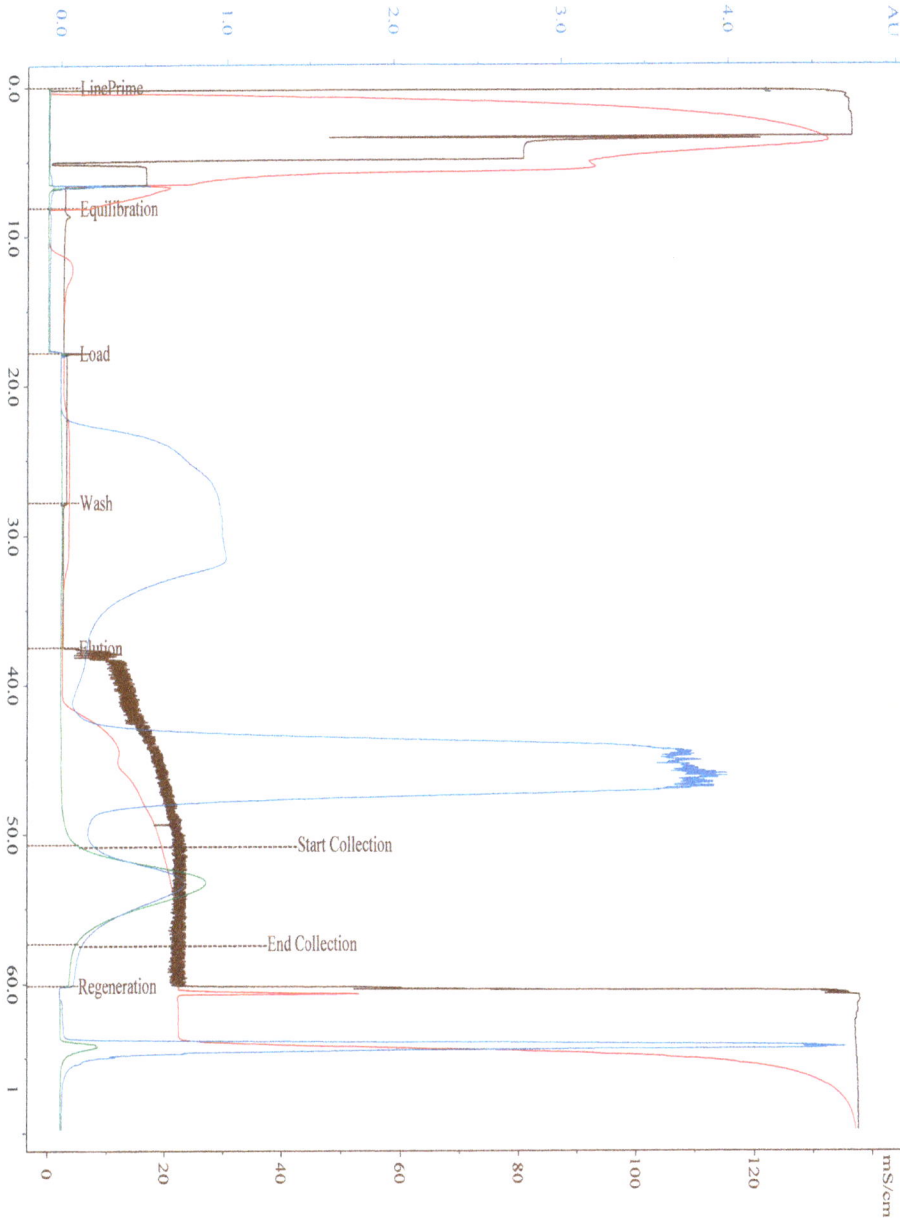

Figure 8.7: A chromatogram for purification of GFP from an *E. coli* lysate using AEC. The y-axis shows precolumn conductivity (brown), postcolumn conductivity (red), and postcolumn UV absorbance at 280 nm (blue) and 395 nm (green) wavelengths. Cumulative volume, in L, is shown on the x-axis. Note that the chromatogram goes only through the regeneration phase and does not include the sanitization, buffer rinse, or storage phases. Image © NC State University; reprinted with permission.

Common methods for packing resin into chromatography columns can be found in a number of references [226] and are not covered here. However, packing can be challenging, and a number of defects can occur in a column – such as channeling, trapped air, clogged bed supports – that result in flow disturbances, which can compromise separation in the column and therefore product quality. Thus, it is common to ensure packing is of high quality just after packing, prior to its first use, and even between runs. Evaluation of column packing is most often accomplished by measuring the HETP, which was discussed previously, and the asymmetry for the packed bed. One procedure is used to determine both values and involves injecting a pulse of tracer into the packed bed (or membrane adsorber or monolith) and then comparing the calculated HETP and asymmetry values to acceptable ranges. The procedure is summarized below [227]:

1. Inject a small volume – usually less than one percent of the bed volume – of a tracer to the column. The tracer should be small so that it can easily access resin pores, nonbinding so that binding effects are eliminated from consideration, and detectable at the column outlet. Commonly used tracers are acetone (1–2% acetone in water with water as the mobile phase to push the acetone pulse through) and concentrated NaCl solutions (e.g., 2 M NaCl).

2. Push the tracer pulse through the column using an appropriate mobile phase. The broadened pulse results in a chromatographic peak as shown in Figure 8.8.

3. Calculate the asymmetry, A_s, of the peak that elutes using the following equation:

$$A_s = \frac{b}{a}$$ (8.17)

where a is the width of the first half of the peak at 10% peak height and b is the width of the second half of the peak 10% peak height, as shown in Figure 8.8. For $A_s > 1$, the peak is referred to as a tailing peak. $A_s < 1$ is a fronting peak. An A_s value of one is ideal, although a range of 0.8 to 1.8 may be considered acceptable [227].

4. Calculate the HETP from equation (8.9) as the height of the bed divided by the number of plates. The number of plates is calculated as:

$$N_{plates} = 5.54 \left(\frac{V_r}{W_h} \right)^2$$ (8.18)

where V_r is the volume of mobile phase required to move the tracer, once injected to a column, to the column outlet, and W_h is the width of the peak at half height. Both are illustrated in Figure 8.8. Note that this equation is derived from geometric properties of a Gaussian peak that are related to the variance and mean of the peak [210]. HETP is often expressed as a reduced plate height, H, defined as the ratio of the HETP to diameter of particles (d_p) packed into the column:

$$H = \frac{\text{HETP}}{d_p} \qquad (8.19)$$

$H \leq 3$ is considered optimal [227] but assumes that optimal test conditions (such as mobile phase velocity as discussed previously for HETP vs velocity curves) are utilized, which may not always be the case in an industrial setting.

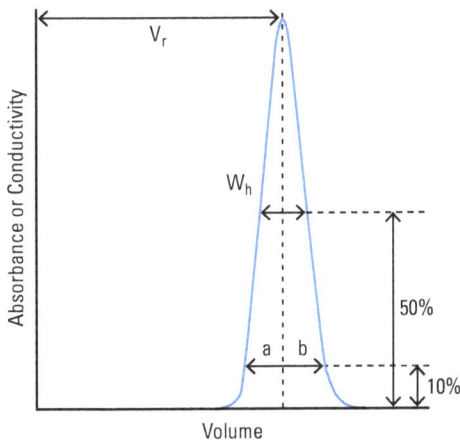

Figure 8.8: Chromatogram showing the output (e.g., UV absorbance or conductivity measured in the column effluent) from a pulse tracer test to determine the HETP and asymmetry of a packed column. All variables shown are used in the equations used to calculate HETP and asymmetry. a and b are the distances shown at 10% of peak height. V_r is the volume of mobile phase required to move the tracer, once injected to a column, to the column outlet, and W_h is the width of the peak at half height. Image © NC State University; reprinted with permission.

We'll wrap up this section on chromatography systems with a few brief comments on single-use systems. There are a number of them on the market, based on ready-to-use, disposable flow paths fitted with the necessary sensors. These include systems from Cytiva, Sartorius, and MilliporeSigma for operation in batch mode [228–230]. Furthermore, continuous chromatography systems are commercially available. For example, Sartorius offers the Resolute® BioSMB 80 and BioSMB 350 Single-Use Multi-Column Chromatography systems, which allow for continuous operation with single-use components [292]. The BioSMB 350 can achieve flow rates up to 350 L/h and process bioreactor runs as large as 4,000 L according to the manufacturer. Continuous operation is achieved through a multi-column system in which multiple, relatively small columns undergo loading, washing, elution, and regeneration phases, with different columns at different phases of the cycle at any given time. Operation in this way allows for uninterrupted feed introduction. A key advantage of this method of operation is increased resin utilization compared to batch chromatography, which enables the processing of a given product volume with less resin.

Single-use stationary phases come in two formats: membrane adsorbers, discussed previously, and prepacked columns. Both are generally reusable, and, in fact, using a prepacked column only once may be too expensive to be considered truly single-use [231]. Membrane adsorbers and prepacked columns can be purchased from a number of vendors. Both types of devices offer the advantage of avoiding challenges with column packing. Worth noting is that Repligen offers Opus® prepacked columns in diameters up to 80 cm [232].

8.8 Process and performance parameters for chromatography

Numerous process parameters can impact chromatography performance. We have already discussed a number of these. Table 8.4 provides a summary of process parameters and their potential impact on the following performance parameters and critical quality attributes (CQAs): step yield, productivity, binding capacity, column lifetime, and purity. A number of the parameters are written generally to apply to most interaction modes, and specifics will vary according to mode. For example, the impact of "load conditions," such as the composition, pH, and conductivity of the buffer in which the product is dissolved, differ according to whether IEC, AC, or HIC is being used. In Table 8.2, we see that for IEC, low conductivity of the load buffer favors product binding by ensuring few ions are present to compete for binding sites on the stationary phase. Thus conductivity impacts binding capacity and, potentially, step yield for IEC. Further, the pH of the load buffer in IEC must be such to provide a charge on the target solute opposite to that of the stationary phase. Therefore, pH of the load buffer also has the potential to impact binding capacity and step yield. For HIC, the presence of certain salts in the load buffer, such as $(NH_4)_2SO_4$, encourages hydrophobic interaction between the target solute and stationary phase. As a result, the type and concentration of these salts impact binding capacity and, potentially, step yield.

Note that the list presented in Table 8.4 makes no assessment as to whether the process parameters are critical, key, or non-key (see Chapter 2). Their classification is determined as part of process development studies, along with establishing the specific relationship between these inputs and the outputs (performance parameters and CQAs). These parameters must then be controlled to these ranges through production control mechanisms such as batch records and automation.

Table 8.4: Common process parameters for a chromatography step and their potential impact on performance.

Process parameter or material attribute	Performance parameters and CQAs potentially impacted
Resin	Binding capacity, purity, step yield, productivity, and column lifetime
Packed bed (column) HETP/asymmetry	Binding capacity, purity, step yield, and productivity
Bed height	Binding capacity, purity, step yield, and productivity
Equilibration conditions (e.g., buffer composition, pH, conductivity)	Binding capacity, purity, and step yield
Equilibration flow rate and CVs	Productivity
Load conditions (e.g., buffer composition, pH, conductivity)	Binding capacity, purity, step yield, and productivity
Product load (e.g., g product/V_{bed})	Purity, step yield, and productivity
Load flow rate	Binding capacity, purity, step yield, and productivity
Wash conditions (e.g., buffer composition, pH, and conductivity)	Purity, step yield, and productivity
Wash flow rate and CVs	Productivity
Elution conditions (e.g., buffer composition, pH, and conductivity)	Purity, step yield, and productivity
Elution gradient slope	Purity, step yield, and productivity
Elution flow rate	Purity, step yield, and productivity
UV gates (start and end collection points for product)	Purity, step yield, and productivity
Cleaning, sanitization, and storage solutions	Purity and column lifetime
Flow rate and CVs for all other phases not mentioned	Productivity
Temperature	Binding capacity, purity, step yield, and productivity

8.9 Chromatography process design

Let's bring together the information we have covered thus far by considering an example that illustrates an important design strategy for the purification stage of a biopharmaceutical process: three or more chromatography steps, each with a different interaction mode, are often necessary to achieve the purity required in the final drug product. The first chromatography step is referred to generally as the capture step, and the second and third as the intermediate and polishing steps, respectively. Data from a purification stage designed with this approach for production of a mAb in Chinese hamster ovary (CHO) cells is shown in Table 8.5 [233]. Specifically, step yields and impurity levels for each of the three chromatography steps used – protein A chromatography, CEC, and AEC – are shown. Review of these data lead to the following observations:

- Protein A chromatography run in bind-and-elute mode – a commonly used AC step as shown in Table 8.2 – is very selective for mAbs as shown by the data in Table 8.5. Significant amounts of important process-related impurities such as HCPs, host cell DNA, and endotoxin (more appropriately classified as a contaminant for a CHO-based process) are removed from the harvested cell culture fluid (HCCF) by the protein A step, while the step yield for the mAb product remains high. However, the protein A step does add leached protein A (the ligand from protein A resin) into the product.
- Cation exchange run in bind-and-elute mode provides significant clearance of HCPs, protein A, and mAb aggregates. The majority of HCPs, leached protein A, and aggregate were eluted from the column after the elution step, during the regeneration phase, which suggests they were more strongly bound to the CEC resin (i.e., more positively charged) than the mAb product.
- Anion exchange run in flow-through mode provides removal of residual HCPs, many of which have a negative charge at the pH of the load and therefore bind to the AEC column. To create flow-through mode, the pH of the buffer in which the mAb resides during loading to the AEC column was adjusted to 0.5–1 pH unit below the mAb pI.

Turning our attention to the design of individual chromatography steps, we recall that the goal of process development activities is to define the relationship between process inputs (e.g., process parameters) and process outputs (e.g., performance parameters and CQAs). Applied specifically to chromatography, process development seeks to provide the necessary purity to meet CQAs of drug substance and optimal performance of each chromatography step by providing high step yield, high productivity, high binding capacity, and low cost, all while maintaining product quality. Thus, the specific relationships between the process parameters and material attributes listed in Table 8.4 to the various performance parameters and CQAs needs to be established. As discussed in Chapter 2, process design requires an understanding of the CQAs for a product, as well as a general understanding of processing, which in the case of chromatography is required to make initial selection of the type of chromatography media to use – e.g., ion exchange or affinity or hydrophobic interaction. Once the initial choice of stationary phases has been made, development activities take place. For chromatography, development relies heavily on experimental studies given the complexity of the product intermediates involved (i.e., product with numerous, often poorly characterized impurities). However, the introduction of mechanistic models – that is, mathematical models that describe the physical processes taking place, rather than statistical models limited to describing data – into design activities is becoming more common [234].

Table 8.5: Typical step yield and purity values for a three-column purification process, similar to that shown in Figure 2.3, for a mAb produced in CHO as reported by Fahrner et al [233].

Process intermediate	Step yield (%)	HCP (ng HCP/mg product)	DNA (pg DNA/mg product)	Endotoxin (EU/mg product)	Protein A (ng protein A/mg product)	Aggregate (%)
Harvested cell culture fluid (HCCF)	NA	250,000–1,000,000	100,000–1,500,000	5–100	NA	5–15
Protein A pool	>95	200–3,000	100–1,000	<0.005	3–35	5–15
Cation exchange pool	75–90	25–150	<10	<0.005	<2	<0.5
Anion exchange pool	>95	<5	<10	<0.005	<2	<0.5

Studies typically begin with screening of both resins and mobile phase conditions, in which the capacity of resins for a solute and selectivity of resins for product relative to key impurities are assessed to select resins that optimize binding capacity (minimizing the volume of resin required, thereby minimizing cost) and ensure acceptable selectivity (maximizing purity). Even if the mode of interaction for a chromatography step has been narrowed to one, there will still be numerous options among that one interaction mode available from the different resin suppliers. The specific design of these screening studies varies according to product type. An excellent example describing screening of CEC resins used to remove aggregates from an Fc fusion protein following a protein A capture step and low pH viral inactivation step has been presented by Shukla [235].

Modern process development tools for chromatography steps include miniaturized systems that make for rapid testing and low sample requirements. Options include 96-well filter plates filled with resin, such as PreDictor™ plates from Cytiva [236] and miniature columns, such as Opus® RoboColumns® from Repligen [237]. Both allow for multiple studies conducted in parallel with automation to conduct screening studies. The Opus® RoboColumns® are small (50–600 μL volume), prepacked columns that come in a 96-well plate format and can be used in combination with a liquid handler, such as the Tecan Freedom EVO®, for automated, parallel execution of studies screening studies [237]. The combined Tecan/Opus® RoboColumns® systems are capable of running step elutions and collecting fractions for analysis. High-throughput methods to analyze effluents from these high-throughput screening studies are desirable as well. Combined with design of experiment (DoE), these methods are now referred to as high throughput process development. A nice example of implementing 96-well filter plates, miniature columns, and lab columns packed to a typical production height for development of a step to separate mAb product from aggregates, testing CEC and MMC resins, is provided by Welsh et al. [238].

Following screening studies, which if successful lead to selection of resin and an initial set of load, wash, and elution conditions, method optimization/characterization typically takes place in which load, wash, and elution conditions are finalized and the remaining phases (see Table 8.3) specified as well. These bench-scale studies may be conducted with a column packed to the same bed height as used in production. Peak pooling criteria – that is, when to start product collection and end production collection based on UV absorbance values from the product peak – are a key outcome from these studies as well.

Once a chromatography step is designed at bench scale, it needs to be scaled up. In addition, chromatography steps may need to be scaled down from production for the purpose of conducting studies, such as those required to demonstrate viral clearance. The objective of scale-up or scale-down is to keep the performance, as measured by purity (or impurity profile), step yield, and other performance parameters, consistent at each scale. We focus on chromatography column scale-up in what follows, although the concepts apply to scale-down as well. Scale-up could theoretically be carried out by replicating enough lab-scale columns, each running in parallel (i.e., the feed is equally distributed to each) at optimized process parameters, to achieve the desired production throughput. However, this approach is often impractical. A more common approach involves increasing the column diameter while maintaining constant column height and mobile phase superficial velocity. This strategy aims to keep the mobile phase residence time constant across different scales. Specifically, the following basic rules can be applied to column scale-up:

1. Optimize chromatography conditions at lab scale.
2. Use the same mobile and stationary phases at both scales; for example, use the same resin, with same particle diameter and the same solutions for each of the phases that make up the chromatography step.
3. Maintain the product load – the ratio of the amount of product loaded (typically in grams) to bed volume – between scales.
4. Keep mobile phase residence time constant between scales (designated as 1 and 2) constant. To do this, maintain bed height between scales, $h_{bed1} = h_{bed2}$, and maintain superficial velocity between scales, $v_{s1} = v_{s2}$. Subscripts 1 and 2 refer to development and manufacturing scales, respectively.
5. For gradients, keep the slopes constant between scales. The gradient slope is $(C_{hi} - C_{low})/(V_{gradient}/V_{bed})$, where C_{hi} and C_{low} are the high and low concentration of the component impacting binding of product to the stationary phase (e.g., the high and low concentration of NaCl in a linear gradient for IEC).

Scaling up using the rules described above results in a column with a larger diameter. The larger diameter allows for processing a larger amount of product at a higher volumetric flow rate. Consequently the process time for the chromatography step remains the same as at development scale, despite the increased product volume. The example below demonstrates how this scale-up approach is applied.

Example: Scaling up a 1.5 cm diameter column to production scale
A chromatography step is optimized at bench scale using a 1.5 cm diameter column packed to a bed height of 25 cm. The flow rate for the load phase is 15 mL/min, and the load volume is 260 mL. (Note that the residence time of fluid in that column was calculated in a previous example in Section 8.4). You will scale up to an estimated load volume of 200 L in production. Answer the following.
(a) What is the bed height of the production-scale column?
(b) How large a column (expressed as the column diameter in cm) is required for production?

Solutions
(a) The bed height at each scale is kept the same ($h_{bed1} = h_{bed2}$); therefore, the bed height at production scale is 25 cm.
(b) Next, calculate the bed volume required for the separation by keeping product load equal at both scales. Assuming that the concentration of product in the load solution is the same at both scales (and designating bench and production scale by subscripts 1 and 2, respectively), results in:

$$\frac{V_{load1}}{V_{bed1}} = \frac{V_{load2}}{V_{bed2}}$$

where the V_{bed} is given by equation (8.1). Solving for V_{bed2} gives:

$$V_{bed2} = \frac{V_{load2}}{V_{load1}} \times V_{bed1} = \frac{200\,L}{0.260\,L} \times \pi \times \left(\frac{1.5}{2}\right)^2 \times 25\,cm$$

$$= 33,984\,cm^3 \text{ or} \approx 34\,L$$

Solving equation (8.1) for D_{bed2} gives:

$$D_{bed2} = \left(\frac{4V_{bed2}}{\pi h_{bed2}}\right)^{1/2} = \left(\frac{4 \times 33,984\,cm^3}{\pi \times 25\,cm}\right)^{1/2}$$

$$= 41.6\,cm$$

But columns are available in discrete diameters, so if you are purchasing a new column, a diameter of 45 cm is a likely choice. Using a 45 cm column results in packing slightly more than the necessary amount of resin if the bed height is kept at 25 cm; i.e., the column is packed with more binding capacity than is actually needed.

From the example above, it is clear that keeping both the amount of product loaded per volume of bed and bed height constant upon scale-up fixes the diameter of the column; that is, the column diameter cannot be independently selected. If you do not have a column of the calculated diameter and do not want to buy one, it is possible to scale up to a particular column diameter by removing the constraint of maintaining bed height between scales. Like the column scale-up method presented previously, the method relies on keeping the mobile phase residence time constant between scales. But rather than holding residence time constant by keeping bed height and superficial velocity constant, the volumetric flow rate, in units of CVs per time (e.g., CVs/h), is held constant. In addition, the amount of product loaded per volume of bed is held constant between scales [239]. This volumetric flow-based method allows the flexibility of using an existing column by adjusting the bed height to provide the same mobile phase residence time between scales. Of course, changing the bed height leads to a different number of theoretical plates in the columns at each scale; however, it has been shown that increasing column length relative to that used at development scale by this procedure will result in equal or better separation [239]. Thus, this volumetric flow-based method of scale-up should be satisfactory when a chromatography step has been optimized at bench scale using a bed height no greater than that used in production.

8.10 Validation considerations for chromatography steps

Like all equipment, a chromatography system – both skid and column – must be qualified prior to use, as described in Chapter 3, to provide assurance that the equipment has been installed properly and operates through the intended ranges. Consistency of

the chromatography steps that are part of a process is demonstrated as part of process performance qualification (PPQ) runs for process validation. Recall from Chapter 3 that PPQ includes execution of multiple batches, usually a minimum of three, at
full manufacturing scale and with more sampling and testing than during routine
commercial production to sufficiently establish consistency and product quality
throughout the process. For a chromatography step designed to remove impurities X,
Y, and Z, sampling and testing during PPQ would be performed to demonstrate clearance of X, Y, and Z. A multitude of other data are also typically examined when assessing consistency of chromatography steps during PPQ. For example, UV elution profiles
related to the product and impurities for each run may be reviewed for consistency.
HETP/asymmetry values may be evaluated after each PPQ run for consistency, and so
on. Following successful completion of the PPQ runs, this additional testing and data
review may be removed because the consistent performance of the chromatography
step has been demonstrated.

In addition to equipment qualification, PPQ runs, and routine monitoring of processing steps following the process qualification stage of process validation, a variety
of studies specific to chromatography systems must be conducted as part of process
validation. These may be viewed as an extension of process design activities (stage 1
of process validation) and include chromatography media (which encompasses
packed resin beds in a column, membrane adsorbers, and monoliths) lifetime studies,
viral clearance studies, impurity removal studies, and chromatography media storage
studies. Note that if chromatography media are not reused, then the need for lifetime
and storage studies is eliminated.

As mentioned previously, it is common to reuse chromatography media in
manufacturing for batches of the same product. Because of the importance of chromatography steps to product purity and overall product quality, data must be obtained to demonstrate that stationary phase can maintain performance over a certain number of runs – that is, the lifetime of the stationary phase expressed as a
number of allowable runs (or cycles) must be determined. A lifetime study must be
performed for each chromatography step used in a process. Studies performed to
generate these data are typically conducted at both small and production scales,
with small-scale studies providing an initial estimate of lifetime, in what are often
referred to as cycling studies, and production-scale studies providing confirmation
of the lifetime. Deterioration of the chromatography stationary phase can result in a
decrease in product purity (i.e., an increase in impurities), increase in pressure drop
across the column, variation in elution profiles, and decreasing step yields. Consequently, these are all parameters that may be monitored as part of bench-scale cycling studies and production-scale runs to confirm lifetime.

An important component of establishing a chromatography media lifetime is assurance of minimal carryover of product/impurities from one batch to the next
throughout the lifetime of the media being used. Minimal carryover is typically demonstrated by performing blank runs at production scale. These runs involve the usual

sequence of phases, except that the load is replaced by a buffer – typically the buffer in which the product is dissolved – so that any components that elute during the blank run, whether product or impurities, have come from the previous product run and indicate undesirable carryover. Blank runs are executed on a single column (or membrane adsorber or monolith) periodically throughout its lifetime, importantly including a blank run at the end of its lifetime. Once the media lifetime is established, blank runs are no longer executed. The lifetime of chromatography media established in CGMP manufacturing can be significant. For example, lifetimes for columns packed with protein A resin for mAb purification used in CGMP manufacturing are commonly 50–200 cycles [240].

As discussed in Chapter 2, processes that pose a risk of introduction of unwanted virus into the product require viral clearance steps. Viral clearance refers to either physical removal or inactivation of the virus. Viral clearance studies assess individual process steps designed to remove or inactivate virus, evaluating their effectiveness in reducing viral load. They are performed using an appropriately scaled-down version of the process step and involve spiking a panel of relevant viruses into the process intermediate that feeds the unit operation being evaluated. The clearance is typically quantified as a log reduction value, defined as the base ten log of the ratio of total virus loaded to the step to the total virus in the product (i.e., output) from the step. Because chromatography steps can provide viral clearance, they are often part of the scaled-down viral clearance studies for a process. For example, protein A chromatography [241] and AEC [242] used in mAb production have been shown to provide robust viral clearance. For chromatography steps, these studies should demonstrate the ability of a stationary phase to clear virus across the process parameter ranges identified in the design of the steps, including media at the end of its expected lifetime, to ensure robust viral clearance throughout its lifetime. Note that the overall goal of these clearance studies is to demonstrate that a process has more than enough viral clearance capacity to ensure drug product safety [243].

In addition to chromatography media lifetime and viral clearance studies, impurity removal studies and storage studies are conducted. Impurity removal studies typically gather bench-scale data that demonstrate consistent removal of impurities. These studies can be conducted as part of the media lifetime evaluation. Storage studies demonstrate that the solution the chromatography media is stored in effectively inhibits microbial growth and does not degrade the stationary phase – that is, does not cause ligand leaching or cause degradation of the support material – over the expected storage times. The storage study should also determine conditions required for complete removal of storage solution.

8.11 Summary

Numerous soluble process- and product-related impurities are present in biopharmaceutical product intermediates that must be removed to produce drug product that is safe and effective. Host cell proteins, endotoxin, leachables from product-contact materials, virus, and product aggregates are some examples. Chromatography is a common and effective operation for separating product from impurities. It separates by preferential interaction of certain solutes (product or impurities) in the mobile phase (the flowing liquid) with the stationary phase (e.g., resin packed into a column, membrane adsorbers, monoliths). For instance, if product binds more strongly to a resin than impurities, it will move through the column more slowly. This difference in speed in which product and impurities move through the column leads to the separation.

Interaction between mobile phase solutes and the stationary phase is influenced by a combination of solute properties, mobile phase conditions, and stationary phase properties. Stationary phases come in a variety of chemistries classified according to their interaction mode: ion exchange, affinity, hydrophobic interaction, reversed phase, and mixed mode. Size exclusion chromatography also exists, although separation is not driven by interaction with stationary phase chemistry but rather occurs because small molecules are able to diffuse into resin pores that are not accessible to larger molecules. The physical process involved in binding of solutes to resin beads includes transfer of the binding solute from the bulk mobile phase to the resin bead, diffusion of the solute into the pores of the bead where most binding sites reside, and finally adsorption to the resin. Intraparticle diffusion is particularly slow and results in broadening of solute bands as they move through a packed column.

Chromatography steps for biopharmaceutical processes are commonly run in bind-and-elute mode, in which product binds to the stationary phase and impurities may or may not bind. Impurities that do bind are separated from product during elution. Flow through is another common mode, in which impurities bind to stationary phase and product flows through. Each mode of operation comprises a number of phases. For example, bind-and-elute operation using reusable columns requires equilibration, load, wash, elution, regeneration, sanitization, and storage phases. Equipment used to carry out the chromatography procedure may be either reusable or single use. Skids with disposable flow paths are available from a number of suppliers; however, single-use chromatography media, such as prepacked columns and membrane adsorbers, may be too expensive to be used only once in routine, commercial manufacturing. Systems that enable chromatography to be run as a continuous rather than batch operation are also available.

Biopharmaceutical processes often consist of three or more different chromatography steps (e.g., anion exchange, cation exchange, and hydrophobic interaction) to achieve the high purity levels required for drug product. Developing chromatography steps is a complex endeavor. It requires a thorough understanding of how numerous process parameters (e.g., equilibration buffer composition, load flow rate, and elution conditions) impact chromatography performance and critical quality attributes (CQAs) of the product. Empirical studies are typically required due to the complexity of the intermediates being purified. Miniaturized automated equipment is often used to perform multiple studies in parallel. In addition to these development studies relating process parameters to chromatography performance, a number of specific studies are conducted to support validation of chromatography steps, including column lifetime, viral clearance, column storage, and impurity profile studies.

Finally, we started off the chapter by saying that chromatography is a big topic, and we conclude this chapter with the admission that there are many important topics related to chromatography that could not be covered in this single chapter. For example, continuous chromatography could have been covered with more depth and no details around mechanistic modeling of chromatography steps, which will undoubtedly play a larger role in chromatography development and process control in the years to come, were provided. But we'll have to end it here.

8.12 Review questions

1. Proteins X, Y, and Z are separated by AEC. Protein X does not bind under the loading conditions (i.e., it elutes in the load flow through), but proteins Y and Z do. Y and Z are eluted from the column using a linear NaCl gradient, with Z eluting first. On a single graph, qualitatively draw charge vs. pH curves for each of the proteins. Draw a vertical line indicating the pH at which the separation takes place.

2. Derive an expression for the superficial velocity in a membrane adsorber with a cylindrical geometry in which mobile phase flows from the exterior of the cylinder formed by the membranes, through the membrane layers, and out through an internal flow channel. Note that at a constant volumetric feed rate, Q, the superficial velocity increases as fluid moves from the exterior of the membrane cylinder to the interior.

3. A solute is continuously fed to a packed column and binds to the resin. Do the following:
   (a) Draw (qualitatively) the resulting breakthrough curve.
   (b) Using the breakthrough curve, graphically indicate the area that corresponds to the total mass of solute adsorbed onto the column. To do this, perform a solute balance over the column from the beginning to end of the run, and use the result to determine the appropriate area.
   (c) On the same curve, show the area that corresponds exactly to $m_{ads,10\%}$.
   (d) On the same curve, show the area corresponding to the mass $m_{ads,10\%}$ estimated by equation (8.11) – specifically by $C_f \times V_{loaded,10\%}$.

4. A study is conducted to generate the adsorption isotherm for a mAb binding to protein A resin. A number of conditions are tested to determine the EBC over a range of mAb concentrations in a mobile phase. In the study, 1.00 mL of resin slurry, which equates to 0.25 mL of resin, is added to a 2.00 mL vial and incubated for 48 h with 1.00 mL of mAb solution. At the end of 48 h, the concentration of mAb in the liquid phase is measured. If the concentration of mAb in the liquid phase added to the vial is 13.0 mg/mL and the concentration of mAb in the liquid phase after incubation is 0.3 mg/mL, (a) what is the value of the EBC at these conditions, and (b) given that the adsorption isotherm is a plot of EBC against the solute concentration with which the stationary phase is in equilibrium, what are the x and y isotherm coordinates that this data represents?

5. The data in Table 8.6 apply to the chromatography step represented in the chromatogram shown Figure 8.7. Based on these data, calculate (a) the % purity for the column load (clarified lysate) and the product (i.e., eluate), and (b) the % step yield for the run.

Table 8.6: Data for chromatography run represented in Figure 8.7.

Process intermediate	GFP concentration (mg/mL)	Total protein concentration (mg/mL)	Phase start volume (L)	Phase end volume (L)
Load (clarified lysate)	1.19	1.90	17.77	27.77
Eluate	1.68	1.70	50.67 (start collection)	57.31 (end collection)

6. A column with an inner diameter of 60 cm is packed for a CEC step to a bed height of 20 cm. The equilibration and wash phases each use 3 CVs of 50 mM MES, pH 6.0 buffer. The elution phase is a linear gradient from 0% to 50% pump B (one of two inlet pumps on the system) over 4 CVs of elution buffer, with the 50 mM MES buffer connected to pump A (the other inlet pump on the system). Estimate the following: (a) the total volume (in L) of MES buffer that will have to be prepared for the step and (b) the amount of MES powder, in kg, that would be required to prepare this total volume of solution.

7. Draw a flow diagram representing the phases involved in a flow-through chromatography step using a single-use chromatography skid and column.

8. Read the article "Defining Process Design Space for a Hydrophobic Interaction Chromatography (HIC) Purification Step: Application of Quality by Design (QbD) Principles" by Jiang et al [78]. Answer the following questions:
   (a) What specific HIC resin is used in this study?
   (b) What type of biopharmaceutical is being purified?
   (c) What mode of operation is used to run the HIC step?
   (d) Which process parameters are considered in the characterization study, and how were these chosen from a much larger list?
   (e) Among the process parameters considered in the characterization study, which were found to have a significant impact on step yield, aggregates (HMW) levels, and HCP levels?
   (f) For the two process parameters that had the most significant impact on step yield, explain the reason for their impact.

9. Development activities for a chromatography step used in the production of a vaccine are conducted at bench scale, using a chromatography system whose in-line UV detector has a path length of 2 mm. The chromatography system used at production scale has a UV path length of 10 mm. If the UV gates (measured at 280 nm wavelength) based on development studies were set to start collecting when the product peak hits 0.1 absorbance units (AU) and to stop collecting on the downward slope of the product peak at 0.2 AU, what UV gate values should be transferred to production to ensure that the start and stop of peak collection occurs at the same product concentration at both scales?

10. As a follow up to the example presented on scaling up a 1.5 cm column used for bench-scale studies to production scale in section 8.9, answer the following questions. (a) At what load flow rate will the 45 cm diameter column be operated? (b) If the sequence of phases used for the step is the same as shown in Table 8.3, estimate the amount of time required for the entire step if the flow rate remains the same at each phase and 3 CVs of solution flow to the column at each phase.

11. A chromatography step for a biopharmaceutical process has been developed using a column with a height of 5 cm, an internal diameter of 1 cm, and a load flow rate of 1.30 mL/min. During the development studies, 100.0 mg of target protein was loaded to the column, which corresponds to 77.0 mL of load material. At production scale, you estimate that you will load 400 L of the same material. You have a 30 cm diameter column available for production scale. (a) Estimate the bed height of the production column based on the volumetric flow scale up method described in the chapter. (b) Estimate the flow rate, in L/h, you will use to load the column.

12. You have been tasked with designing a test for an OQ protocol to check that the pumps on a production chromatography system can deliver the flow rates required for production of a viral vector for gene therapy. The chromatography system is automated and controls flow rate based on a set point value in L/h. Describe the OQ test design, including the following information: (i) the parameter(s) you would set to execute the test, (ii) any measurements you would need to make as part of the test, and (iii) how you would demonstrate whether the chromatography system passes or fails the test. Use the flow meter calibration example in Section 3.3.4 as a guide.

Chapter 9
Formulation operations: ultrafiltration

Ultrafiltration (UF) is an important membrane-based operation common to the formulation/fill stage of most biopharmaceutical processes. Following the final chromatography step in a process, the product is in a clear – that is, essentially particle-free – liquid solution and contains low levels of the process- and product-related impurities discussed in the last chapter. However, it is also at a concentration and buffer composition dictated by that last chromatography step and likely different from what is required for the drug product. Consequently, an adjustment to the product concentration and buffer composition is needed, which can be accomplished by UF. This adjustment produces bulk drug substance, which is used to create the drug product as described in Chapter 2. UF steps are also used for other purposes in biopharmaceutical processes, as discussed shortly. Note that we use the term *product* here to mean the active molecule at any step in the process and the term *product solution* or *product intermediate* to refer to the solution in which product is dissolved.

In manufacturing processes, UF is run in tangential-flow mode and is closely related to tangential-flow microfiltration (MF), another membrane-based operation that was introduced in Chapter 7. The main difference between the two is membrane pore size, with pores in an MF membrane larger than those in a UF membrane. The pore size in MF membranes makes them well suited for solid-liquid separations, including clarification steps in which solids are retained and product solution permeates the membrane. The smaller pores in UF membranes allow them to be used for solute-solvent separation, including concentration steps in which product (the solute) is retained by the membrane while water (the solvent in biopharmaceutical applications) and other small molecular weight components permeate. Thus, in MF steps, product is often in the permeate stream, while in UF steps, product is retained by the membrane.

This chapter begins by introducing UF and answering the following questions:
– What is the basis of separation in UF?
– In addition to formulation of drug substance, what are applications of UF in biopharmaceutical processes?
– What equipment is used to execute a UF step?

As the chapter continues, we take advantage of the similarities in the two types of tangential-flow filtration (TFF) by bringing tangential-flow MF back into the discussion to complete the coverage started in Chapter 7. Questions to be addressed include the following:
– What materials are UF and MF membranes made from, and what modular formats are commonly used to house them in biopharmaceutical production?
– What is a typical TFF procedure for a current good manufacturing practice (CGMP) manufacturing environment?

https://doi.org/10.1515/9783111112459-009

- What are the process and performance parameters for TFF steps?
- How are TFF steps developed and designed for CGMP use?
- How are TFF steps scaled up from bench-scale development studies?

9.1 Basis of separation in UF and applications

Recall from Chapter 7 that our definition of filtration included not only the separation of *solid particles* in a fluid suspension according to their size by flowing under a pressure differential through a porous medium, but it also included separation of *components in solution* by size. While the former applies to MF, the latter part of the definition describes UF.

Like MF, UF steps rely on a porous membrane to create a size-based, pressure-driven separation, as shown in Figure 9.1. Small molecules, such as water, salts, and buffering agents, permeate the UF membrane while larger molecules or components are retained. As discussed in Chapter 7, UF membranes are generally designed to retain solutes with diameters ranging from 0.005 to 0.15 μm [190]. However, their pore sizes are typically described using a molecular weight cutoff (MWCO), which generally ranges from 1 to 1,000 kDa [244] and is discussed in more detail shortly.

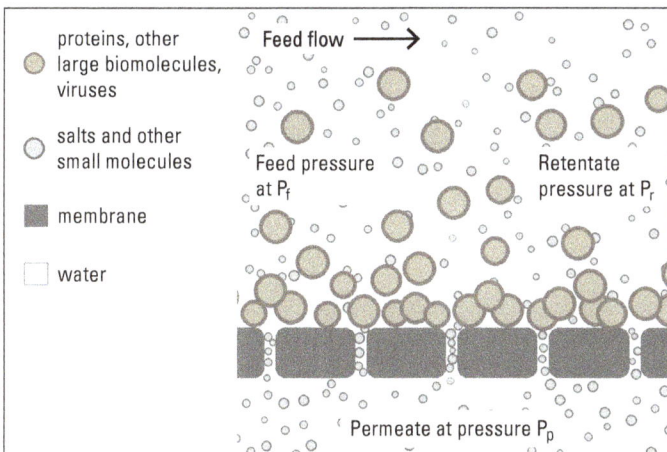

Figure 9.1: Illustration of UF. Small molecules like water, salts, and buffer components permeate the membrane, while larger components like proteins and viruses are retained. Note that in tangential-flow mode, $P_f > P_r$. Image © NC State University; reprinted with permission.

UF membranes are most commonly used to concentrate product (i.e., increase the concentration of product in solution by removing solvent, in our case water) or for diafiltration (i.e., removing or exchanging solutes, such as buffer or salts). We use the terms *diafiltration (DF)* and *buffer exchange* interchangeably throughout this chapter.

Buffer exchange, in this context, refers not only to the actual exchange of buffering agents, such as tris, but also solutes that are not true buffers, such as NaCl. Concentration and DF are required near the end of the drug substance process for most biopharmaceuticals. Note that the term *UFDF* is commonly used to describe both the concentration and diafiltration steps performed with UF membranes. During concentration, product solution is fed to the membrane. Water and other small molecules permeate, while product is retained. As water continues to permeate, the product becomes more concentrated and its volume is reduced. During DF, as the product solution is fed to the membrane, a buffer solution is added, typically at the same rate that water permeates the membrane. In a UF system operated in this configuration, the new buffer system washes out the old, which results in an exchange of buffers or other small molecular weight components. We discuss each of these operations in more detail shortly, but first let's consider the questions below to illustrate these concepts.

Example: Calculating target volume during a concentration step

Green fluorescent protein in 50 mM tris is introduced to a UF device at a concentration of 1.50 mg/mL and a volume of 300 L. Assuming that no GFP permeates the membrane but that water, tris, and all other small molecular weight components freely permeate, calculate the following: (a) the retentate volume for a final GFP concentration of 20.0 mg/mL, and (b) the amount of water that has been removed from the product in the permeate stream.

Solution

(a) If no product is lost in the permeate, then the mass of GFP remains unchanged from beginning to end of the concentration step, and the following relationship applies:

$$C_{initial} V_{initial} = C_{final} V_{final} \tag{9.1}$$

where C refers to the concentration of product and V to volume. So the final volume is given by

$$V_{final} = \frac{C_{initial} V_{initial}}{C_{final}} = \frac{1.50\,\frac{g}{L} \times 300\,L}{20.0\,\frac{g}{L}}$$

$$= 22.5\,L$$

Note that in a production environment, a product concentration target is typically met by concentrating to a target volume measured by a level sensor on a process vessel or using a scale, rather than by measuring a concentration value inline.

(b) During concentration, water, tris, and any salts present permeate the membrane, with most of the volume being water. Consequently, the volume of water removed is calculated by $V_{initial} - V_{final}$:

$$V_{final} - V_{initial} = 300\,L - 22.5\,L = 277.5\,L$$

Example: Product volume and buffer composition after diafiltration

In a separate step, a phosphate buffered saline (PBS) solution is fed to the concentrated GFP solution at the same rate that permeate is removed from the system. If the volume of PBS fed is 10 times that of the concentrated product solution (i.e., GFP solution at 20 mg/mL), determine the final volume of product and the composition.

Solution
Because PBS solution is fed to the membrane at the same rate that permeate is withdrawn, the product volume remains unchanged at 22.5 L. Both PBS components and tris freely permeate the UF membrane; however, PBS is replenished during the UF step, while tris is not. As a result, tris is washed out of the product solution and replaced with PBS. This leaves the product at its starting concentration in the PBS solution, with only trace amounts of the original tris buffer remaining.

UF used for product concentration and buffer exchange is designed and implemented with a number of objectives in mind. These include consistently achieving the target product concentration, consistently reaching the desired extent of buffer exchange, maintaining product integrity, delivering consistently high step yield, and achieving reproducible and high permeate flux.

We mentioned that UF may be used for more than just concentration and buffer exchange to formulate drug product. Other applications in biopharmaceutical processes include:

– Concentrating and/or diafiltering an intermediate product stream prior to a chromatography step. Concentration reduces the volume loaded to a chromatography system, and DF adjusts the product matrix (e.g., removes salts or exchanges buffer components) to optimize product binding to the resin and/or selectivity of the resin for the product.
– Purifying viruses as part of a vaccine or gene therapy viral vector process in which the large virus particle is retained while smaller impurities such as host cell proteins permeate the membrane.
– Removing unreacted reagents from chemically modified biopharmaceuticals, such as antibody-drug conjugates discussed in Chapter 1, and PEGylated proteins, which are therapeutic proteins to which polyethylene glycol has been attached to increase retention time in the blood or to stabilize the protein [245].

9.2 UF equipment and process configurations for production

At production scale, UF is typically operated in tangential-flow mode (refer to Chapter 7) and the equipment used is similar to the setup shown in Figure 7.12 for tangential-flow MF. In fact, systems for tangential-flow UF and MF can often be used interchangeably, with only minor modifications. For example, MF setups are usually equipped with a pump on the permeate to keep permeate flux at an optimal level. A permeate pump may be used in UF applications in which the pore size is relatively large (100 kDa or greater) [190] but seldom for UF applications with tighter pores. Diagrams showing a basic UF system for product concentration and buffer exchange are shown in Figure 9.2.

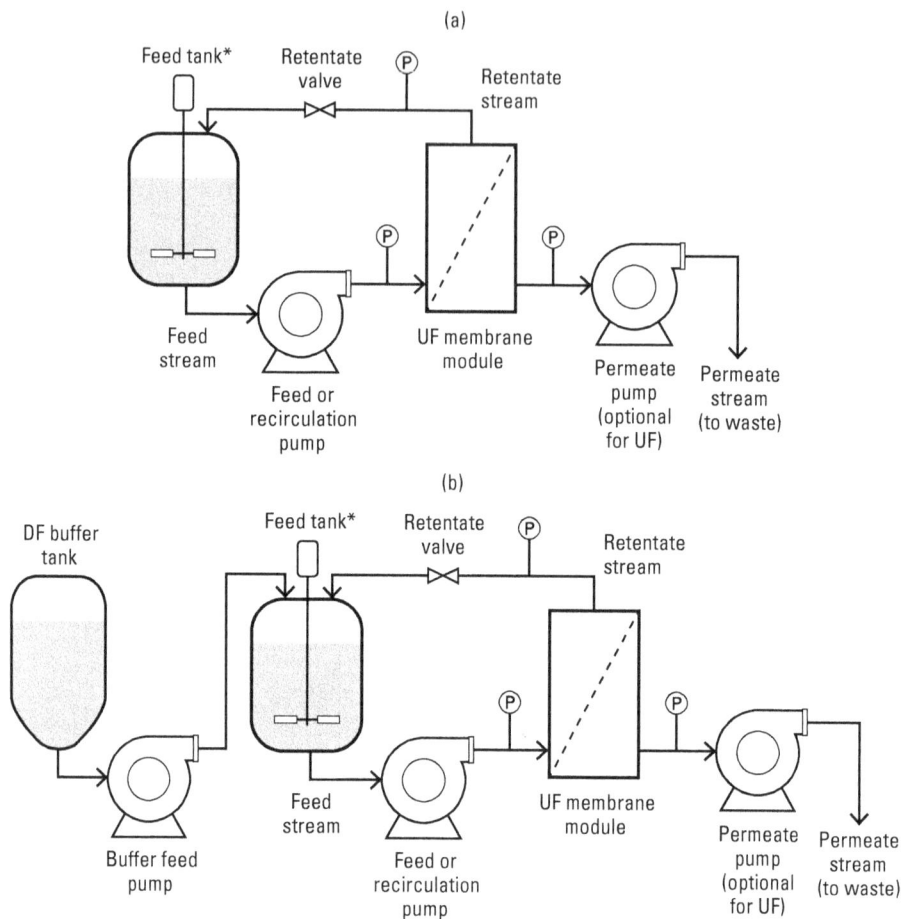

*Also referred to as retentate, recycle, or recirculation tank

Figure 9.2: UF system examples. (a) The setup used for batch concentration. (b) The setup typically used for buffer exchange (i.e., diafiltration). Note that a system used for fed-batch concentration would be similar to that for DF, except the DF buffer tank is replaced by a feed tank, and the feed tank as labeled in Figure 9.2(b) would be referred to as the recirculation (or recycle or retentate) tank. Image © NC State University; reprinted with permission.

The main components of these basic UF systems are:

- A feed tank that holds the product solution being processed.
- A feed pump that moves the product tangentially over the membrane surface and pressurizes the feed stream. It is often a rotary lobe pump [190]. As discussed in Chapter 7, the portion of the feed that does not permeate the membrane is referred to as retentate. The portion that permeates is referred to as permeate or filtrate. Consequently,

$$Q_f = Q_r + Q_p \qquad\qquad (9.2)$$

where Q represents volumetric flow rate and subscripts f, r, and p refer to feed, retentate, and permeate streams, respectively.

- Module(s) that contain the UF membrane required for separation. We discuss membranes and modules in greater detail in the next section of this chapter.
- A retentate valve that along with the feed pump establishes transmembrane pressure (TMP), which drives permeate flow through the membrane. Recall from equation (7.2) that TMP = $(P_f + P_r)/2 - P_p$.
- Pressure sensors, used to measure the feed, retentate, and permeate pressures. These values are used to determine the TMP and enable control to a TMP set point (for systems operated under TMP control).
- A DF tank and pump used to feed buffer into the product solution for performing buffer exchange. This pump is often a peristaltic type [190].

In addition to pressure gauges on the feed, retentate, and permeate streams, a number of other sensors may be found on a UF system. These include flow meters to measure flow rate through the feed or retentate line and the permeate line; conductivity and pH sensors on the permeate line to monitor the progress of the DF step (as DF progresses, the conductivity of the permeate will shift to that of the DF buffer); a UV detector on the permeate line to detect product loss to the permeate; a level sensor on the feed vessel for volume measurement; and temperature sensors on the feed or retentate line to detect temperature increase.

Concentration is commonly carried out in batch mode. During concentration only a portion of the permeable components (e.g., water, buffer components) permeate the membrane after a single pass. Retentate is recycled to the feed tank until the concentration target, usually measured by a volume reduction, is met; thus, feed/retentate undergoes multiple passes through the system shown in Figure 9.2(a). Permeate is directed to waste.

During DF, which typically follows concentration, retentate recirculates through system and permeate is sent to drain, as with concentration. However, a new buffer solution from the DF tank is also added, often at the same rate that permeate is withdrawn, to accomplish buffer exchange. Recirculation of product and addition of new buffer continues until the desired extent of exchange is achieved. When the rate of buffer addition matches the permeate rate, the volume of product solution in the feed vessel remains constant as does the product concentration. What does change is the composition of buffer components in the product. This method of operation is referred to as constant-volume DF. As already discussed, DF can be viewed simply as washing out the old buffer solution and replacing it with a new one.

Concentration by UF can also be carried out in a fed-batch configuration, which is particularly useful when a large volume reduction is needed for concentration (e.g., when the final volume is too small to be measured in a large feed vessel). Referring to Figure 9.2(a), in fed-batch operation, product intermediate from a separate vessel would be pumped to the feed tank, better referred to as the retentate vessel in this case, at the same rate that permeate is withdrawn. Because the feed rate to the retentate vessel and the permeate rate are the same, the volume of solution in the retentate vessel remains constant. Once all product from the separate vessel is fed to the retentate vessel, concentration of product can continue as a typical batch concentration step.

An example of a pilot-scale TFF unit is shown in Figure 9.3. This system can accommodate either UF or MF membrane devices. The system in Figure 9.3 is skid mounted and sits on wheels, which is common for TFF systems. However, there are a range of TFF systems in use for CGMP manufacturing – from relatively simple benchtop units operated manually that might be used for small-volume clinical lot manufacturing to much more complex, highly automated permanent units mounted to the floor.

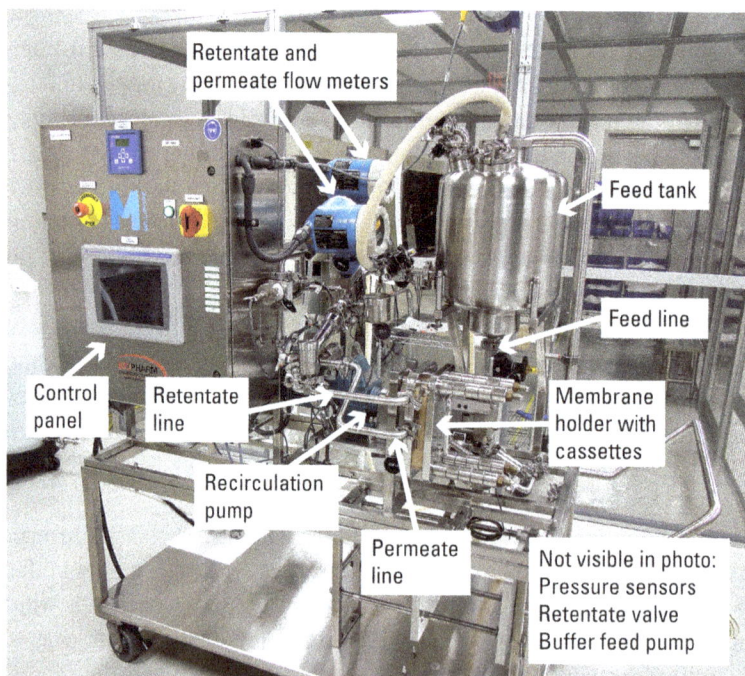

Figure 9.3: Photo of a pilot-scale UF system, with major components labeled. Photo © NC State University; reprinted with permission.

The photo in Figure 9.3 shows a stainless steel TFF skid that is cleaned, sanitized, and reused between batches. Single-use TFF skids are also available. An example is the Sartoflow® 4500 single-use TFF system from Sartorius. It uses a single-use flow kit and various single-use sensors. The Sartoflow® 4500 can accommodate between 3 and

10 m^2 of membrane area (cassette format) and deliver feed flow rates up to 4,500 L/h [246]. Note that this is one of a number of single-use TFF systems offered by Sartorius. Other single-use TFF systems on the market include the Mobius® FlexReady Solution for TFF from MilliporeSigma [247] and the KrosFlo® KTF TFF systems from Repligen [248].

9.3 TFF membranes and modules

As mentioned previously, UF membranes are characterized by a molecular weight cutoff (MWCO), also referred to as a nominal molecular weight limit (NMWL). These membranes are available with fixed MWCOs ranging from 1 kDa to 1,000 kDa. The MWCO represents the smallest molecular weight (in daltons) of a component that the membrane can retain. However, the retention or passage of components across the membrane depends on more than just their molecular weight relative to the MWCO. Several factors contribute to this, including:
- UF membrane pores are not all the same size – there is a distribution.
- Membrane surfaces and pores can become blocked by the retained components, effectively lowering the MWCO of the membrane.
- The shape and flexibility of the molecule may impact its retention [190, 249]. For example, a solute that is long, thin, and/or flexible may align and even elongate under the influence of flow through the membrane pore in such a way that solute enters the pore despite having a molecular weight greater than the MWCO.

Recall that in Chapter 7, we discussed that MF membranes have pores with sizes expressed as a μm rating. For MF, the range of pore sizes is as small as 0.1 μm or as large as 10 μm [173].

The images in Figure 9.4 show electron micrographs of UF membranes made from two of the most common materials: modified polyethersulfone on the left and regenerated cellulose on the right. Their structure is asymmetric, which is typical of UF and MF membranes: a thin top layer provides the separation on top of a substrate with larger pores [250]. The porous substrate provides mechanical strength for the thin ultrafiltration layer while not significantly impeding permeate flow through the membrane.

In general, regenerated cellulose is hydrophilic, exhibits low protein adsorption and low fouling, and is more compatible with organic solvents than polyethersulfone-based membranes. However, regenerated cellulose is less tolerant of pH extremes and oxidants, such as hypochlorite, a key component in bleach. Consequently, cleaning and sanitization of regenerated cellulose membranes with solutions having a high concentration of NaOH (high pH) or including bleach (an oxidant) may be problematic, although these more stringent cleaning solutions may not be required due to the low protein binding and fouling exhibited by regenerated cellulose membranes. Polyethersulfone tends to adsorb pro-

(a)   (b)

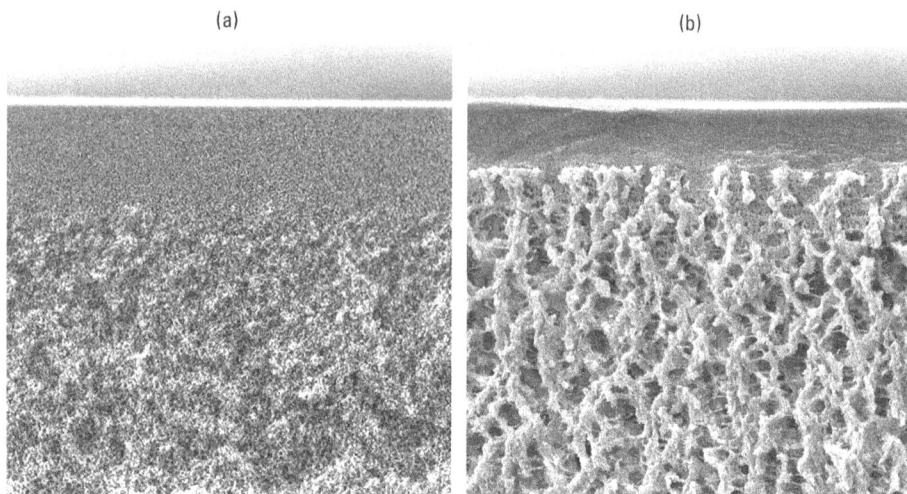

Figure 9.4: Scanning electron micrograph images of two membranes: (a) a 10 kDa MWCO Biomax®
modified polyethersulfone membrane and (b) a 10 kDa MWCO Ultracel® regenerated cellulose
membrane. Images from MilliporeSigma Technical Brief "Protein Concentration and Diafiltration by
Tangential Flow Filtration," page 5. © 2003. Reproduced with permission from Merck KGaA, Darmstadt,
Germany and/or its affiliates [244].

teins and other biological components but is stable over a wide range of pH values and
to oxidants, which may allow for more effective cleaning and sanitization. For bioprocessing applications, it is often modified to be more hydrophilic, which reduces protein
binding. In addition to polyethersulfone and regenerated cellulose, other materials are
used for UF and MF membranes for biopharmaceutical processing. An Internet search of
the websites of membrane vendors such as MilliporeSigma, Repligen, Sartorius, and Cytiva shows that polyvinylidene fluoride, mixed cellulose ester, and polysulfone are also
in use. An exhaustive review of membrane materials is beyond the scope of this book;
however, vendors can provide information to help decide on the type of membrane suitable for a particular application.

For ease of use, UF and MF membranes are assembled into modules that provide
(1) a high surface-area-to-volume ratio, (2) physical separation of the retentate and
permeate streams, (3) scalability, (4) cleanability/sanitizability, and (5) flow characteristics required for suitable performance. The two most common module geometries in
biopharmaceutical manufacturing are flat-sheet cassettes and hollow-fiber modules.
These are shown in Figures 9.5 and 9.6. A hollow-fiber module contains a bundle of
hollow, parallel fibers, with inner diameters typically ranging from 0.25 mm to 1 mm
based on a review of offerings from membrane vendors [251, 252]. The fibers are
housed in a cylindrical shell. The interior of the fiber is referred to as the lumen, and
the space outside the fiber is the shell. The thin retentive layer of the membrane is
usually on the lumen side of the fiber, in which case feed flows through the lumen,

while permeate flows radially through the fiber and to the shell side. Hollow-fiber modules are typically equipped with a single feed inlet, a single outlet for the retentate, and two ports for permeate.

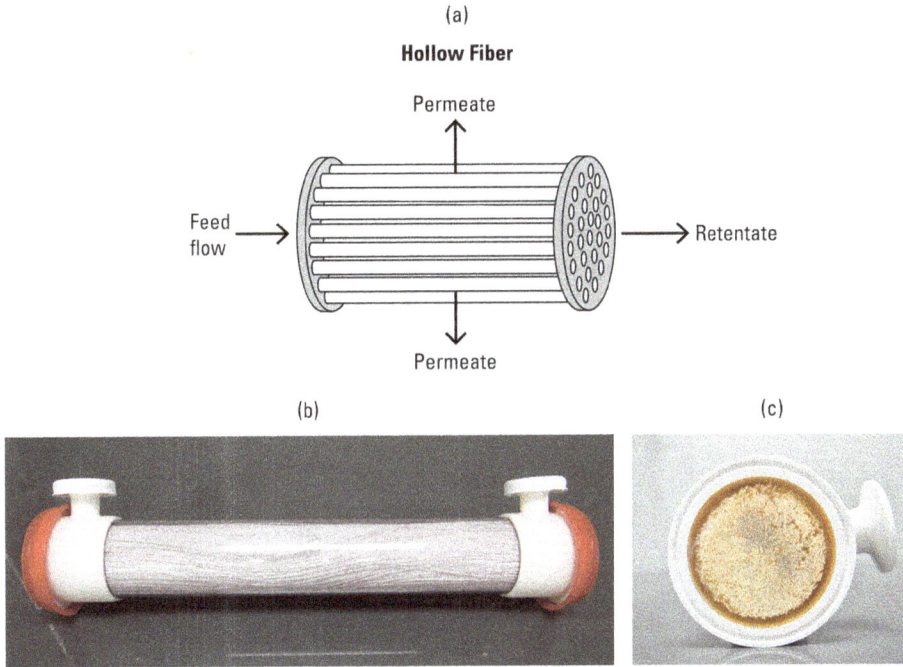

(a)

Hollow Fiber

Permeate

Feed flow → Retentate

Permeate

(b) (c)

Figure 9.5: Hollow-fiber modules for TFF applications. (a) A schematic showing feed, retentate, and permeate flow through the module. (b) A photo of a Spectrum (now Repligen) KrosFlo® hollow-fiber module. It is 50 cm in length, has modified polyethersulfone fibers of 0.5 mm inner diameter and a pore size of 0.2 μm, and contains 2.60 m² of membrane area. This is an MF membrane, but similar devices are available with UF membranes. (c) shows the front of the module, to which feed flows and enters into the fiber lumen. Images © NC State University; reprinted with permission.

Cassettes are the most commonly used membrane module type for TFF in bioprocessing [190]. Figure 9.6(a) provides a view of the interior of a cassette and the flow through it. The cassette has multiple layers of membrane across and through which feed flows. (Note that the number of membrane layers shown in that drawing is not necessarily representative of an actual cassette.) The tight retentive side of the membrane (refer to Figure 9.4) is on the feed side. Retentate flow from the cassette is separated from permeate. Screens may be inserted into the feed channels to promote turbulence on the feed side of the membrane, which serves to better sweep dissolved components (in the case of UF) or particles (in the case of MF) away from the surface. This sweeping enables high permeate rates to be achieved at lower tangential flow rates, which results in a reduced pumping requirement and the use of less membrane

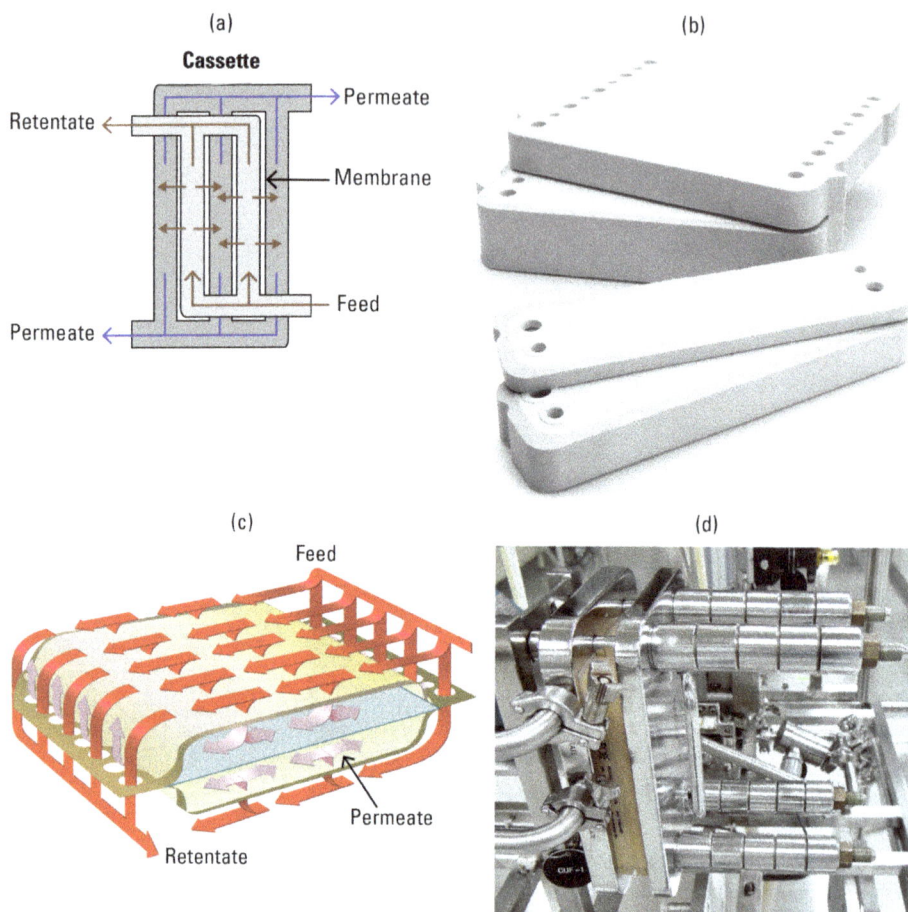

Figure 9.6: Images of cassettes and holders for TFF applications. (a) Flows through a cassette. Image adapted and reproduced with permission from Merck KGaA, Darmstadt, Germany and/or its affiliates. (b) Actual TFF cassettes. These are MilliporeSigma Pellicon® 3 UF cassettes with an area of 88 cm², 0.11 m², 0.57 m², and 1.14 m². Reprinted from "Pellicon® 3 Cassettes with Ultracel® Membrane," MilliporeSigma Data Sheet, 2018, Lit. No.: DS1209EN001. © Merck KGaA, Darmstadt, Germany and/or its affiliates, and reproduced with permission [253]. (c) Flow through a single "packet" of membranes within a cassette. A cassette is made up of a stack of multiple packets in parallel. Image from "Integrity Testing of Ultrafiltration Systems for Biopharmaceutical Applications," by S. Yee Lau, P. Pattnaik, and B. Raghunath, 2012, *BioProcess International, 10*(9), 54. Reprinted with permission [254]. (d) Two 0.5 m² Pellicon® 2 cassettes, 10 kDa MWCO, placed in a stainless steel holder. Photo © NC State University; reprinted with permission.

area. Details on screen designs appropriate for different applications can be obtained from the supplier.

A more detailed view of the flow of feed, retentate, and permeate in a cassette is shown in Figure 9.6(c), which illustrates flow in a membrane "packet" – defined here as two membrane sheets and a permeate spacer. The holes shown on each side of the packet in Figure 9.6(c) are specific to the feed, retentate, and permeate streams, and their circumference is sealed in such a way as to prevent flow from feed/retentate holes to permeate holes and vice versa. Feed is pumped to a holder, which contains the membrane cassettes. The cassette holder directs the feed to the feed holes on one side of the cassette. The feed stream flows through its designated holes (every other hole, as illustrated in Figure 9.6(c)) throughout the depth of the cassette, which is made up of many packets. That is, feed flows parallel to each membrane packet within the cassette through the alternating feed holes. The resulting retentate flows to the other side of the cassette, where it accesses only the retentate holes, combines with retentate from other membrane packets, and is directed by the holder out of the cassette to the retentate line. Permeate flows through the membrane to the permeate spacer and then out the permeate holes, where it combines with permeate from other packets and is directed to the system permeate line by the holder.

To increase the membrane surface area within a cassette for scale-up, the feed flow channel is widened and/or more packets are stacked in parallel within a cassette. Note that widening the flow channel does not mean that the space between membrane layers is made greater; instead, the membranes themselves are widened giving flow more area to "spread out" so that larger volumetric flow rates can be accommodated. In addition, individual cassettes can be installed in parallel if even more surface area is required. With hollow-fiber modules, area is increased by using modules of the same length but with more hollow fibers included in the module. And like cassettes, more modules can be installed in parallel if needed. We discuss scale-up in more detail shortly.

Table 9.1 contains information on various commercially available UF and MF membranes for commercial-scale biopharmaceutical applications.

Table 9.1: Various commercially available UF and MF membranes, and their properties, for use in biopharmaceutical manufacturing.

Membrane	References	Manufacturer	Module format	Membrane material	MWCO or pore size	Reusable or single-use module
Polysulfone	[255, 256]	Cytiva	Hollow fiber	Polysulfone	3–750 kDa 0.1–0.65 μm	Reusable and single use
Ultracel®	[257, 258]	MilliporeSigma	Cassette	Composite RC	3–1,000 kDa	Reusable cassettes and single-use Pellicon® capsules available

Table 9.1 (continued)

Membrane	References	Manufacturer	Module format	Membrane material	MWCO or pore size	Reusable or single-use module
Biomax®	[259]	MilliporeSigma	Cassette	Modified PES	5–1,000 kDa	Reusable
Durapore®	[260]	MilliporeSigma	Cassette	PVDF	200 kDa 0.1–0.65 μm	Reusable
Supor®	[261]	Pall (now part of Cytiva)	Cassette	Modified PES	0.1–0.65 μm	Reusable
Omega™	[262]	Pall (now part of Cytiva)	Cassette	Modified PES	1–300 kDa	Reusable
Delta	[263]	Pall (now part of Cytiva)	Cassette	RC	10–100 kDa	Reusable
ProStream	[264]	Repligen	Cassette	Modified PES	1–300 kDa	Both single-use and reusable cassettes available
HyStream	[264]	Repligen	Cassette	Modified PES	5–300 kDa, 0.1–0.65 μm	Both single-use and reusable cassettes available
Modified PES	[251]	Repligen	Hollow fiber	Modified PES	3–750 kDa, 0.65 μm	Single use
Mixed cellulose ester	[251]	Repligen	Hollow fiber	Mixed cellulose ester	0.1, 0.2 μm	Single use
Polysulfone	[251]	Repligen	Hollow fiber	Polysulfone	10–500 kDa, 0.05 μm	Single use
PES	[251]	Repligen	Hollow fiber	PES	0.2 μm	Single use
Hydrosart®	[265]	Sartorius	Cassette	Stabilized cellulose	2–300 kDa, 0.2, 0.45 μm	Reusable
PESU	[265]	Sartorius	Cassette	PES	1–300 kDa, 0.1 μm	Reusable

Note that RC is regenerated cellulose, PES is polyethersulfone, and PVDF is polyvinylidene difluoride.

9.4 Typical TFF procedures for CGMP manufacturing

In describing a typical TFF procedure, we start with concentration and DF by UF, similar to what might be used for product formulation just prior to bulk fill. This procedure can be readily adapted to UF steps at other points in a biopharmaceutical process and to MF steps as well. More on that at the end of this section. A diagram of the individual steps that make up an overall UF step for a biopharmaceutical process is shown in Figure 9.7.

The first step is installation of the membrane module(s). Specific instructions for installation are provided by the supplier. Modules from the supplier are often shipped in a storage solution, such as glycerin or a low concentration of NaOH that keeps the membrane wet – so it doesn't dry out and collapse – and inhibits microbiological growth. Modules are flushed with water after installation to remove storage solution. Referring to Figure 9.2, the term *flush* means a process configuration in which both the permeate and retentate lines are directed to drain, with some backpressure on the system. The flush configuration enables removal of material held up in the UF system. We also use the term *recirculate*, which refers to a process configuration in which liquid in the retentate, permeate, or both is returned to the feed vessel and back to the membranes. Following the water flush, the membrane modules are cleaned/sanitized. We cover cleaning and sanitization in more detail shortly.

After the initial cleaning and sanitization step, modules are again flushed with water to remove cleaning and sanitization agents. To ensure that the membrane has not been damaged during shipping and installation, an integrity test is performed. A membrane without integrity is one that does not perform as designed; in particular, for the case of a UF step designed to concentrate and diafilter a product, a lack of integrity would be any defect that allows product that should be retained to be lost to permeate. A hole in the membrane or improper sealing of the permeate and feed streams within the module are two causes of loss of integrity. An air diffusion test is often used for UF membranes to assess integrity. This test involves wetting the membrane with water then flowing air to the feed side of the membrane at a specified pressure. The only way for air to be transported to the permeate side in a completely wetted membrane with integrity is by the very slow process of diffusion. Thus, air flow rates in excess of what would be measured by diffusion alone suggest loss of membrane integrity. Suppliers provide test conditions and a specification for the largest air flow rate allowable for a passing test. For example, a Pellicon® 2 UF cassette containing Biomax® membrane with a MWCO of 10 kDa and area of 0.5 m^2, like the cassettes in the holder shown in Figure 9.6(d), must have an air flow (from feed side to permeate) of ≤18 cm^3/min at an air pressure of 10 psig [266].

Following integrity testing, the normalized water permeability (NWP) is measured. NWP serves as a measure of membrane cleanliness. It is performed by feeding high purity water (HPW) or water for injection (WFI) to the UF membrane at the same flow rate and TMP each time, then measuring the rate at which water perme-

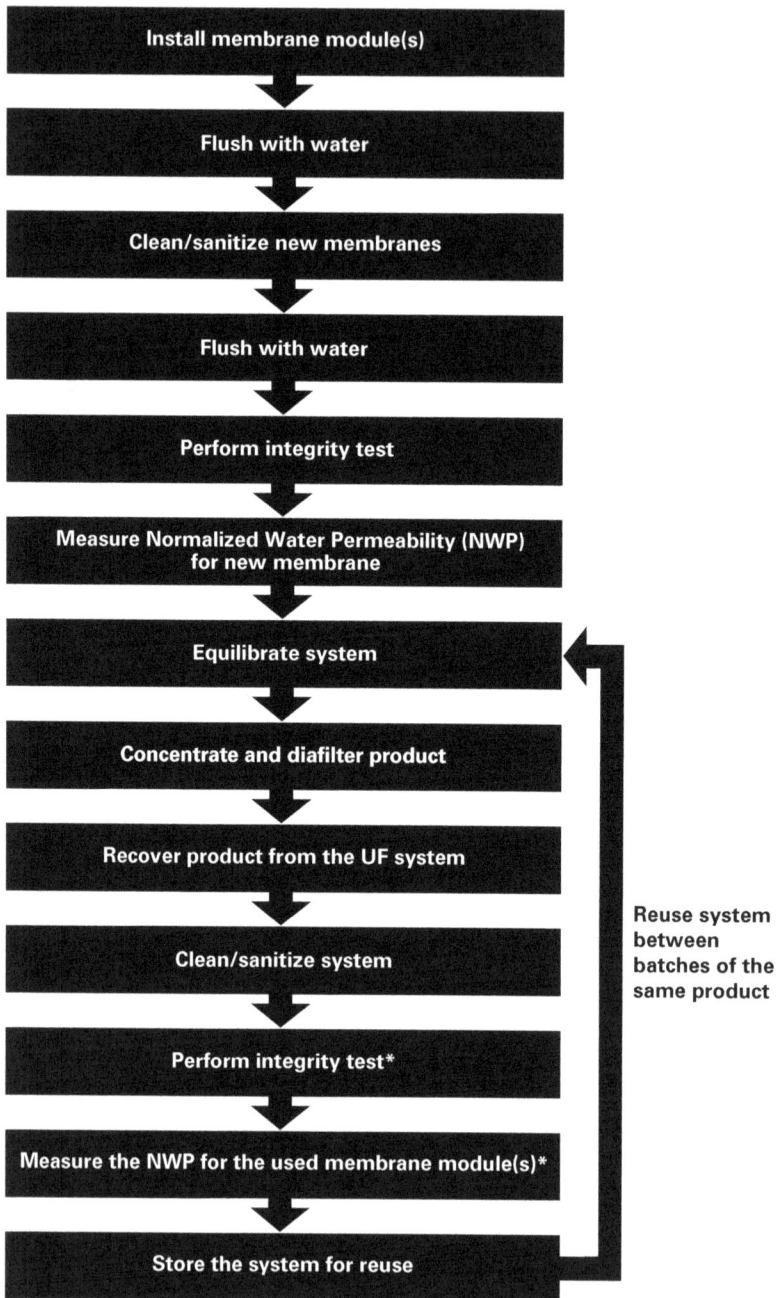

Figure 9.7: A typical procedure for tangential-flow UF used for concentration and buffer exchange of product. The steps shown apply to the case in which membranes are reused. Note that steps with an asterisk (*) may be executed prior to the following run, just before the system equilibration. Image © NC State University; reprinted with permission.

ates. Simply put, the permeate rate for water remains relatively constant if the membrane is cleaned effectively (i.e., no fouling from run to run) and declines if the membrane becomes fouled.

The equation used to calculate NWP is as follows:

$$NWP = \frac{Q_p \times TCF}{TMP \times A_m} \tag{9.3}$$

where Q_p is the volumetric flow rate of the permeate water typically measured in L/h, TCF is a dimensionless temperature correction factor (values are usually available from the membrane supplier), TMP is the transmembrane pressure as defined previously by equation (7.2) and typically measured in psig, and A_m is the membrane area in m². Note that the ratio Q_p/A_m is the permeate flux of the water, which we designate J_w. Referring to equation (7.7), the flux of water in an NWP measurement with an unused membrane can be written $J_w = TMP/(\mu R_m)$. Note that R_{total} in equation (7.7) has been replaced by R_m, the intrinsic membrane resistance, because resistance from a filter cake and internal membrane fouling would be nonexistent for a water flux measurement. The flux of water becomes dependent on temperature through the viscosity, μ. Higher temperatures reduce the viscosity of water thereby increasing permeate flux. For example, the viscosity of water at 25 °C is 0.8903 cp, while the viscosity of water at 18 °C is 1.0532 cp [267]. Based on these viscosity values, we would expect J_w at 25 °C to be 1.1830 times higher than J_w at 18 °C ($J_{w,25C}/J_{w,18C} = \mu_{18C}/\mu_{25C} = 1.1830$), all other conditions being the same. The TCF corrects for differences in permeate flow rates due to temperature by converting water flux values measured at any temperature to a standard reference temperature, typically 25 °C; thus, the TCF at 18 °C is 1.1830.

So if NWP is a measure of membrane cleanliness, why measure it on a new, unused membrane? Because NWP is often reported as a percent of the original value, with the value measured immediately after installation/initial cleaning serving as the original value:

$$\% \text{ of original NWP} = \frac{NWP_{after\ run}}{NWP_{original}} \times 100 \tag{9.4}$$

The percent of NWP recovered is a parameter typically trended for a campaign using the same set of membrane modules. For cassettes, it is common to see a decline in NWP after their first use. This initial decline has been attributed to the cassette screen compressing into the membrane and obstructing the area for permeation [190]. Beyond this initial decline, a steady decrease in NWP from run to run suggests ineffective membrane cleaning. Relatively constant NWP suggests effective cleaning and that process permeate flux values will not decline significantly from run to run. NWP as a percentage of the original value is typically used to define the membrane module lifetime. The exact criteria differ from company to company, but once the NWP falls below a predetermined percent of original value (e.g., NWP < 70% of the original), existing membrane modules are replaced with new modules.

Prior to exposing the UF system to product during concentration and DF, the system is equilibrated by flushing and recirculating buffer. Equilibration is necessary to remove water from the system (for the case in which new membranes are in use) or storage solution from the previous run. Exposure of product to either water or storage solution can lead to precipitation and/or denaturation, which is to be avoided. It is worth noting that the pre-concentration/DF steps just described may be greatly reduced by using a pre-sterilized single-use, self-contained TFF module, which may only need to be installed and flushed with equilibration buffer before the concentration/ DF steps [268].

Following equilibration, concentration and/or DF take place. We discuss these steps in more detail in the sections on parameters and process development. Following the concentration and DF steps, product must be recovered from the UF system. Given the value of biopharmaceutical products, it is critical to maximize their recovery while ensuring product integrity is maintained and final concentration targets are met. Product recovery begins by depolarizing the membrane. Depolarization involves recirculating the retentate, which resides in the feed tank, with the permeate valve closed. This procedure allows product that is concentrated at the membrane surface to diffuse back into the bulk retentate. Next, the system contents are collected in a separate vessel. This can be accomplished by collecting the retentate from a drain located at a low point on the system. We refer to product collected as the recovered retentate.

At this point in the procedure, most of the product has been removed from the system, but residual product may remain in the system hold-up volume. A number of different methods can be used to recover this residual product, including blowing down the system with air to push remaining product out of the system, flushing buffer through the feed/retentate to displace and collect the residual product, and recirculating buffer through the system and collecting the resulting solution. Residual product recovered by air blowdown should be at the same concentration as the recovered retentate and can be added back to that retentate directly. However, the residual product collected from the buffer flush or recirculation steps contains product at a concentration less than in the recovered retentate, which presents a problem: how can that product be returned to the collected retentate, so that it is not wasted, while also achieving the target concentration for the step? This can be accomplished by concentrating to slightly above the target – while ensuring no impact on product quality – then using the recovered product-containing buffer (from the flush or recirculation step) to dilute recovered retentate to the target concentration. For example, if the target concentration for final formulation is 10 mg of product/mL solution, then the product might be concentrated to 12 mg/mL and subsequently diluted to 10 mg/mL with the buffer recovered from the flush or recirculation steps. The following equation can be used to calculate the amount of recovered buffer, V_b, at product concentration C_b, that can be added to an overconcentrated recovered retentate to achieve the target final product concentration, C_{final}:

$$V_b = \frac{(C_r - C_{final})V_r}{C_{final} - C_b} \tag{9.5}$$

C_r is the product concentration in the recovered overconcentrated retentate and V_r is the collected volume of the recovered retentate. Note that C_r and C_b must be measured (e.g., in the quality control laboratory) to use equation (9.5). In practice, it is desirable to minimize overconcentration, while at the same time adding as much of the recovered buffer flush as possible back to the recovered retentate to maximize step yield. To write a procedure that consistently achieves the target concentration and maximizes step yield, a typical value for C_b is required. This value depends on the amount of buffer used for the flush or recirculation steps. Knowledge of the typical value for C_b along with the buffer volume used for flush or recirculation will determine the appropriate product concentration value to target in the overconcentrated retentate. That value is necessary for the UF procedure in a CGMP process. At-scale, engineering runs that precede CGMP runs provide a good opportunity to define this overconcentration procedure. Note that in a well-design UF process, step yields of >95% are readily achievable.

After product recovery, membranes are cleaned and sanitized to minimize carryover of the product from one batch into the next, to keep process permeate flux values high upon reuse, and to kill microbial contaminants. Cleaning and sanitization of membranes can be conducted as a single step or two different steps, and the solutions used depend on the membrane type. Cleaning typically involves flushing and recirculating cleaning/sanitization solution; the sanitization step may also involve a hold. Membrane manufacturers typically offer recommendations for cleaning and sanitization procedures and solutions. This information can be found in the user's guide for the cassette or hollow-fiber module. For example, for Pellicon® 2 cassettes with Biomax® membranes, MilliporeSigma recommends 0.1 M NaOH/250 ppm NaOCl – 0. 5 M NaOH/600 ppm NaOCl at 20–50 °C for cleaning, among a number of other possible cleaning agents. They also recommend sanitization be performed as a separate step after cleaning, using NaOH between 0. 5 M and 1.0 M at 20–50 °C, among a number of other options [266]. Of course, for a commercial biopharmaceutical product, the cleaning procedure must be validated as described in Chapter 3. Following cleaning and sanitization, the system is flushed with water to remove residual cleaning and sanitization agents, and a post-use integrity test and NWP measurement are performed.

Reusable membranes are typically stored in a liquid solution to keep the membrane wet and inhibit microbial growth until their next use. The storage step should include a flushing procedure to remove the cleaning/sanitization solution or water (if a water flush is performed) introduced in the previous step followed by recirculation. As with cleaning and sanitization, manufacturers often recommend a storage procedure and solutions. For the Pellicon® 2 cassettes with Biomax® membranes mentioned previously, 0.1 M NaOH is recommended [266].

It is important to note that reusable membrane modules are typically used for only a single product. If modules were used for multiple products, then there is the risk that Product A adsorbs on the polymeric membrane and is released into a subsequent batch of Product B.

Before concluding this section, let's briefly summarize how a tangential-flow MF procedure differs from the typical UF procedure depicted in Figure 9.7:

- For an MF step, the concentration/DF portion of the procedure is replaced by a solid-liquid separation such as clarification of a cell culture broth containing extracellular product. For this example, product permeates; therefore the permeate stream is collected rather than being sent to drain as in a UF step used for concentration and DF. Note that DF may be part of a tangential-flow MF step used for clarification, but the objective is enhanced product recovery from the retentate, once most of the product has permeated, rather than buffer exchange.
- The step yield for a tangential-flow MF step used for clarification is calculated as follows because the product permeates:

$$\% \text{ step yield} = \frac{C_p V_p}{C_f V_f} \times 100 \tag{9.6}$$

where C_p is the measured product concentration in the permeate, V_p is the volume of permeate collected, and C_f and V_f are the concentration of product in and volume of the initial feed.

- Whereas the integrity of a UF membrane is typically measured by an air diffusion test, for an MF membrane, a bubble point test may be used. Bubble point refers to the pressure at which a wetting liquid, such as water, is pushed out of the pores of a membrane. Suppliers specify the minimum bubble point value for a particular wetting agent that demonstrates membrane integrity. Because bubble point pressure is inversely proportional to pore size, the larger pores of MF membranes allow for a lower bubble point pressure that stays within the membrane module's pressure rating. In contrast, the smaller pores of UF membranes would result in high bubble point pressures that might exceed the module's pressure rating; therefore, lower pressure air diffusion testing is used to assess integrity.

9.5 Process and performance parameters for tangential-flow UF and MF

The purpose of this section is to discuss process and performance parameters that are of particular importance to tangential-flow UF and MF steps. We actually started this discussion in the previous section by defining a couple of performance parameters for both UF and MF: step yield and NWP. Let's continue the discussion of parameters by considering the diagram in Figure 9.8, which represents flow of retained product between two UF membranes, similar to the feed channel of a cassette. As discussed, the tangential flow on the feed side of the membrane is driven by a pump. The system is operated so that the feed-side pressure is higher than the permeate-side pressure, which forces liquid to permeate the membrane. P_f is greater than P_r, and the average pressure on the feed side of the membrane, $(P_f + P_r)/2$, is greater than the permeate-side pressure, P_p; that is, the TMP > 0. As liquid permeates the membrane, the concentration of the product at the membrane surface, C_w, increases because the permeating liquid carries dissolved product to the membrane surface, where it is retained. Thus, $C_w \gg C_{feed\ channel}$. Elevated concentration at the membrane surface is referred to as polarization, a term used previously in our discussion of the product recovery procedure. Note that the same polarization occurs in a tangential-flow MF system, except that it is solids that are accumulating at the membrane surface rather than dissolved components (i.e., product).

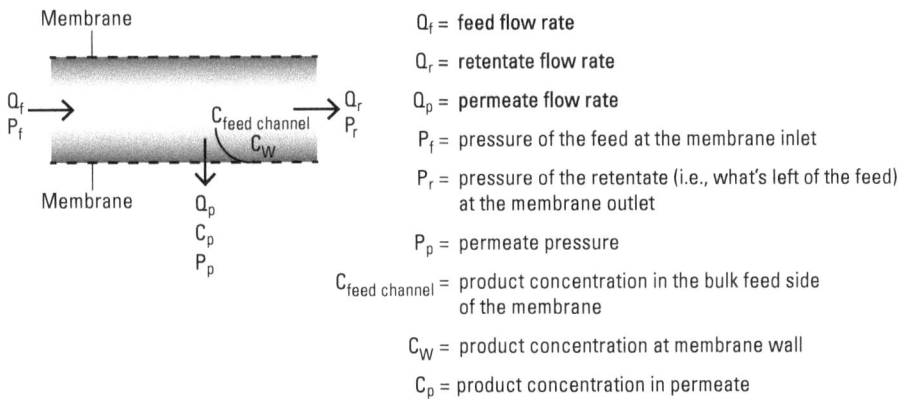

Q_f = feed flow rate
Q_r = retentate flow rate
Q_p = permeate flow rate
P_f = pressure of the feed at the membrane inlet
P_r = pressure of the retentate (i.e., what's left of the feed) at the membrane outlet
P_p = permeate pressure
$C_{feed\ channel}$ = product concentration in the bulk feed side of the membrane
C_W = product concentration at membrane wall
C_p = product concentration in permeate

Figure 9.8: Schematic of flows, pressures, and product concentrations in flow between two UF membranes. Image © NC State University; reprinted with permission.

UF is typically operated under TMP control, meaning TMP is a process parameter with a specified range in the batch record and controlled either manually or through process automation. Under TMP control, the permeate flux, designated J_p, is a response to TMP (along with other parameters). It is desirable to maximize the permeate flux because high flux leads to short processing times and/or lower membrane area requirements as the example below illustrates.

> **Example: Estimating the time to concentrate a solution of viral vectors for a gene therapy product**
> A bench-scale UF system is used to concentrate a feed of adeno-associated virus serotype 2 (AAV2) vectors, used in gene therapies, from 8.5×10^{10} vg/mL to 8.5×10^{11} vg/mL. (Note: vg stands for vector genomes, which is the number of AAV2 capsids that contain the gene of interest). The membrane is 100 kDa MWCO regenerated cellulose with an area of 88 cm². If the starting feed is 2 L and the average permeate flux during concentration is 150 L/m²/h, how long will it take to complete the concentration step? Assume no loss of AAV2 in the step.

Solution

First, determine how much feed must be removed as permeate to meet the concentration target of 8.5×10^{11} vg/mL. Once the volume permeated is determined, the process time can be calculated using equation (6.7) ($t_{process} = V_{feed}/Q$), rewritten for the special case of a membrane separation as follows:

$$t_{process} = \frac{V_p}{J_p A_m} \tag{9.7}$$

where V_p is the volume permeated, J_p is the permeate flux, and A_m is the membrane area. The retentate volume at the end of the concentration step is calculated from equation (9.1) as $V_r = (8.5 \times 10^{10}$ vg/mL $\times$ 2,000 mL)/ 8.5×10^{11} vg/mL = 200 mL. So the volume permeated is 2,000 mL − 200 mL, or 1,800 mL. Substituting these values into equation (9.7) results in:

$$= \frac{1.8\,L}{150\,L/(m^2 h) \times 0.0088\,m^2}$$

$$= 1.4\,h$$

Obviously if the permeate flux is <150 L/m²/h, the time required for the UF step increases.

We've already discussed permeate flux in tangential-flow MF in Chapter 7 (refer to equation (7.7)). For UF, the permeate flux during a concentration or DF step is strongly dependent on the extent of polarization, among a number of other variables. A polarization layer adds resistance to permeate flow, similar to what was discussed for accumulation of particles at the surface of an MF membrane. In addition, the accumulation of solute at the membrane wall creates an osmotic pressure gradient that opposes the TMP driving force. Osmotic pressure refers to the pressure that would have to be applied to prevent a solvent, in our case water, from moving from a solution of low solute concentration (permeate side) to high solute concentration (feed side). In the case of UF, because solute concentration is significantly higher on the feed side of the membrane than the permeate side, pressure must be applied just to prevent water from flowing in the opposite of the desired direction – from the permeate to the feed side of the membrane. With these polarization effects in mind, an equation for the permeate flux, J_p, for UF can be written similar to equation (7.7) for MF [269]

$$J_p = \frac{Q_p}{A_m} = \frac{TMP - \Delta\pi}{\mu R_{total}} \tag{9.8}$$

where the variables are defined as previously in equation (7.7), with the exception of $\Delta\pi$, which is the osmotic pressure difference for the solute concentration at the feed side and permeate side of the membrane surface. Its value can be quite high. For example, monoclonal antibodies are often formulated at high concentration (e.g., 200 g/L) so that only small volumes are required for subcutaneous administration. Binabaje et al. measured the osmotic pressure of a 200 g/L monoclonal antibody solution to be as high as 7 psi [270]. In the presence of concentration polarization, the osmotic pressure of a 200 g/L monoclonal antibody solution will be even higher, leading to a significant pressure difference that must be overcome to generate permeate flux. Note also that in the case of UF, R_{total} is the sum of resistances from the membrane and from any fouling that occurs, such as adsorption of solutes to the pore walls and gel layer formation.

Three input parameters that impact the process permeate flux, and therefore the process time, in tangential-flow UF are TMP, crossflow rate, and product concentration. Note that crossflow rate refers to the flow rate at which solution flows through the feed channel of (i.e., over) the membrane. To understand the impact of each of these process parameters on permeate flux, consider the plots in Figure 9.9. As discussed previously, the flux of water across a membrane is given by the equation $J_w =$ TMP/(μR_m), which shows that permeate flux varies linearly with TMP. This relationship is shown in Figure 9.9. When a solute is present that is retained by the membrane, such as a biopharmaceutical product, the permeate flux initially increases with TMP and then levels off, also shown in Figure 9.9. The leveling off occurs because of concentration polarization, just discussed, which leads to a high osmotic pressure gradient and greater resistance to permeation. Further increases in TMP simply increase polarization, which counteracts the additional driving force for permeate flow from the increased TMP.

Figure 9.9 also shows the impact of crossflow rate and product concentration in the feed on permeate flux. Higher feed rates result in higher permeate flux values than lower feed rates as a result of the enhanced sweeping action across the surface of the membrane, which minimizes polarization. Higher product concentration, on the other hand, leads to lower permeate flux because polarization effects are more significant. For concentration/DF steps under TMP control, the TMP target value is usually chosen at a point between the pressure-dependent and pressure-independent portion of the curve, as shown in Figure 9.9.

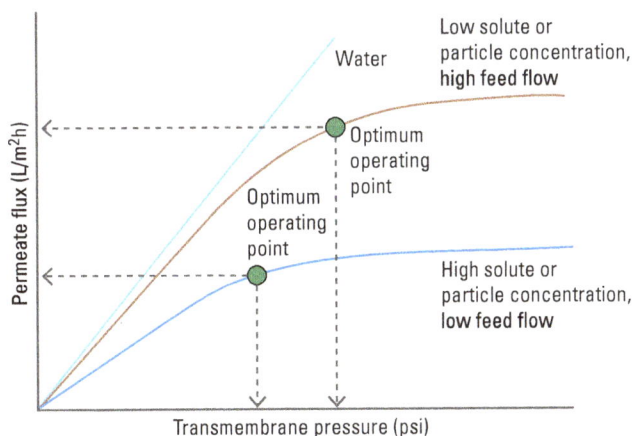

Figure 9.9: Typical permeate flux versus TMP curves for TFF operations. Note that low or high solute concentration applies to UF, and low or high particle concentration to MF. Image © NC State University; reprinted with permission.

In addition to TMP and crossflow rate, a number of other process parameters are controlled for the concentration and DF steps. Two important ones, concentration factor (CF) and DF volumes, summarized in Table 9.2, determine when concentration and diafiltration steps conclude.

Table 9.2: Additional process parameters that are controlled during concentration and DF steps.

Parameter	Definition		Comments
Concentration factor (CF) and Volume concentration factor (VCF)	**CF**: C_r/C_f (9.9) **VCF**: V_f/V_r (9.10) CF = VCF, assuming no loss of product to the permeate. Note that in these equations, C_f and C_r refer to the product concentration in the initial feed and final retentate, respectively. Likewise, V_f and V_r are the volumes of the initial feed and final retentate.	– –	For final formulation, the final product concentration in the UF system, C_r, is dictated by what is required to formulate drug substance. A concentration target is usually met by targeting a final retentate volume, which is determined by assuming VCF = CF.
DF volumes or diavolumes (DVs)	**DV**: $V_{df\ buffer}/V_r$ (9.11) In this equation, $V_{df\ buffer}$ is the volume of DF buffer fed and V_r is the retentate volume at the time of DF.	– –	In DF, the number of DVs dictates the extent of buffer exchange. For buffer components that freely permeate the membrane, almost 95% of buffer exchange occurs after only 3 DVs. For "complete" exchange, 7–12 DVs are common. For constant-volume DF, DV is often monitored and controlled by measuring V_p collected during DF and assuming $V_{df\ buffer} = V_p$.

For final formulation steps, the final production concentration, and therefore CF, is dictated by the product concentration required for formulation of drug product. Typically in a manufacturing process, the target concentration is achieved via the volume concentration factor (VCF) in equation (9.10) (see Table 9.2). The VCF is assumed to equal the required CF, and from the VCF, a final retentate volume is calculated and targeted to end the concentration step. This strategy is used because volume (or weight) of retentate is more readily measured inline (and in real time) than is product concentration in the retentate. Instructions for calculating the VCF from the target concentration value would be included in the batch record.

During DF, the extent of buffer exchange depends on the number of DF volumes (DVs) used, defined by equation (9.11) in Table 9.2. In a production process using constant-volume DF, DVs, or alternatively the volume of buffer added based on the number of DVs required, are often monitored by measuring the value of V_p during DF and recognizing it is equal to the volume of buffer added in a constant-volume DF setup. So how many DVs are required for complete buffer exchange? For constant-volume DF, the theoretical extent to which a solute targeted for removal, such as a buffer component or a salt, is remaining or removed is described by the following equations:

$$\% \text{ solute remaining} = \frac{C_{\text{solute,rf}}}{C_{\text{solute,ro}}} \times 100 = \exp[(R-1)DV] \times 100 \tag{9.12}$$

$$\% \text{ solute removed} = [1 - \exp(R-1)DV] \times 100 \tag{9.13}$$

where $C_{\text{solute,rf}}$ and $C_{\text{solute,ro}}$ are the solute concentrations in the final and initial retentate during DF, respectively, and R is the rejection (also referred to as retention) of the solute. Note that we use the term initial retentate rather than initial feed here assuming that the product (i.e., initial feed) has undergone a concentration step prior to diafiltration. Rejection is defined as

$$R = 1 - \frac{C_{\text{solute,p}}}{C_{\text{solute,r}}} \tag{9.14}$$

where $C_{\text{solute,p}}$ and $C_{\text{solute,r}}$ are the concentrations of the solute in the permeate and the retentate being fed to the membrane, respectively. To avoid ambiguity, we note the assumption that in the following analysis, the solute concentration in the feed tank, feed line, and retentate returning to the feed tank is similar enough to be considered the same and is designated $C_{\text{solute,r}}$. If a buffer component or salt passes through the membrane unhindered, its concentration in the permeate and retentate will be the same, and $R = 0$. If a solute is completely retained by the membrane, its concentration in the permeate is zero, and $R = 1$.

Equation (9.12) (and by extension, equation (9.13)) is derived by performing a mass balance on the solute that is removed during DF, such as a salt or buffer component, over the DF system, as shown in Figure 9.10(a). That is, the accumulation of the solute ($=d(V_r C_{\text{solute,r}})/dt$) during DF equals its input (in the DF buffer) minus its output (in the permeate). In the case of DF, the rate of input of a solute to be removed is zero, and the output is $C_{\text{solute,p}} \times Q_p$, which can be rewritten as $(1–R) \times C_{\text{solute,r}} \times Q_p$. For a constant-volume DF, in which the rate of buffer addition is equal to the permeate flow rate, the resulting differential equation can be solved to obtain equation (9.12). Additional details on deriving equations (9.12) and (9.13) are included in the questions at the end of this chapter.

Figure 9.10(b) shows plots of the percent of a solute retained versus diafiltration volumes for various R values as calculated by equation (9.12). As you can see, for a solute that freely permeates the membrane ($R = 0$), only 5% of that solute remains in the feed after 3 DVs; that is, 95% of the solute has been removed. And after 12 DVs, >99.999% of the solute has been removed. When performing buffer exchange for final formulation of product, it is common to use 7–12 DVs to ensure as complete a buffer exchange as possible. The required number of DVs and instructions for monitoring DVs are included in the batch record for the process.

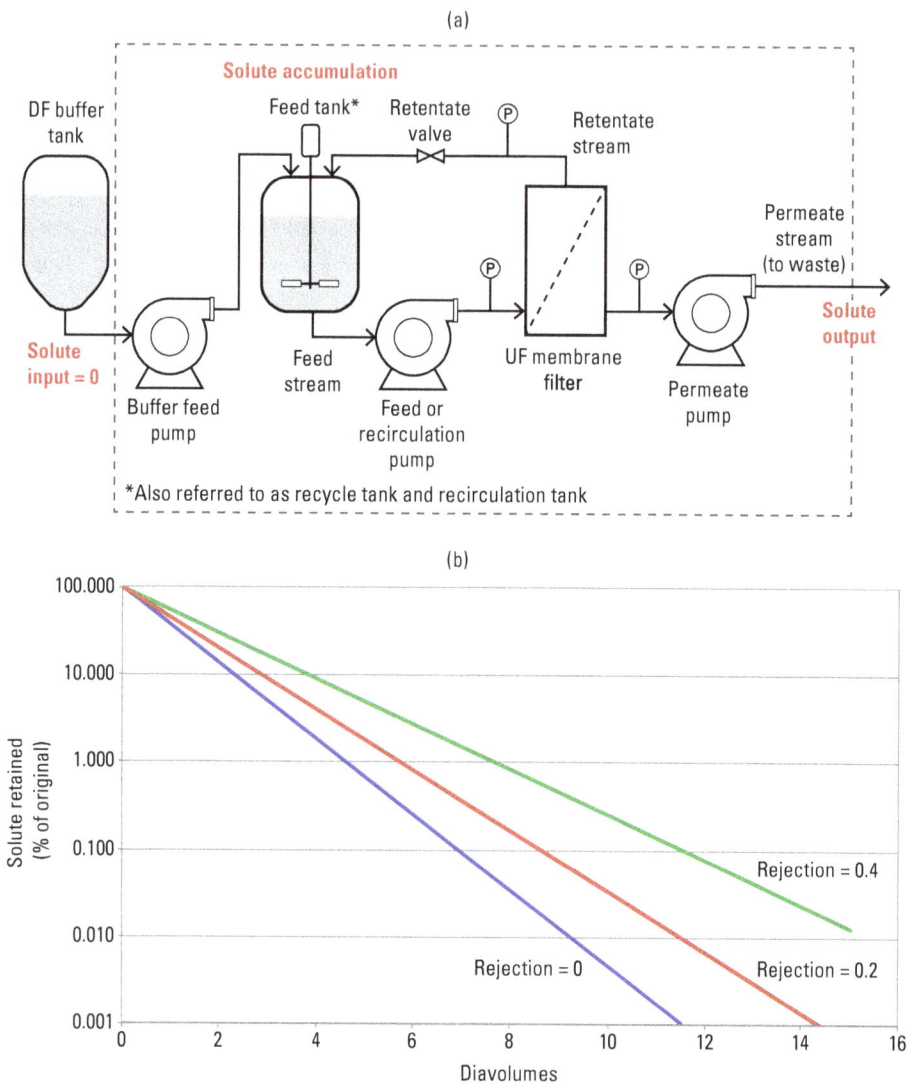

(a)

(b)

Figure 9.10: (a) The boundaries of the component balance used to derive equation (9.12). (b) Graphical results of equation (9.12). Images © NC State University; reprinted with permission.

Example: Estimating the concentration of a buffering agent at the end of a DF step

50 L of a solution containing a protein therapeutic in 50 mM tris is diafiltered with a buffer that contains no tris. If 200 L of DF buffer is used, estimate the concentration of tris at the end of the DF step. Assume that tris freely permeates the UF membrane being used.

Solution

To begin, solve equation (9.12) for $C_{solute,rf}$, which represents the concentration of tris at the end of DF:

$$C_{solute,rf} = \exp[(R-1)DV] \times C_{solute,ro}$$

Recognize that because tris freely permeates the membrane, $R = 0$. Further, 200 L of DF buffer added against 50 L of protein solution equates to 4 DVs of tris-free buffer added as calculated by equation (9.11). Substituting these values along with $C_{solute,ro} = 50$ mM leads to

$$C_{solute,rf} = \exp[(0-1) \times 4] \times 50 \text{ mM tris}$$

$$= 0.92 \text{ mM tris}$$

This value represents a significant reduction from the 50 mM starting value.

Note that throughout this chapter, we have indicated that concentrating the product commonly precedes DF. But the reverse sequence, in which DF precedes concentration, is also an option. In fact, there is an optimal product concentration at which to perform buffer exchange that will minimize the DF process time; specifically, it is desirable to minimize the ratio of permeate volume (thus minimizing the amount of DF buffer that must be fed during the DF step) to permeate flux (thus maximizing the rate of permeation during DF). There are both empirical and theoretical methods for determining this optimal product concentration for minimizing DF time. Those details can be found elsewhere [190, 244]. To take advantage of this optimal product concentration, a UF step may be designed to perform an initial concentration to achieve the optimal product concentration for DF, followed by DF for buffer exchange, followed by a final concentration step. As mentioned previously, this final concentration step may be designed to "overconcentrate" – that is, concentrate product to a value slightly greater than the target for bulk drug substance – so that the recovered product in the buffer from the flush or recirculation steps may be added back to the recovered retentate. As discussed, adding the recovered buffer dilutes the overconcentrated retentate to the target product concentration while returning residual product from the UF system to the product stream, thereby maximizing step yield.

Much of the previous discussion on process and performance parameters for tangential-flow UF applies to tangential-flow MF as well, with a few differences. These differences are considered in the following points:

- The schematic in Figure 9.8 applies to tangential-flow MF, except that the profile for product concentration (i.e., $C_{feed\ channel}$ and C_w) in the UF scenario applies to particle (not dissolved product) concentration in the MF scenario.
- The permeate flux versus TMP curves in Figure 9.9 apply to tangential-flow MF, with the exception that product concentration would be replaced by particle concentration.

- For tangential-flow MF, the CF defined in Table 9.2 applies to particulate. For MF steps used to clarify an extracellular product, the product is not retained but should freely permeate the MF membrane. Therefore, CF and VCF are not used as process parameters. Note, however, that dilution of the permeated product from DF buffer during an MF step does occur, and there is a tradeoff between step yield and product concentration; that is, the more DVs used for product recovery in the permeate, the more dilute the product becomes.
- Equations (9.12) and (9.13) apply to tangential-flow MF, but with a twist. For clarification of extracellular product, the solute being removed is the product. So equation (9.13) can be used to predict the percent of product remaining in the feed vessel just prior to DF that will be recovered as a function of DVs (i.e., the % solute removed in equation (9.13) for UF = % of product recovered for MF of extracellular product).

It is also important to note that while feed flow rate and TMP are frequently controlled parameters that are specified in a batch record and/or under automated control in a UF step for concentration/DF, tangential-flow MF may rely on permeate flow control. Because of the relatively large pore size of MF membranes, permeability is higher than with UF membranes, which results in a high permeate flux at the outset of an MF run under TMP control. In other words, most of the feed is converted to permeate. This high initial permeate flux pushes particles onto the surface of the membrane and into the pores, resulting in significant fouling. Consequently, rather than control TMP in MF steps, permeate flow rate is often controlled by using a pump on the permeate line to maintain flux at a level that avoids significant membrane fouling. In this approach, TMP responds to the permeate flow rate and is not an independently controlled parameter.

9.6 TFF development and scale-up

The goal of most tangential-flow UF and MF development efforts for biopharmaceutical processing is to design a step that consistently meets the target concentration of product and the correct extent of buffer exchange (for UF), or an acceptable product clarity (for MF), and that results in acceptable product quality, high product recovery (i.e., step yield), and high permeate flux with reproducible process times. Of course for UF, high product retention is needed for high step yields, while for clarification using MF, low product retention is needed for high step yields. To produce acceptable performance, an appropriate membrane and module must be selected. In addition, process parameters and their ranges must be identified and implemented that lead to acceptable values for performance parameters such as step yield, process time, and various quality attributes.

For TFF steps, development studies are likely to begin by screening different membranes. In the particular case of concentration/DF of proteins by UF, a rule of thumb is to select a membrane with a MWCO that is one-third to one-fifth of the molecular weight of the target protein [244]. This guideline should lead to good product retention while simultaneously allowing a high permeate flux. For clarification by MF, select a membrane pore size that maximizes particle retention, minimizes product retention, and maximizes process permeate flux. If using a membrane cassette with a turbulence-promoting screen on the feed side, the type of screen to be used is another choice that must be made. Also, the membrane chemistry must be chosen as well. Advantages and disadvantages of two of the most common membrane materials, regenerated cellulose and modified polyethersulfone, were presented previously. Clearly, there are a number of attributes to consider when choosing the right membrane (MWCO, pore size, membrane chemistry, module type, and turbulence promoting screens), and there is value in experimentally screening different membranes and modules to select the one that is best suited for a particular tangential-flow UF or MF application.

Following membrane screening, additional studies and information gathering are required to identify and determine ranges for process parameters that lead to acceptable values for performance parameters and quality attributes. Guidelines on development activities for the various steps involved in a typical UF procedure for concentration and DF of product are detailed in Tables 9.3 and 9.4. These tables present an expanded list of process parameters, relative to what was presented in the previous section, that should be considered. They also include performance parameters and quality attributes that are potentially impacted and strategies for identifying "ranges" for these parameters. These development studies, including the membrane screening previously described, are typically conducted at bench scale using systems that can reliably be scaled to manufacturing. Most of the membranes and modules listed in Table 9.1 come in sizes suitable for these bench-scale studies, and there are a number of TFF units commercially available that are likewise suitable for use with the smaller volumes typically used to conduct these bench-scale studies (<1 L). Suppliers can provide significant input on membrane screening and parameter setting for each of the procedural steps via vendor-performed studies for a specific process, user manuals, technical briefs, and other sources of information.

It is worth noting that many of the process parameters listed in Tables 9.3 and 9.4 likely will have no effect on product quality. They are in place primarily to ensure processing consistency. Process parameters in Tables 9.3 and 9.4 that may be critical – that is, must be controlled tightly to ensure drug substance critical quality attributes meet their specifications – include the CF and DVs.

Because development studies are typically done at bench scale, scale-up is often required for execution at production scale. Matching the optimized performance developed at bench scale is a primary objective of scale-up. To properly scale up, the pressure, flow, and product concentration profiles (or particulate concentration for MF)

Table 9.3: Process development activities by step in a UF procedure for concentration and DF, from membrane installation to equilibration.

Step in UF procedure	Process parameter or material attribute	Performance parameter(s) or quality attribute(s) potentially affected	Source of process parameter, material attribute information
Installation	– Membrane/module type – Module area – Torquing requirements for cassettes	Step yield and process time (for membrane/module type and area) Sealing and integrity	Membrane screening studies Scale-up calculations Supplier instructions
Cleaning/ sanitization	– Cleaning/sanitization agents – Cleaning/sanitization agent concentration – Cleaning solution temperature – Recirculation time – Flush volume	For all process parameters listed, corresponding performance parameters are process time, membrane lifetime, impurities from previous batches, and bioburden	Supplier recommendations; studies at small scale to demonstrate effectiveness, assessed by NWP measurement, product carryover measurement, sanitization studies, etc.
Integrity test	– Test pressure (for air diffusion test) – Allowable air flow rate (for air diffusion test)	Process parameters in place to ensure consistent and accurate measurement of integrity	Supplier instructions
NWP	– Feed rate – TMP – Pass/fail criterion (% of original)	Process parameters in place to ensure consistent and accurate measurement of NWP	Supplier instructions, with the exception of pass/fail criterion, which is established by the user
Equilibration	– Buffer – Feed flow rate – TMP – Recirculation time and flush volume	For all process parameters listed, corresponding quality attribute is product-related impurities. Removal of previous system contents (e.g., water and storage solution) during equilibration minimizes the formation of these impurities.	User chooses buffer solution, typically the buffer in which the product is in or DF buffer; supplier recommendations

Refer to Figure 9.7 for the procedure on which this table is based (adapted from Lutz, Herb, 2015, Ultrafiltration for Bioprocessing) [190].

Table 9.4: Process development activities by step in a UF procedure for concentration and DF, from the concentration step to system storage.

Step in UF procedure	Process parameter/performance parameter or quality attribute potentially affected	Source of process parameter information
Concentration/DF	– TMP (for TMP control)/process time – Permeate flow rate (for permeate flux control)/process time – Crossflow rate (Q_f)/process time – Concentration factor (CF) or concentration end point (C_r)/product concentration, process time, and aggregates – DF buffer/drug substance formulation (pH, osmolality, and conductivity) – Diavolumes/extent of buffer exchange (pH, osmolality, and conductivity), process time, and aggregates	For concentration, perform TMP scouting studies (J_p vs. TMP at various flow rates as shown in Figure 9.9) at initial and final product concentration to choose TMP and Q_f. These flux studies are conducted with permeate recycled to the feed vessel. For concentration, concentrate product to final target value. Measure J_p vs. time (or C_r), C_p to ensure acceptable rejection of product and to ensure final target concentration can be consistently met. Also ensure that step yield and product quality are acceptable. For DF, perform TMP scouting studies (J_p vs. TMP at various flow rates as shown in Figure 9.9) at product concentration during DF to choose TMP and Q_f. Diafilter product with buffer to determine DVs that produce acceptable buffer exchange with no loss of product or product quality.
Product recovery	– Depolarization flow rate and TMP/step yield – Buffer flush volume or recirculation volume and flow rate/product recovery	Bench studies to determine parameter values that maximize % recovery of product while minimizing product dilution
Cleaning/ sanitization, NWP, integrity test	See Table 9.3	See Table 9.3

Table 9.4 (continued)

Step in UF procedure	Process parameter/performance parameter or quality attribute potentially affected	Source of process parameter information
Storage	– Storage solution and concentration/ membrane lifetime and bioburden – Flow rate/membrane lifetime and bioburden – Recirculation time/membrane lifetime and bioburden – Flush volume/membrane lifetime and bioburden – Allowable hold time between runs/ membrane lifetime and bioburden	Supplier instructions

Refer to Figure 9.7 for the procedure on which this table is based (adapted from Lutz, Herb, 2015, Ultrafiltration for Bioprocessing) [190].

along the length of the membrane module must be maintained across scales during concentration and DF [190, 271]. To maintain these profiles across scales, a scale-up method that maintains feed-side velocity and channel length on scale-up can be used. The following steps apply in general:

1. Optimize the UF or MF step at lab scale by performing the studies described previously.
2. Use the same membrane (chemistry, MWCO, or pore size) and similar module (geometry, feed channel or lumen dimensions, turbulence-promoting screen if applicable) as used in bench-scale development/optimization studies. The TMP is scale independent and should be transferred directly if operating under TMP control.
3. Maintain V_f/A_m between scales, where V_f is the initial volume of material fed to the membrane unit and A_m is the membrane area.
4. Maintain the length of the flow path between scales. Suppliers offer cassettes and hollow-fiber modules with small surface areas suitable for bench-scale studies and much larger areas suited to production applications (as has been described previously) while keeping the feed path length constant. Figure 9.6(b) shows how this is done for Pellicon® 3 cassettes. It is clear in that photo that membrane area is increased by adding feed channels (resulting in thicker cassettes) and widening feed channel width (resulting in wider cassettes) while keeping the length of the flow path the same.
5. Maintain the volumetric feed rate per membrane area (e.g., L/m²/h) between scales. As a result, the feed flow rate (L/h) (obviously) increases with increased membrane area.

Membrane area requirements for CGMP production applications vary depending on the product. There are UF systems in use for commercial manufacturing with <1 m^2 of membrane area and much larger systems with up to 120 m^2 of UF membrane area in operation commercially [190].

9.7 Validation considerations

TFF systems must go through installation qualification (IQ) and operational qualification (OQ) as a prerequisite to process performance qualification (PPQ) runs as described in Chapter 3. There are numerous tests to be conducted during IQ and OQ, and a detailed listing of these tests is beyond the scope of this book. However, it is worth noting that during OQ, it is particularly important to check the following for UF and MF systems: that the feed tank on the system provides good mixing (this is particularly important to ensure adequate buffer exchange), and that product can be readily drained from the TFF system. In addition, the minimum and maximum operating volumes of the equipment must align with what is required for the process. It is particularly important to confirm that a UF system is capable of accurately measuring the volume that will be targeted for a concentration step. PDA Technical Report No. 15 recommends these and a number of tests to consider as part of a TFF system OQ [272].

Also as discussed in Chapter 3, PPQ runs combine the actual facility, utilities, qualified equipment, and trained personnel to produce commercial batches. However, enhanced sampling, testing, and monitoring of process and performance parameters occur during a PPQ run compared to a normal commercial batch. For example, you might also test for aggregates during PPQ to ensure that the scaled-up version of the TFF step is not adversely impacting product quality. This test would be in addition to monitoring performance parameters such as process time and step yield, as well as drug substance quality attributes such as product concentration, extent of buffer exchange (e.g., pH and conductivity), and bioburden as would be monitored in a routine commercial batch. Upon demonstrating that aggregate levels remain within expected limits during PPQ, you may choose to stop testing for aggregates at the UF step during routine commercial production.

Cleaning validation for TFF systems typically takes place concurrently with PPQ runs and demonstrates that the cleaning/sanitization procedure for the system (including both membrane modules(s) and skid) ensures that at process scale there is minimal carryover of product from one batch to another using methods described previously in Chapter 3; that bioburden is controlled within acceptable limits; that the process permeate flux is recoverable after each run; and that cleaning/sanitizing agents are removed. Note that an effective cleaning/sanitization procedure would have been developed at bench-scale prior to the validation runs. That bench-scale evaluation would include an assessment of the compatibility of cleaning/sanitization agents with the membrane as well as sanitization studies that look at the effectiveness of the sanitizing agent, often the same as the cleaning agent, for reducing microbial growth.

For membranes that are used for multiple runs of the same product, validation studies are also required to establish a membrane lifetime (i.e., the maximum number of runs). The evaluation may start with bench-scale studies aimed at assessing how reuse affects membrane performance. Starting at bench scale offers the advantage of allowing for adjustment of cleaning/sanitization and storage procedures if they are leading to membrane degradation. The maximum time that membranes can be exposed to cleaning/sanitization and storage solutions should also be determined as part of a membrane lifetime study. The supplier should be able to supply data to support setting these limits. Ultimately, the membrane lifetime should be validated under protocol at process scale by closely monitoring appropriate performance parameters and ensuring they remain within applicable ranges throughout the membrane lifetime. The example below considers the design of such a protocol.

Example: Designing a UF membrane lifetime protocol

A validation engineer is writing a lifetime protocol to validate the maximum number of uses for a UF membrane in a process for a new vaccine. The protocol will be executed concurrently with production of commercial batches. Suggest parameters to be monitored at production scale to establish the lifetime.

Solution

The goal of the study is to demonstrate that performance of the membrane remains consistent throughout its lifetime. To that end, the following performance parameters may be monitored for each batch processed up to the maximum number of runs being validated. Note that acceptance criteria would have to be established for each:

- NWP
- Permeate flux during concentration and DF (and/or process time)
- Membrane integrity
- Step yield

In addition, monitoring bioburden and endotoxin as part of the UF step is useful to assess the ability of the sanitization procedure to control microbial contamination throughout the lifetime of the membrane. Finally, it is also recommended to assess carryover of product from batch to batch to ensure it remains within acceptable limits. One method to accomplish the carryover assessment is by performing blank runs at some pre-determined frequency, including a blank run at the end of membrane lifetime. Similar to the blank runs described in Chapter 8 for chromatography, these runs would use the routine CGMP procedure except that product replaced by a buffer. The presence of product or impurities in the resulting "buffer product" suggests carryover from one batch to the next.

9.8 Single-pass TFF

In the UF process configurations we have discussed previously (batch concentration, constant-volume DF, and fed-batch concentration), retentate makes multiple passes through the setup as described for Figure 9.2. If instead of recycling retentate to the system feed vessel, it is discharged, as shown in Figure 9.11(a), a single-pass TFF (SPTFF) system configuration is created. It is a method that has gained popularity over the last few years, with the first patent being awarded in 2008 [273].

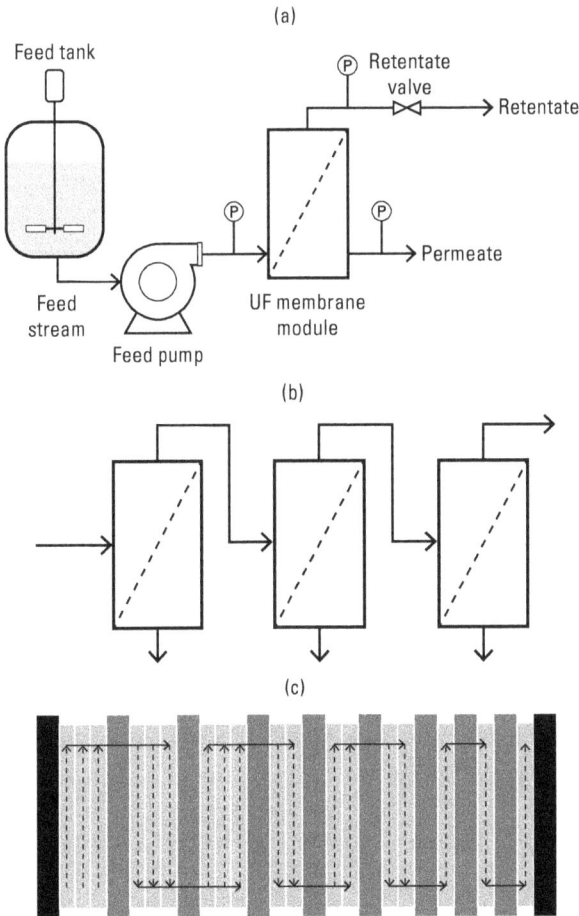

Figure 9.11: Single-pass TFF. (a) A UF system configured for single-pass operation. Image © NC State University; reprinted with permission. (b) Conventional UF membranes placed in series to create an SPTFF membrane stack. Image © NC State University; reprinted with permission. (c) The diagram for an SPTFF module. Adapted from Pall Life Sciences' "Application Note: Cadence™ Systems Employ New Single-Pass TFF Technology to Simplify Processes and Lower Costs," 2011, 2. Copyright 2011 Pall Corporation [274].

A parameter often used to characterize the performance of SPTFF is conversion, defined as the ratio of the permeate flow rate, Q_p, to the feed flow rate, Q_f:

$$\text{Conversion} = \frac{Q_p}{Q_f} \tag{9.15}$$

To achieve the desired concentration target in SPTFF, high conversion values are required. High conversion can be achieved by operating at low flow rates through the module and/or by increasing the path length to increase residence time. There are a number of ways to implement a single-pass configuration with increased path length. Conventional UF cassettes can be placed in series by the end user, similar to the configuration shown in Figure 9.11(b). Membrane providers like Pall and MilliporeSigma have created special manifolds and gaskets to enable placement of conventional UF cassettes in series within a single existing holder [275, 276]. In addition, SPTFF cassettes, comprising an internally manifolded series of UF cassette stages, can be used. Pall's Cadence SPTFF modules with Delta membrane are an example [274]. The flow path for these cassettes is shown in Figure 9.11(c). The number of flow channels through each stage decreases so that flow velocity remains high as the feed stream is converted to permeate. Note that unlike conventional batch concentration by UF, in which product concentration in the retentate changes with time, product concentration in the retentate exiting the SPTFF device of a given length reaches a steady-state value.

One of the advantages of recirculating feed as described for Figure 9.2 is the possibility of minimizing the membrane area for a concentration or DF step. So why use SPTFF? There are several potential advantages versus the traditional batch concentration configuration shown previously:

- A process stream can be fed continuously, while retentate and permeate are continuously withdrawn, making SPTFF well suited for continuous processing.
- Unlike batch concentration by UF, product in single-pass UF sees only a single pass through a pump or, alternatively, the process stream can be moved by pressurizing the feed tank. Reduced pump passes relative to batch concentration makes SPTFF a gentler method of processing and useful for fragile molecules.
- The smaller hold-up volume in an SPTFF system can translate to better product recovery.
- The equipment footprint for single-pass UF is smaller than for the batch concentration configuration.
- Product concentration in the retentate from the single-pass UF setup reaches a steady state, unlike batch concentration in which product concentration continuously changes.

SPTFF is well suited to a number of applications in biopharmaceutical processes. As we've mentioned, it is a preferable option for continuous processing. Even in a batch process, it is a good option for inline concentration (i.e., volume reduction); for example, it can be used to easily reduce the load volume for a chromatography step, which can shorten the time for processing. It is also a good option for inline desalting, which can be accomplished by concentrating via SPTFF then diluting with a no- or low-salt buffer.

More recently, inline cassettes for DF have become available. These address a gap in single-pass UF technology by enabling inline buffer exchange, to complement the concentration capabilities of the SPTFF systems just described [277]. Few details of the module configuration are provided; however, presumably it is configured to allow DF buffer to be fed to stages within the SPTFF module while simultaneously allowing permeate to be withdrawn from each stage, thereby enabling buffer exchange.

9.9 Summary

Tangential-flow UF and MF are widely used membrane separation techniques for biopharmaceutical processes. Separation in both is size based. UF is used for concentrating product and performing buffer exchange in a step referred to as diafiltration, often as a formulation step just before bulk filling. Both concentration and diafiltration are typically performed in batch mode. In batch concentration, retentate is recycled to the feed tank until the target concentration value is achieved. Retentate is also recycled to the feed tank during diafiltration; however, a buffer is simultaneously added to the feed vessel, often at the same rate that fluid permeates the membrane. The result is a constant volume of solution in the feed vessel. MF, which uses membranes with slightly larger pores than UF, is used for solid-liquid separation, such as clarifying a cell culture broth.

Most membranes are made from modified polyethersulfone or regenerated cellulose, and numerous suppliers offer UF and MF modules for biopharmaceutical applications. For biopharmaceutical applications, UF and MF membranes are packaged as modules in either a cassette or hollow-fiber format. To accommodate larger scale applications that require higher flow rates, membrane area in a cassette is increased by widening the membrane or adding more membrane layers in parallel within the cassette while keeping the path length the same. With hollow-fiber modules, area is increased by adding more fibers of the same length within a module. To achieve even greater surface area for large-scale applications, multiple cassettes or hollow-fiber modules are installed in a parallel configuration.

Most UF and MF membrane modules are reusable, and a procedure similar to that shown in Figure 9.7 is used for CGMP processing. Design of a UF or MF step is typically done at bench-scale, using small membrane modules that are readily scaled up. There are numerous process parameters that apply to UF and MF steps, such as TMP, crossflow rate, concentration factor, and diafiltration volumes. Ranges must be set appropriately through development studies, with the goal of designing a step that meets either the target concentration of product and the correct extent of buffer exchange (for UF) or an acceptable product clarity (for MF), while resulting in acceptable product quality, high step yield, and high permeate flux with reproducible process times. Once the UF or MF step has been optimized at bench scale, it can scaled up using a linear scale-up methodology in which the length of the flow path, the volumetric flow rate per unit area of membrane, and the feed volume per unit membrane area are held constant between scales. Among a number of expectations for validation of UF and MF steps is the execution of a study, done under protocol, to establish a membrane lifetime for a specific process.

Single-pass tangential-flow filtration is a method that has gained in popularity in recent years, particularly for concentration of products by UF. Rather than recirculating retentate through the system as is done in batch concentration or constant-volume diafiltration, the retentate is discharged from the system after a single pass. This process configuration may be particularly useful for concentration and buffer exchange steps in continuous processes.

9.10 Review questions

1. As mentioned earlier in the chapter, an in-line UV sensor (set to a wavelength of 280 nm) is often included on the permeate line of a tangential-flow UF system to detect product loss to the permeate. Why not also include a UV sensor in the feed tank of a UF system to provide a direct measure that a concentration target has been met?

2. You are working on a procedure to measure the NWP for a UF step to be used as part of a process to produce a protein therapeutic for clinical trials. That process uses 2.5 m^2 of membrane area, and you know from bench-scale studies that the NWP for the membrane is 23.7 LMH/psi. The measurement of NWP in the scaled up process is manual. Estimate the amount of HPW that will be collected as permeate over a 1-minute period at a TMP of 5 psi. The HPW temperature is 18 °C.

3. Below is a graph showing NWP data for 15 UF runs for a validated process.

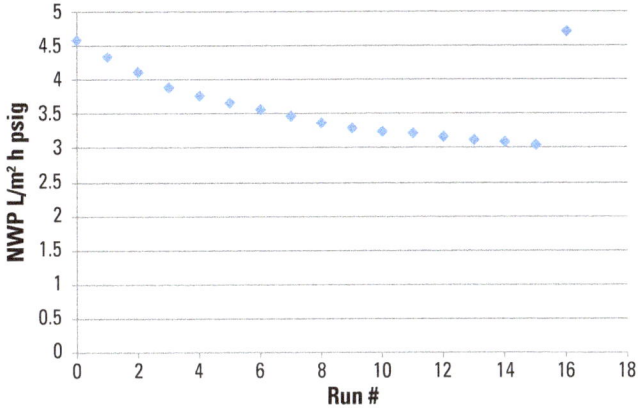

Figure 9.12: NWP data for 15 UF runs for a validated process. Image © NC State University; reprinted with permission.

Give three possible reasons why the NWP value suddenly increased at run #16.

4. A solution containing a 27 kDa protein was concentrated using a 10 kDa MWCO UF membrane with an area of 50 cm² at a TMP of 22 psig. The starting concentration was 1 mg/mL. Data showing the retentate volume and time for the run are shown below. Answer the following: (a) what is the protein concentration of the final retentate? (Assume no protein is lost in the permeate); (b) what is the average permeate flux for the run in LMH?; (c) plot the permeate flux in $L/m^2/h$ (i.e., LMH) versus the protein concentration in the retentate and explain the trend you see.

Table 9.5: Retentate volume versus time data for concentration of a protein by UF.

Time (s)	V_r (mL)	Time (s)	V_r (mL)
0	500	2,072	200
121	475	2,285	175
262	450	2,506	150
416	425	2,735	125
576	400	2,970	100
746	375	3,217	75
917	350	3,474	50
1,097	325	3,742	25
1,283	300	3,797	20

Table 9.5 (continued)

Time (s)	V_r (mL)	Time (s)	V_r (mL)
1,473	275	3,855	15
1,666	250	3,915	10
1,868	225	–	–

5. A protein therapeutic is to be concentrated to a target of 7.00 ± 0.20 mg/mL. To achieve this concentration, the product is initially overconcentrated and then buffer flush from the product recovery step is added back to the retentate. The overconcentrated retentate is drained from the pilot-scale UF skid and collected into a vessel with a tare weight of 1.34 kg. The gross weight of the vessel following collection is 6.43 kg, and the product concentration is 9.42 mg/mL. The system is flushed with 3 L buffer, of which 2.89 L is recovered at a product concentration of 0.49 mg/mL.

   a) How much of the recovered flush can be added to achieve the final target concentration?

   b) What is the step yield achieved? The initial feed to the UF step had a product concentration of 2.00 mg/mL and a volume of 25.0 L.

6. It was shown that for batch concentration, VCF = CF assuming no loss of product to the permeate. Derive the more general expression $V_r = V_f \times (C_r/C_f)^{-1/R}$ that accounts for the possibility of product permeating the membrane. To do this, (i) perform a total mass balance (in the form: accumulation of mass = total mass input - total mass output) on Figure 9.2(a) operated in batch concentration mode and (ii) perform a product mass balance in the form: accumulation of product = product mass input – product mass output. Recall that $R = 1 - C_p/C_r$ and solve the three equations for V_r.

7. Referring to the above problem, if a protein therapeutic permeates a UF membrane with a rejection value of 0.8, what final retentate volume would be targeted for a concentration factor of 10 (in batch concentration configuration) if the initial feed volume is 100 L? How does this value compare to the final volume required for a product that is completely retained by the membrane?

8. Derive equation (9.12) for the percent of a solute retained in the feed, with no solute added in the DF buffer, during a constant-volume batch DF step. To do this, (i) perform an overall mass balance (in the form: accumulation of mass = total mass input - total mass output) on Figure 9.10(a) operated in constant-volume DF mode

and (ii) perform a solute mass balance in the form accumulation of solute = solute mass input − solute mass output.

9. For a solute that is completely permeable to a UF membrane, the solute concentration is reduced by a factor of 10 in a constant-volume DF after every _____ DVs.

10. 500 L of product intermediate in a buffer solution of 50 mM tris and 0.1 M NaCl is being diafiltered (by constant-volume DF) with 50 mM tris to reduce NaCl concentration to below 0.005 M. For NaCl, $R = 0$, and for product $R = 1$ for the membrane in use. If the average permeate flux is 40 LMH and 25 m^2 of membrane area are used, answer the following questions.
    a) What are the minimum number of DVs of DF buffer required to reduce the NaCl concentration to the desired value?
    b) How much permeate is generated during the DF step?
    c) How much time is required for the DF step?

11. A concentration/DF step using UF for formulation of drug substance has an expected step yield of 93–100%. After 15 runs of the same product using the same membrane modules, the step yield measures 75%. A deviation is initiated and root cause must be determined. Give eight possible reasons for the out-of-range step yield value.

12. For an SPTFF system in which product is completely retained by the membrane, what conversion value is required to produce a product concentration of 15 mg/mL from a feed at 4 mg/mL?

Chapter 10
Summary and trends in biopharmaceutical processing

Rather than providing a traditional conclusion to the book, we close by sharing an imagined conversation between a new employee, fresh out of school, at a biopharmaceutical company and a thoughtful manager at the same company who has significant experience in the field. The new employee has little hands-on experience in making biopharmaceuticals and only rudimentary book knowledge. The experienced manager has significant knowledge of the biopharmaceutical field but only 15 minutes before having to depart for an important meeting. This conversation is meant to summarize salient points made in the book, to acknowledge some topics that we were unable to cover (due to space limitations) but that deserve more in-depth discussion, and to highlight trends in biopharmaceutical processing.

New employee: I would benefit from a better understanding of biopharmaceuticals. I know you need to get to a meeting, but can you give me a quick synopsis of biopharmaceutical products?

Experienced (and thoughtful) manager: You will soon discover that the biopharmaceutical field is an excellent way to apply the science and engineering courses you took in college to solve real-world problems. You are probably aware that biopharmaceuticals are medicines manufactured by or from living organisms such as bacteria like *E. coli*, mammalian cells like Chinese hamster ovary (CHO) cells, or even human beings. The active ingredients in these medicines are so large and complex that they are difficult to produce by chemical synthesis, which is used to manufacture many traditional pharmaceuticals. Cells have the ability to produce what the chemist cannot. Some familiar examples include insulin, which is used to treat diabetes and is produced through genetic engineering of *E. coli* or yeast cells. Another is Humira®, a monoclonal antibody produced in CHO cells and used to treat rheumatoid arthritis among a number of other diseases. For many years, Humira® was the biggest-selling drug (by revenue) in the world. Vaccines, such as those used to prevent COVID-19 or the flu, are biopharmaceuticals as well. Many newer types of biopharmaceuticals are coming on the market too, such as gene therapy products.

The final dosage form of a biopharmaceutical product is not a pill or tablet. It is usually a liquid solution or freeze-dried product that is reconstituted to a liquid solution before being administered. Biopharmaceuticals are delivered in liquid state because they are injected or infused. If they were administered orally, they would break down in the stomach and not be effective.

https://doi.org/10.1515/9783111112459-010

New employee: How are biopharmaceutical products made? Even a brief explanation would help me as I begin working on the projects assigned to me.

Experienced manager: Processes for producing biopharmaceutical drug products typically start with steps necessary to grow the cells to high concentrations and large volumes. Large-scale cell growth requires a bioreactor. Once the cells have produced the active ingredient, the product is separated from the culture broth that results. This separation requires a series of steps – referred to as the harvest process – that may include cell lysis, centrifugation, depth filtration, and tangential-flow microfiltration. Those steps are capable of separating product from cells, but they typically cannot separate the product from the many soluble impurities that are generated during processing. The concentration of these impurities should be minimized to ensure the safety and efficacy of the medicine. A clarified (i.e., particle-free) product intermediate, in which the active ingredient is dissolved, is produced by the harvest process and is loaded to a chromatography column. In fact, product is likely to go through several independent chromatography (or other purification) steps to separate the product from impurities such as host cell proteins, endotoxin, host cell DNA, and product aggregates. These chromatography steps are essential to ensuring that critical quality attributes (CQAs) related to purity, potency, and safety are met.

Once the product has been purified, the resulting product intermediate – the product dissolved in a buffer – is formulated. Formulation involves steps like ultrafiltration, which are performed to put the product into the proper buffer system with all necessary excipients added. The formulated product can then be filled into one of a variety of devices – such as vials, prefilled syringes, or autoinjectors – to create the drug product that goes to a patient.

I would be remiss if I led you to believe that this is all there is to manufacturing a biopharmaceutical product. There are also requirements, known as current good manufacturing practice (CGMP), that must be followed to ensure that products are consistently produced with the required quality. The manufacturing process, facility, and CGMP requirements all come together to ensure product quality. People are central to these activities as well. It is people who design and execute the manufacturing processes, design and build facilities, test the product, and implement systems in place to ensure quality. I've given you quite a bit of information and need to get going, but I know of a good book that tells *most* of the "story" of biopharmaceutical manufacturing. It's called Biopharmaceutical Manufacturing: Principles, Processes, and Practices.

New employee: This information is useful and has increased my interest. I know your meeting is getting ready to start, but I have one more question. You said the book you mentioned covers most of the "story" of biopharmaceutical manufacturing. What doesn't it cover?

Experienced manager: The book provides the details of most steps required to produce final drug product. But there are a few operations that you may want to know more about. For example, viral clearance steps, which are briefly discussed in Chapters 2 and 8, could be the topic of a separate chapter. If you are not familiar with viral clearance, it is a term used to describe steps required to remove or inactivate unwanted viruses that can make their way into a process, particularly those processes that use mammalian-derived components such as CHO cells. The book also gives only brief mention of final fill-finish operations in Chapter 2. Fill-finish refers to the steps needed to produce final drug product from the purified active ingredient. It, too, could have been the topic of a separate chapter. The book also only touches on the different testing – of raw materials, in-process samples (i.e., samples of product inter-mediate), drug substance, and drug product – required as part of a biopharmaceutical process. That topic is large enough to deserve its own book.

Remember, though, that biopharmaceutical manufacturing is a big and evolving field. There are many advances being made in a number of areas within biopharmaceutical manufacturing.

New employee: I know you have to go to your meeting, like yesterday, but do you have any information on these trends? I'm interested in everything you've said so far and would like to do additional reading.

Experienced manager: In fact, I keep a running tab of trends based on my discussions with colleagues, seminars I attend, and articles that I read. The list includes:

– Growth in use of single-use technologies. Single-use technologies are already widely adopted in processes to manufacture biopharmaceuticals for pre-clinical and clinical trial studies. It is expected that adoption for commercial processing will increase [278]. Single-use technologies offer a number of advantages, including the flexibility offered by rapid changeover between products. Single-use technologies have been and will continue to be particularly important to enabling personalized cell therapy processes.

– More processes designed to operate continuously. By creating more continuous processing options, potential advantages such as smaller equipment, a smaller manufacturing facility, and less cost for equipment and facilities may be realized.

– Development and implementation of intensified processes, in addition to development of continuous process steps. There is no single accepted definition of process intensification, but we use it to refer to "technologies to increase the amount of intermediate or final product manufactured per volume, footprint, unit of time, or expense by increasing efficiency in one or more unit operations" [279]. There are numerous examples of advances that fit this definition. One method of intensifying is to integrate multiple processing steps into a single unit operation, resulting in less capital cost and high process yields. The Emphaze™ AEX Hybrid Purifier from 3 M is an example. It combines the ability to clarify a process stream while simultaneously clearing soluble impurities, objectives that would typically be met by the use of two separate unit operations rather than just one. Intensified processes will lead to greater productivity, less equipment in a process, smaller facilities, and less cost for equipment and facilities.

– Advances in process analytical technology (PAT) will lead to greater implementation. Process analytical technology refers to a system for designing, analyzing, and controlling manufacturing through timely measurements (i.e., during processing) of critical quality and performance attributes of raw and in-process materials and processes, with the goal of ensuring final product quality [280]. It is most often associated with real-time inline measurement (sensor placed directly in a process vessel or process stream) and online measurement (sensor placed in a sample stream diverted from the manufacturing process and then returned) of critical quality attributes or performance parameters, with adjustments made to process parameters to keep the CQA or performance parameter within range. Central to PAT is sensor technology, and while inline and online sensors are common in biopharmaceutical processes, there remain many measurements that require a sample to be pulled and tested in a lab, away from the process. This takes time and makes it difficult to control the process using the measured value. Consequently, development of inline/online sensors is widely considered essential to advancing biopharmaceutical processing by enabling real-time testing (samples won't have to be sent to the QC lab for testing) and even real-time release [281]. Jiang et al. [282] provide a nice summary of analytical technologies that may significantly advance PAT implementation.

– More sophisticated automation and process control. These changes will reduce manual interventions in processes and one day lead to "lights-out manufacturing" in which no human presence is required. Interest in improving automation and control in biopharmaceutical processes is high. BioPhorum, a consortium in which member companies can collaborate to accelerate their progress, published the results of a technology roadmapping exercise a few years back for biopharmaceutical processing [281]. Process automation was identified as one of six areas in which industry would be investing innovation efforts over the next decade.

– Use of artificial intelligence (AI). AI is already being used for drug discovery, which refers to the steps involved in identifying new medicines by searching for molecules that can safely interact with a specific biological target to treat a disease. In the coming years, it will likely play an increasingly important role in the design and optimization of biopharmaceutical processes and the control of those processes. Specifically, control systems that use AI may enable real-time, adaptive control of biomanufacturing processes, allowing for improved responsiveness to changing process conditions.

And now back to the readers of the book: thank you for your enthusiasm and interest. We expect you will have a fulfilling and engaging experience in biopharmaceutical manufacturing for many years to come.

Glossary

Adsorption Attachment of a molecule, particle, etc. to a solid surface.

Affinity chromatography (AC) A type of chromatography in which a target solute, such as a protein therapeutic, has a very specific affinity for a stationary phase, such as a resin.

Air diffusion test A type of filter integrity test that challenges the feed side of a wetted membrane with pressurized air and measures the rate that air permeates. If the flow rate of air that permeates is less than or equal to the maximum that would be expected by diffusion alone, the membrane has integrity.

Air lock A room that typically has two doors in series to separate a cleanroom from a less clean environment, such as a corridor. The two doors are typically interlocked to avoid being opened at the same time.

Allogeneic cell therapy A type of cell therapy in which cells are derived from a donor (or donors) other than the patient who will receive the cells.

Antigen Any substance (e.g., molecule) that induces the body to make an immune response.

Aseptic Describes a process and/or facility designed to keep a product sterile.

Aseptic filling Filling of drug product under aseptic conditions.

Aseptic technique Practices for working in a cleanroom that minimize the risk of contaminating the environment and the product.

Assay A lab procedure for qualitatively or quantitatively measuring the presence, identity, concentration, activity, purity, or some other property of specific biological substances within a sample.

Asymmetric membrane Membrane consisting of a thin, dense layer that provides the desired separation on top of a thicker, more porous bottom layer, which provides mechanical strength without significantly hindering permeate flow.

Asymmetry A parameter that provides a measure of packing quality; it is typically calculated from the peak resulting from a pulse injection of a tracer to a packed column. Specifically, asymmetry is a measure of the symmetry of the elution peak that results from the pulse tracer test.

Autologous cell therapy A type of cell therapy in which cells are taken from and administered to the same individual.

Auxotroph A strain that has been genetically modified to be deficient in production of an essential metabolite.

Axenic Free from living organisms other than the cells required for the process. When growth of cells is carried out in a bioreactor, the fermentation or cell culture should be axenic, not sterile.

Band broadening (or zone broadening) The widening of a solute band moving through a chromatographic stationary phase. It results from hydrodynamic dispersion, axial diffusion of the solute, and resistance to transport of the solute from the mobile phase to the surface of the stationary phase. Band broadening dilutes the solute and reduces resolution between components being separated.

Batch A defined quantity of product processed in one process or series of processes so that it could be expected to be homogeneous. Definition adapted from EudraLex – Volume 4 – Good Manufacturing Practice (GMP) guidelines [283].

https://doi.org/10.1515/9783111112459-011

Batch disposition The process of assigning a specific status or product usage to a batch of product or product intermediate. Disposition status may include released/approved, conditionally released/approved, rejected.

Batch production record An approved document that provides instructions on and generally requires data be recorded from an operation associated with the production of drug substance and drug product.

Batch release Approval of a batch of product or product intermediate for further processing or commercial distribution.

Bind-and-elute chromatography A mode of chromatography in which product binds to the resin during the load step, and impurities may or may not bind. The product is recovered during an elution step by making a change in mobile-phase (elution buffer) composition that disrupts interaction between the stationary and mobile phase.

Binding capacity (equilibrium and dynamic) The amount of solute that can bind to a stationary phase per unit volume of stationary phase. When measured under static conditions in which the mobile phase and stationary phase are allowed to come to equilibrium, the term *equilibrium binding capacity* is used. Measured under flow conditions, binding capacity is referred to as dynamic binding capacity.

Bioavailability The fraction of drug that enters systemic circulation.

Bioburden The amount and type of microorganisms that can be present in raw materials, process intermediates or drug substance.

Biological product In this text, synonymous with biopharmaceutical.

Biopharmaceutical A medicine that is inherently biological in nature and manufactured by or from living organisms, including cells from living organisms.

Biosafety The discipline addressing the safe handling and containment of infectious microorganisms and hazardous biological materials [89].

Biosafety cabinet (BSC) A type of enclosure for working with potentially infectious biological agents or materials when protection for the worker, environment, and material is needed. The BSC creates a controlled airflow that captures and removes airborne contaminants generated during manipulation of hazardous materials within the cabinet.

Biosimilar A biological product highly similar to and interchangeable with an approved biological product.

Bubble point test A type of filter integrity test that measures the minimum air pressure required to force liquid, such as water, from the pores of a filter or membrane. Measured bubble point pressures equal to or greater than the minimum bubble point value specified by the vendor for a particular wetting agent/ gas (typically air) combination demonstrates membrane integrity.

Buffer A solution that resists changes in pH when acid or base is added to it. Buffers often include a weak acid or weak base together with one of its salts. They are used to avoid shifts in pH in a solution. In biopharmaceutical processing, the term is often used to refer to solutions – even those that do not buffer against pH changes – in general.

Buffer exchange Replacing the existing buffer solution surrounding a protein or other biomolecule with a different buffer solution. The term is also used (some might say incorrectly) when a salt that is not a buffer is removed from a process fluid. When performed using ultrafiltration, this process step is called diafiltration.

Buoyancy The force experienced by a particle submerged in a fluid that results from the pressure at the bottom of the particle being greater than that at the top. In the case of gravity settling, the force is upward; in the case of centrifugation, the force acts toward the axis of rotation.

Calibration Comparison of the results from a particular instrument or device with results produced by a reference or traceable standard over an appropriate range of measurements. The comparison must fall within specified limits [61].

Cassette A tangential-flow filtration module, usually with a rectangular geometry, containing multiple layers of flat sheet UF or MF membranes and separate channels for feed, retentate and permeate.

Cavitation Phenomenon in which a sudden drop in the pressure of a liquid leads to the formation of small vapor-filled cavities in a liquid. When these bubble collapse, shock waves are formed.

Cell culture The growth of animal cells, including mammalian cells like Chinese hamster ovary (CHO) and insect cells like Sf9 (cells from ovaries of a fall armyworm).

Cell lysis (also referred to as cell disruption) Breaking open of cells, usually to recover intracellular components. There are numerous methods used to lyse cells, some of which are more scalable than others.

Cell membrane (also referred to as the plasma membrane or cytoplasmic membrane) A semipermeable membrane that surrounds the cytoplasm of every cell. It is composed of lipids and proteins. Generally speaking, the cell membrane serves as a barrier keeping the cell components in and unwanted substances out. It also regulates transport of essential nutrients into the cell and removal from the cell of waste products.

Cell therapy The use of living, whole cells for the treatment of a disease.

Cell wall A flexible layer that surrounds the plasma membrane of the cells of bacteria, fungi such as yeast, and plants. The composition of the cell wall varies by cell type. It provides cells with structural support, which is particularly needed as water enters the cell due to the osmotic pressure difference between the cytoplasm and surrounding fluid.

Centrifugal force An apparent force that acts on an object moving in a circular path that causes movement outward from the center of rotation. When a slurry consisting of particles dispersed in a less dense liquid is exposed to a centrifugal force, the particles move away from the axis of rotation. This movement is the basis of centrifugation for solid-liquid separation.

Centrifugation A unit operation that uses centrifugal force to separate components based on differences in density and size. In biopharmaceutical manufacturing, centrifugation is typically used to separate denser solids, such as host cells or cell debris, from less dense surrounding liquid in which they are suspended.

Centrifuge A machine used for centrifugation. A centrifuge sets immiscible phases in rotation around a fixed axis, which applies a centrifugal force that enables separation of the two phases by causing the denser phase to move outward in the radial direction. Common centrifuge types used in biopharmaceutical production include the disc-stack centrifuge and tubular-bowl centrifuge.

Change control A formal system by which qualified representatives of appropriate disciplines review proposed or actual changes that might affect the validated status of facilities, systems, equipment or processes. The intent is to determine the need for action to ensure and document that the system is maintained in a validated state [88].

Changeover The replacement of wetted soft parts on equipment – such as gaskets, O-rings, and valve diaphragms – to avoid the possibility of contamination of one product that may have been absorbed into the soft parts into the next product.

Chromatogram A plot of a sensor response (e.g., UV absorbance, conductivity, and pH) from a chromatography system versus volume or time.

Chromatography column The tube (and other associated hardware) that holds chromatography resin in place.

Chromatography support The nonfunctionalized base for the stationary phase medium.

Clarification The process of making a liquid clear by removing suspended solids.

Cleaning The removal of residues (e.g., proteins and buffer components), particulates, and cleaning/sanitizing agents from equipment surfaces.

Clean-in-place (CIP) A cleaning mode in which removal of soil from product contact surfaces is carried out with equipment in its process position by flowing and/or spraying cleaning agents and water rinses over the surfaces to be cleaned.

Clean-out-of-place (COP) A cleaning mode in which equipment is disassembled and manually cleaned in a location away from the processing area.

Cleanroom (also referred to as a classified area) A room in which the concentration of airborne particles is controlled and classified (e.g., ISO 8) and which is constructed and used in a manner to minimize the introduction, generation, and retention of particles inside the room [85]. Cleanrooms are monitored to demonstrate that the classification is met [284].

Clonality The condition of being genetically identical.

Closed system A system (process or process steps) designed and operated so that product is not exposed to the surrounding environment during processing [84].

Column volumes (CVs) Term used to describe the amount of a mobile phase fed to a chromatography column relative to the volume of the column. For example, 600 L of equilibration buffer fed to a column packed to a volume of 200 L corresponds to 3 CVs of equilibration buffer.

Complex component A chemical component (e.g., cell culture medium) whose constituents are not completely defined. All constituents may not be known and their relative abundance difficult or impossible to discern.

Component From 21CFR210, any ingredient intended for use in the manufacture of a drug product, including those that may not appear in such drug product [53]. Additionally, a component can refer to constituent part of something (for example, a component of a piece of equipment).

Concentration Generally refers to the amount of a component per unit volume or mass. As a biopharmaceutical processing step, the term refers to increasing the mass of product per unit volume of solution by removing solvent, which is typically water in biopharmaceutical processes.

Concentration factor (CF) Ratio of the final concentration of the product to its starting concentration during a UF concentration step.

Concentration polarization Elevated solute (usually product) concentration at the surface of a UF membrane that results from permeating liquid carrying dissolved solutes, such as product, to the membrane surface where they are retained.

Contaminants An external component introduced into the process that is not part of the process [199].

Controlled not classified (CNC) Space in which the particulate levels in the room air are controlled, but the rooms is not necessarily monitored and therefore cannot be classified (as ISO 8, for example).

Corrective action An action taken to correct and/or eliminate the cause of a detected non-conformity or other undesired situation to prevent recurrence.

Critical quality attribute (CQA) "A physical, chemical, biological or microbiological property or characteristic that should be within an appropriate limit, range, or distribution to ensure the desired product quality. CQAs are generally associated with the drug substance, excipients, intermediates (in-process materials), and drug product" [52].

Crossflow rate Flow rate on the feed side of a membrane.

Current good manufacturing practice (CGMP) a system for ensuring that processes and facilities produce products of consistent quality. CGMP applies to industries such as pharmaceuticals (and biopharmaceuticals), food, cosmetics, and medical devices. The "C" stands for "current," reminding manufacturers that they must employ technologies and systems that are up to date in order to comply with the CGMP regulations. Note that the term is also commonly abbreviated as cGMP or just GMP [55, 285].

Dead leg A portion of a piping or distribution systems through which fluid cannot flow when connections are removed and capped or valves are closed.

Death phase The terminal phase of growth where cellular stresses due to starvation or toxic compounds causes cell death.

Defined component A chemical component (e.g., cell culture medium) with clearly defined chemical formulae. All chemical constituents and their abundance are known.

Denature To change the three-dimensional structure (i.e. secondary, tertiary, or quaternary) of proteins or other biomolecules by application of an external stress such as heat, strong acid or base, or an organic solvent.

Depth filtration A type of filtration that uses a porous filtration medium to capture particles throughout the depth of the filter rather than just at the surface.

Design of Experiments (DoE) A statistical approach for planning and executing defined experiments and analyzing the generated results to provide insight into factors that significantly affect the outcome of a process.

Design qualification (DQ) The documented verification that the proposed design of the facilities, systems and equipment is suitable for the intended purpose [88].

Design space The combination of process parameters and material attributes that ensures biopharmaceutical product quality [52].

Deviation An occurrence, problem, or other undesirable event that represents a departure from an approved process or procedure or from an established standard or requirement.

Diafiltration (DF) A processing step that uses UF membranes to remove solutes or exchange solutes (e.g., buffers and salts) for others by adding a new solution and simultaneously permeating out the old.

Diavolume (DV) The ratio of the volume of diafiltration buffer fed in diafiltration step to the volume of retentate being diafiltered. For example, 500 L of a diafiltration buffer added to 100 L of product solution (retentate) is five DVs.

Diffusion The movement of solute molecules from areas of high concentration to low concentration driven by the random thermal motion of solute molecules.

Dissolved oxygen (DO) A measure of the amount of oxygen molecules dissolved in a liquid.

DNA replication The cellular process by which genetic material is copied and sorted in preparation for cellular division.

Doubling time The time it takes for cell concentration to double in value.

Downstream (DS) process Later steps in a process for production of biopharmaceuticals. The downstream portion of a process comprises the purification stage and formulation and fill stage. It may also include steps in the harvest stage.

Drug In this text, both small-molecule pharmaceuticals produced by chemical synthesis as well as biopharmaceuticals.

Drug product The finished dosage form – for example, tablet, capsule, or in the case of biopharmaceuticals, commonly a liquid solution or lyophilized product – that contains the active ingredient. It contains the drug substance generally, but not necessarily, in association with one or more other ingredients [43].

Drug substance The active ingredient that is intended to furnish pharmacological activity or other direct effect in the diagnosis, cure, mitigation, treatment, or prevention of disease. It is intended to be incorporated into the finished dosage form [43].

Dynamic binding capacity (DBC) See *Binding capacity*.

Effluent Flow exiting a unit operation, such as the flow from a packed bed during any phase of chromatography.

Eluate Product intermediate from a chromatography step.

Elution Chromatography phase in which product bound to the resin is moved through the stationary phase for recovery of the purified product. For chromatography that involves binding of a mobile phase solute to the stationary phase, elution steps weaken the interaction between a solute and the stationary phase by changing mobile phase composition, typically in a stepwise fashion or linearly. Isocratic elution involves no change in mobile phase composition and is common in size exclusion chromatography.

Endotoxin Literally, a toxin from within. It is most often used in reference to lipopolysaccharides found in the outer membrane of gram-negative bacteria like *E. coli*. It is a common impurity or contaminant in bioprocess streams, and it can produce a variety of reactions in humans, including fever and a lowering of blood pressure.

Equilibrium binding capacity (EBC) See *Binding capacity*.

Eukaryotic cell Cells from animals, plants, fungi, or protists. They contain a membrane-bound nucleus that holds the cellular DNA and numerous membrane-bound organelles.

Excipient Anything in the drug product other than the active ingredient.

Exponential phase The period following the lag phase where cells are rapidly dividing. It is marked by an exponential increase of cell mass with respect to time.

Extracellular Outside the cell.

Factory acceptance test (FAT) Testing performed at an equipment/system manufacturer's facility to demonstrate that a piece of equipment or system performs as expected prior to delivery to the end user.

Feed flux Volumetric feed rate to a filter per unit area. It is often expressed in units of LMH (L of feed per m^2 of filter area per hour).

Feed interval The time between discharges in a disc-stack centrifuge.

Fill-finish Steps involved in putting drug product into its container (e.g., vial or syringe). Some sources consider the labeling, packaging, and QA release of product part of fill-finish operations.

Filter A porous medium used for separation of components (e.g., solids from liquid and solute from solvent) in a fluid stream.

Filter capacity (also referred to as loading or throughput) Maximum volume of suspension that can be filtered – because a maximum pressure difference has been reached, for example – per unit area of filter. It is typically expressed in units of L/m^2.

Filtrate (also referred to as permeate) Portion of the feed that permeates a filter or membrane.

Filtration Any technique used to separate particulate in a fluid suspension or components in solution according to their size by flowing under a pressure differential through a porous medium.

Flow-through chromatography A mode of chromatography in which impurities rather than product bind to the stationary phase. Flow through also refers to material passing through column during the load of bind-and-elute chromatography without being bound.

Formulation Refers to the buffer system and other excipients in which active ingredient is dissolved in drug product. The steps required to place the active ingredient into the proper formulation are part of the formulation and fill stage of a process.

Fouling Plugging of the filter due to accumulation of solids on or within the filter medium that leads to poor filter performance (e.g., reduced filtrate or permeate flux).

Gene therapy A technique that uses genetic material to treat, prevent, or even cure a disease. This can be done by replacing a faulty or missing gene with a healthy copy, inactivating a gene that is functioning improperly, introducing a new gene into the body to help fight a disease, or by editing existing genes within the body.

Glycosylation attachment of sugar molecules to other molecules, such as proteins.

Gram staining A multi-step method of staining bacteria that uses crystal violet as the primary dye. Results from the procedure are used to classify bacteria as gram positive or gram negative. Bacteria with a thick peptidoglycan cell membrane are stained purple by crystal violet and referred to as gram positive. Bacteria with thinner peptidoglycan layers do not retain the crystal violet dye, and are counter-stained pink by safranin. These bacteria are referred to as gram negative.

Harvest The processing stage that separates product from the production system. Also referred to as recovery.

Harvest cell culture fluid (HCCF) Broth resulting from the culturing of cells in a bioreactor.

Height equivalent to a theoretical plate (HETP) The height of one theoretical plate within a packed column (or other stationary phase). HETP is based on the idea that a packed column (or membrane adsorber or monolith) contains theoretical plates that allow for equilibration between the mobile and stationary phases, and the larger the number of plates, or equivalently the smaller the HETP, the better the separation.

High-efficiency particulate air (HEPA) filter A filter that is rated to remove 99.97% of bacteria, molds, virus or other airborne particles of 0.3 μm or less. HEPA filters are found in BSCs and in HVAC systems in a CGMP facility.

High purity water (HPW) In this text, we use the term *high purity water* to refer to a type of purified water that meets the requirements set for Purified Water in Chapter 1231 of the USP-NF. In biopharmaceutical manufacturing, it is often used for equipment cleaning.

Hollow-fiber module A membrane module that contains a bundle of hollow fibers made from UF or MF membranes that are housed in a cylindrical shell and used for tangential-flow applications. The interior of the fiber is referred to as the lumen and has a typical diameter of 0.5–1.0 μm.

Homogenizer Machine that pumps a cell slurry at high pressure through a small orifice created by a specially designed valve resulting in lysis.

Homogenizer passes The number of times a suspension of solids (or insoluble liquid) in liquid (e.g., cells or cell debris in medium or buffer) is passed through a high-pressure homogenizer.

Host cell proteins (HCPs) Native proteins from the host cell that may become part of the product stream. They are a heterogeneous group made up of hundreds and possibly even thousands of different proteins, with widely different properties. They are problematic for a patient because they may cause an unwanted immune response and must be removed from product.

Hydrophobic interaction chromatography (HIC) A type of chromatographic interaction based on hydrophobicity, that is, a hydrophobic solute (such as the product) binds to a hydrophobic stationary phase. The greater the hydrophobicity, the stronger the interaction between the solute and stationary phase. Addition of certain salts is required to adsorb hydrophobic solutes to the resin.

Impurities "Any component present in the drug substance or drug product which is not the desired product, a product-related substance, or excipient including buffer components" [199].

Inclusion body A form of intracellular product (protein) in which the product takes the form of micron-sized solid aggregates that are relatively pure. Because the protein that makes up the inclusion body is misfolded, the inclusion body must be solubilized and then refolded to create a bioactive, soluble biopharmaceutical product.

Installation qualification (IQ) The documented verification that the facilities, systems, and equipment, as installed or modified, comply with the approved design and the manufacturer's recommendations [88].

Intermediate Material produced during processing of drug substance that undergoes further processing before it becomes drug substance.

Intracellular Within the cell.

Ion exchange chromatography (IEC) A chromatographic interaction mode in which adsorption of a component from the mobile phase is based on attraction of opposite charges. For example, a negatively charged solute binds to a positively charged resin bead and vice versa. Separation between components in the mobile phase is based on differences in charge between components. Positively charged resins are used in anion exchange chromatography; negatively charged resins are used in cation exchange chromatography.

Ionic strength A measure of the concentration of ions in a solution.

Isoelectric point (pI) The pH at which the net charge on a molecule (e.g., protein) is zero. The pH of the solution in which a molecule is dissolved determines its charge. If the solution in which a molecule is dissolved is at a pH < pI, the molecule has a net positive charge. If the solution is at a pH value > pI, the net charge on the compound is negative.

Kosmotropic salt Salts such as $(NH_4)_2SO_4$ that lead to protein aggregation and, by extension, binding of hydrophobic solutes such as proteins to hydrophobic ligands in HIC.

Lag phase The period of time following inoculation where cells are becoming acclimatized to their new environment. It is marked by a minimal amount of cell division.

Laminar flow An airflow moving in a single direction and in parallel layers at constant velocity from the beginning to the end of a straight line vector [286].

Leachables Compounds that migrate from any product-contact material under normal process conditions.

Ligand The molecule coupled to a chromatography support that creates interaction between a mobile phase solute and the stationary phase.

Lyophilization (also referred to as freeze drying) A process that removes water from a product by freezing it, then reducing the pressure to allow the frozen water to sublimate directly into vapor, bypassing the liquid state. This technique is commonly used to produce a cake/powder dosage form for biopharmaceuticals and small-molecule drugs.

Lysate The intermediate that results from a cell lysis step. In the case of lysis by homogenization, the intermediate may be referred to as homogenate. Lysate that has had solids removed is often referred to as clarified lysate.

Master cell bank (MCB) Cells that have been grown and dispensed into multiple vials and stored under defined conditions, often cryogenically. The master cell bank is used to produce the working cell bank. It prevents genetic variation by minimizing the number of times a cell line is passaged or handled during the manufacturing process.

Membrane A thin porous medium used for separation of components (e.g., solids from liquid, solute from solvent) in a fluid stream. A membrane is a type of filter, but not all filters would be classified as membranes.

Microbial fermentation The growth of microorganisms, like the bacterium *E. coli* or the yeast *Saccharomyces cerevisiae*.

Microfiltration (MF) A type of filtration that relies on a microfiltration membrane to capture particles at the membrane surface. Microfiltration membranes have pore sizes as small as 0.1 μm or as large as 10 μm.

Mobile phase The liquid phase that flows through a packed chromatography bed or other chromatography device (e.g., membrane adsorber or monolith).

Molecular weight cutoff (MWCO) Also referred to as the nominal molecular weight limit (NMWL), represents the molecular weight of a solute, in daltons, retained 90% by the membrane. MWCO values for UF membranes range from 1 kDa to 1,000 kDa.

Multimodal chromatography (MMC) A chromatography mode in which separation of a solute from other components is based on more than one type of interaction. For example, a mixed mode resin might include a ligand that is both charged and hydrophobic.

Normal-flow filtration (NFF) A mode of filtration in which feed flows perpendicular to the filter medium, in the same direction as filtrate flow.

Normalized water permeability (NWP) A measure of membrane cleanliness in which the permeate rate of water, at a given TMP, temperature, and crossflow rate, is measured and calculated per equation (9.3). Typical units of NWP are $L/(m^2 \text{ h psig})$.

Open system A process that is not closed and therefore must be performed in an environment where the probability of contamination is acceptably low.

Operational qualification (OQ) The documented verification that the facilities, systems, and equipment, as installed or modified, perform as intended throughout the anticipated operating ranges [88].

Osmolality/osmolarity Measures of solute concentration. Osmolarity is the number of osmoles of solute per L of solution, while osmolality is the number of osmoles per kg of solvent (water). Osmoles are the number of moles of solute that contribute to the osmotic pressure of a solution. For example, one mole of NaCl would equate to two osmoles of solute particles, because NaCl dissociates to Na^+ and Cl^- ions in water.

Osmotic pressure Pressure that would have to be applied to prevent a solvent, such as water, from being transported across a semipermeable membrane from low solute concentration (such as UF permeate) to high solute concentration (such as the feed side of a UF membrane).

Parenteral Other than through the gastrointestinal tract. Common parenteral routes of administration for biopharmaceuticals are intravenous, subcutaneous, and intramuscular injection and intravenous infusion.

Passaging The act of transferring growing cells to a freshly prepared medium, typically to increase the volume of growing culture. Each time cells are passaged, the passage number increases by one. The passage number should be tracked.

Peptide A molecule composed of a short chain of amino acids linked by peptide bonds that is generally considered to contain fewer amino acids than a protein. The U.S. FDA specifically defines a peptide as containing 40 or fewer amino acids [9].

Percent yield The ratio of that amount of product at the end of one step or series of steps to the amount of product at the beginning of that step (or series of steps). The term percentage step yield is used when applied to one step in the process and percentage process yield when applied to the entire process.

Performance parameter An output parameter that cannot be directly controlled but is indicative of process performance [80].

Performance qualification (PQ) The documented verification that systems and equipment can perform effectively and reproducibly based on the approved process method and product specification [88].

Perfusion culture A production mode in which fresh culture medium is fed at the same rate that spent medium is removed. Perfusion differs from a simple continuous operation in that cells are retained within the bioreactor vessel as spent medium is removed.

Permeate (also referred to as filtrate) Portion of the feed that permeates a filter or membrane. In this book, we use the term permeate in the context of membrane separations.

Permeate flux Permeate flow rate per unit area of membrane. Typically expressed in units of LMH (L of permeate per m^2 of membrane area per hour).

Platform The use of common methods, procedures, and equipment in the development of processes for or manufacturing of different products. Products with similar active ingredients lend themselves to platform processes.

Precipitation A process step in which a dissolved component – either the product or impurities – is separated from other solutes by changing a property, such as temperature or pH, or adding a precipitant that alters the solubility of the component and causes it to fall out of solution. The precipitated component is a solid phase.

Preventive action Actions taken to eliminate the cause of a potential deviation.

Procedure A document specifying instructions for carrying out an activity or process. Procedures typically do not require recording of information.

Process parameter (also referred to as operational parameter) An input parameter to a unit operation that is directly controlled. Process parameters can be further categorized as critical, key, and non-key. A critical process parameter is one whose variability has impact on a critical quality attribute. A key process parameter is one whose variability is important to process performance, but does not affect a CQA. Non-key process parameters are easily controlled and therefore likely to have no impact on a CQA or performance parameter [80].

Process-related impurity An impurity originating from the manufacturing process [199].

Process residuals Chemicals added to the biopharmaceutical manufacturing process that are not fully consumed and therefore remain in the process stream in small amounts. Process residuals must be removed to safe and acceptable levels within the final drug substance or drug product.

Production In Chapter 2, the term used to refer to the product synthesis stage of a process.

Product-related impurity Molecular variants of the product that may form during manufacture and storage. They are often physically and chemically similar to the product, but not as safe or effective [199].

Prokaryote A structurally simple single-celled organisms. Prokaryotes do not have a nucleus or other organelles within the cytoplasm.

Protein A macromolecule composed of amino acid monomers.

Protein therapeutic Protein used for therapeutic applications, i.e., to treat an existing disease.

Purification Generally refers to separating product from impurities present in a process stream. In Chapter 2, the term refers to the process stage that purifies the product.

Quality assurance (QA) The unit within a biopharmaceutical facility with responsibility over quality systems and ensuring CGMP requirements are fulfilled. QA typically has responsibility for disposition of raw materials, product intermediates, drug substance, and drug product.

Quality control (QC) The group within a biopharmaceutical facility whose primary responsibility is conducting routine analytical testing on raw materials, product intermediates, drug substance, and drug product to ensure acceptance criteria and specifications are met.

Quality target product profile (QTPP) A prospective summary of the quality characteristics of a drug product that ideally will be achieved to ensure the desired quality, taking into account safety and efficacy of the drug product [52].

Raw materials A general term used to denote starting materials, reagents, and solvents intended for use in the production of intermediates or active ingredients [61].

Records and reports Documents stating results achieved or providing evidence of activities performed.

Rejection factor (also referred to as retention factor) A measure of solute retention by a membrane, defined specifically as $1 - C_{solute,p}/C_{solute,r}$, where $C_{solute,p}$ and $C_{solute,r}$ are the measured solute concentrations in the permeate and the retentate. Solutes that are completely retained by a membrane have a rejection factor of 1.

Relative centrifugal force (RCF) A measure of the outward force generated by rotation of a centrifuge relative to the force of gravity. It can be calculated using the following relationship: $RCF \equiv \omega^2 \times R/g$, where ω = angular velocity (in units of rad/s, for example), R = distance from center of rotation, the product $\omega^2 \times R$ is the centrifugal acceleration, and g = acceleration due to gravity (=9.8 m/s^2).

Resin Beads that serve as the stationary phase in a packed chromatography column.

Resolution A measure of the degree of separation between two components in a chromatography step, specifically defined as distance between two neighboring peaks divided by the mean of the peak widths.

Retentate The effluent from the feed side of UF or MF membrane module, after liquid has permeated the membrane, in a tangential-flow filtration operation. UF retentate will be more concentrated in retained components than the feed; likewise MF retentate will be more concentrated in solids (for a clarification application) than the feed.

Reverse osmosis (RO) A filtration technique that uses reverse osmosis membranes, which are able to reject ions, such Na+ and Cl-, from an aqueous solution.

Reversed phase chromatography (RPC) A chromatographic interaction mode in which binding to a stationary phase is based on hydrophobic interaction, similar to HIC, but the binding is stronger and typically requires an organic solvent for elution of bound solutes.

Sanitization A process used to reduce microbial contamination to acceptable levels. In biomanufacturing, sanitization is often performed using chemical solutions such as isopropyl alcohol, sodium hydroxide solutions, or disinfectants.

Selectivity A measure of how much time one solute (e.g., product) spends on the stationary phase relative to another (e.g., impurities). The greater the selectivity, the better the resolution between solutes.

Settling velocity The velocity of a particle as it moves through a fluid under a force, such as gravity or centrifugal force. It increases as the size of the particle increases and the density difference between the particle and surrounding fluid increases. It decreases with increasing fluid viscosity.

Shear stress A force, or more correctly a force per unit area, that acts along the surface of an object, such as a cell, that may cause damage.

Sieving The removal of particles from a fluid stream by size (i.e., a particle is too large to flow through the pores of the filter media).

Sigma (Σ) factor Centrifuge scale-up parameter that corresponds to the area in a gravity settling device that would be required to give the same performance as the centrifuge at equal volumetric flow rates.

Site acceptance test (SAT) Testing performed upon delivery to the end user's facility to verify the system or equipment performs as expected.

Size exclusion chromatography (SEC) A chromatography method in which separation is based on size differences between solute molecules. Larger molecules in the mobile phase cannot diffuse into pores within the resin particles and elute after a volume of buffer equal to the interparticle column void volume (i.e., the volume between resin particles) has passed through the column. Smaller molecules access the resin pores and are held up longer in the column, resulting in separation from the larger molecules.

Skid A collection of components mounted on a frame to support a given activity. Skids are often on wheels so that they are easily moved.

Slurry A mixture of particles suspended in liquid.

Standard Operating Procedure (SOP) An approved written document that provides instructions for performing a task or operation not necessarily specific to a given product.

Stationary phase (related to chromatography) The fixed phase through which the mobile phase flows in a chromatography device and with which solutes interact to create separation. Examples of stationary phases used in biopharmaceutical processing include resin beads packed into a column, the membrane used in a membrane adsorber, or a monolith.

Stationary phase (related to fermentation and cell culture) The phase of cell growth where division is arrested due to the exhaustion of nutrients or accumulation of inhibitory substances.

Steam-in-place (SIP) The use of steam to sterilize or sanitize equipment in place.

Sterile Free from living organisms.

Sterilization A process used to eliminate all microbial contamination, rendering the item sterile. In biomanufacturing, sterilization is often performed using heat (i.e., autoclave), exposure to gamma irradiation or through chemical means.

Superficial velocity The velocity of a mobile phase fluid element in a column or other chromatography device if the stationary phase were not present. In a packed column, it is calculated as the ratio of the volumetric flow rate to cross sectional area of the bed.

Tangential-flow filtration (TFF) (also referred to as crossflow filtration) A mode of filtration in which feed is pumped tangentially along the surface of a filter or membrane with only a portion of the feed permeating the filter.

Terminal sterilization Filling a low-bioburden product into the final container followed by sterilization *after* filling.

Titer The term generally refers to the concentration of a component in solution. In biopharmaceutical manufacturing, it is most often used to refer to the amount of product produced per unit volume of bioreactor media. Units are often g/L.

Total organic carbon (TOC) An analytical method used to determine the amount of organic carbon in a sample. This method is used for detection of product and process residues to verify equipment cleaning procedures are effective.

Transcription Part of the process by which DNA is decoded to generate a functional product, such as a protein. Transcription refers specifically to the process by which a segment of DNA is copied into a new strand of messenger RNA.

Translation Along with transcription, part of the process by which DNA is decoded to generate proteins. Translation specifically refers to the process by which messenger RNA, produced by transcription, is decoded for protein synthesis.

Transmembrane pressure (TMP) Pressure difference across a membrane that drives permeate flow in a TFF step. It is defined as the average pressure on the feed side of the membrane minus the pressure on the permeate side: $TMP = (P_f + P_r)/2 - P_{filtrate}$.

Turbidity Cloudiness of solution due to the presence of particles. It is typically measured using a nephelometer, which measures the scattered light in an illuminated sample by placing a detector at a specific angle (often at a right angle) to the incident light. The more scattered light, the higher the turbidity. It is expressed in nephelometric turbidity units (NTU).

Ultrafiltration (UF) A type of filtration that relies on an ultrafiltration membrane, typically used to separate large molecular weight components in solution from smaller components. UF membranes have molecular weight cutoff values from 1 kDa to 1,000 kDa.

Unit operation Any distinct step in a process that serves a specific function, often associated with a single major piece of equipment.

Upstream (US) process Initial steps in a process involved in the production of a biopharmaceutical product. The upstream portion of a process includes steps in the production stage and may include the harvest stage.

User requirements specification (URS) The set of owner, user and engineering requirements necessary and sufficient to create a feasible design meeting the intended purpose of the system [88].

Vaccine A biopharmaceutical used to prevent a disease by improving immunity to that disease.

Validation Establishing documented evidence that provides a high degree of assurance that a specific process (e.g., a manufacturing process, a cleaning procedure) will consistently produce a product meeting its predetermined specifications and quality attributes [286].

Vessel volumes per minute (vvm) The volume of gas sparged per working volume in the bioreactor per minute. For example, if you were sparging 1 L of air into 30 L of medium per minute, the vvm would equal 0.033.

Viable cell density (VCD) The number of viable cells (that is, cells capable of reproducing and growing) per mL in a cell culture. This is often determined using trypan blue exclusion.

Viral clearance The inactivation or removal of unwanted virus in a biopharmaceutical process.

Water for injection (WFI) A type of purified water that meets the requirements for Water for Injection in Chapter 1231 of the USP-NF or in other pharmacopoeia. WFI is used as an excipient in the production of parenteral drugs, like biopharmaceuticals. HPW and WFI have similar testing requirements, except WFI has a bacterial endotoxin specification that must be met that is not in place for HPW.

Working cell bank (WCB) A (usually) cryogenically preserved vial of cells produced from a master cell bank that is used to initiate a seed train that produces inoculum for a production bioreactor.

Yield of cells on substrate The ratio of cell mass produced to growth substrate consumed during growth.

Yield of product on cells The ratio of product produced to cell mass accumulated during growth and production.

Yield of product on substrate The ratio of product produced to growth substrate consumed during production.

References

[1]   Rader RA. (Re)defining biopharmaceutical. Nature Biotechnology 2008; 26: 743–51.

[2]   Diabetes Complications. 2024. CDC. (Accessed Dec 29, 2024, at https://www.cdc.gov/diabetes/com plications/?CDC_AAref_Val=https://www.cdc.gov/diabetes/managing/problems.html).

[3]   Schmid RD. Pocket Guide to Biotechnology and Genetic Engineering. Weinheim Germany: Wiley-VCH, 2003, 1942.

[4]   Eli Lilly and Company. Humulin [package insert]. 2022.

[5]   Seqirus Inc. Flucelvax Quadrivalent [package insert]. 2023.

[6]   Akazawa-Ogawa Y, Nagai H, Hagihara Y. Heat denaturation of the antibody, a multi-domain protein. Biophysical Reviews 2018; 10: 255–58.

[7]   Maa Y, Hsu CC. Protein denaturation by combined effect of shear and air-liquid interface. Biotechnology and Bioengineering 1997; 54: 503–12.

[8]   Salvi G, De Los Rios P, Vendruscolo M. Effective interactions between chaotropic agents and proteins. Proteins, Structure, Function, and Bioinformatics 2005; 61: 492–99.

[9]   Impact Story: Developing the Tools to Evaluate Complex Drug Products: Peptides 2019. U.S. Food and Drug Administration, Center for Drug Evaluation and Research. (Accessed Dec 29, 2024, at https://www.fda.gov/drugs/regulatory-science-action/impact-story-developing-tools-evaluate-complex-drug-products-peptides).

[10]   What Are: "Biologics" Questions and Answers 2018. Center for Biologics Evaluation and Research. (Accessed Dec 29, 2024, at https://www.fda.gov/about-fda/center-biologics-evaluation-and-research -cber/what-are-biologics-questions-and-answers).

[11]   Becker Z, Liu A, Dunleavy K, Kansteiner F, Sagonowsky E. The top 20 drugs by worldwide sales in 2023. 2024. Fierce Pharma. (Accessed Dec 01, 2024, at https://www.fiercepharma.com/special-re ports/top-20-drugs-worldwide-sales-2023).

[12]   AbbVie Inc. Humira [package insert]. 2023.

[13]   Aspirin. National Library of Medicine. (Accessed Dec 29, 2024, at https://pubchem.ncbi.nlm.nih.gov/ compound/2244).

[14]   Jayapal KP, Wlaschin KF, Hu W, Yap MGS. Recombinant protein therapeutics from CHO cells – 20 years and counting. Chemical Engineering Progress 2007; 103: 40.

[15]   Thomas F. Considering alternative dosage forms in biologics. BioPharm International 2019; 32: 16, 18, 50.

[16]   Mantaj J, Vllasaliu D. Recent advances in the oral delivery of biologics. Pharmaceutical Journal 2020; 304.

[17]   Siew A. Ensuring sterility of parenteral products. Pharmaceutical Technology Europe 2013; 25: 37.

[18]   rAAV Average Molecular Weight – SignaGen Blog. 2018.

[19]   Tasnim H, Fricke GM, Byrum JR, Sotiris JO, Cannon JL, Moses ME. Quantitative measurement of naïve T cell association with dendritic cells, FRCs, and blood vessels in lymph nodes. Frontiers in Immunology 2018; 9: 1571.

[20]   Kim S, Chen J, Cheng T, et al. PubChem 2019 update: improved access to chemical data. Nucleic Acids Research 2018; 47: D1102–9.

[21]   Novo Nordisk Inc. Novoeight [package insert]. 2020.

[22]   Spark Therapeutics Inc. Luxturna [package insert]. 2022.

[23]   Novartis Pharmaceuticals Corp. Kymriah [package insert]. 2024.

[24]   Ponomarenko EA, Poverennaya EV, Ilgisonis EV, et al. The size of the human proteome: the width and depth. International Journal of Analytical Chemistry 2016; 2016: 7436849–6.

[25]   Dimitrov DS. Therapeutic Proteins. In: Voynov V, Caravella J, eds. Therapeutic Proteins. Methods in Molecular Biology. Totowa, NJ: Humana Press, 2012: 1–26.

[26]   Amgen Inc. Neulasta [package insert]. 2021.

https://doi.org/10.1515/9783111112459-012

[27]   Genentech Inc. Avastin [package insert]. 2022.

[28]   Allergan Pharmaceuticals. Botox [package insert]. 2023.

[29]   MedImmune LLC. FluMist Quadrivalent [package insert]. 2023.

[30]   Sanofi Pasteur SA. IPOL [package insert]. 2022.

[31]   PaxVax Bermuda Ltd. Vaxchora [package insert]. 2024.

[32]   Daptacel [package insert]. 202x.

[33]   Merck & Co. Inc. Gardasil 9 [package insert]. 2024.

[34]   Merck & Co. Inc. Recombivax HB [package insert]. 2023.

[35]   Pfizer (Wyeth Pharmaceuticals, Inc.). Prevnar 20 [package insert]. 2023.

[36]   Pfizer. Comirnaty [package insert]. 2024.

[37]   GlaxoSmithKline. Bexsero. [package insert]. 2024.

[38]   Approved Cellular and Gene Therapy Products 2024. U.S. Food and Drug Administration. (Accessed October 26, 2024, at https://www.fda.gov/vaccines-blood-biologics/cellular-gene-therapy-products/approved-cellular-and-gene-therapy-products).

[39]   AveXis Inc. Zolgensma [package insert]. 2024.

[40]   Kite Pharma Inc. Yescarta [package insert]. 2024.

[41]   Walsh G, Walsh E. Biopharmaceutical benchmarks 2022. Nature Biotechnology 2022; 40: 1722–60.

[42]   TEConomy Partners LLC. The Economic Impact of the U.S. Biopharmaceutical Industry: 2022 National and State Estimates, 2024.

[43]   21CFR314.3 – Code of Federal Regulations – Title 21 -- Food And Drugs. Chapter I -- Food And Drug Administration, Department Of Health And Human Services. Subchapter D – Drugs for Human Use. Part 314 -- Applications For FDA Approval to Market a New Drug. Subpart A – General Provisions. Sec. 314.3 Definitions. Current as of Aug 30, 2024. (Accessed Dec 30, 2024, at https://www.accessdata.fda.gov/scripts/cdrh/cfdocs/cfcfr/CFRSearch.cfm?fr=314.3).

[44]   Emami F, Vatanara A, Park E, Na D. Drying technologies for the stability and bioavailability of biopharmaceuticals. Pharmaceutics 2018; 10: 131.

[45]   Bye J, Platts L, Falconer R. Biopharmaceutical liquid formulation: a review of the science of protein stability and solubility in aqueous environments. Biotechnology Letters 2014; 36: 869–75.

[46]   Siew A. Drug delivery systems for biopharmaceuticals. BioPharm International 2015; 28: 14–19.

[47]   Amgen (Mfg. by Immunex Corp.). Enbrel [package insert]. 2024.

[48]   Anselmo AC, Gokarn Y, Mitragotri S. Non-invasive delivery strategies for biologics nature reviews. Drug Discovery 2019; 18: 19–40.

[49]   Gutka H. Rational selection of sugars for biotherapeutic stabilization: a practitioner's perspective. BioProcess International 2018; 16: 40–52.

[50]   Khan TA, Mahler H, Kishore RSK. Key interactions of surfactants in therapeutic protein formulations: a review. European Journal of Pharmaceutics and Biopharmaceutics 2015; 97: 60–67.

[51]   European Medicines Agency. ICH Topic Q6A, Specifications: Test Procedures and Acceptance Criteria for New Drug Substances and New Drug Products: Chemical Substances. 2000.

[52]   US Dept. of Health and Human Services, Food and Drug Administration, Center for Drug Evaluation and Research (CDER), Center for Biologics Evaluation and Research (CBER), adopted from ICH. Guidance for Industry – Q8(R2) Pharmaceutical Development. 2009.

[53]   CFR – Code of Federal Regulations Title 21 – Title 21 -- Food and Drugs, Chapter 1 -- Food and Drug Administration, Department of Health and Human Services. Subchapter C – Drugs: General. Part 210 – Current Good Manufacturing Practice in Manufacturing, Processing, Packing, or Holding of Drugs: General. (Accessed Dec 30, 2024 at https://www.accessdata.fda.gov/scripts/cdrh/cfdocs/cfcfr/CFRSearch.cfm?CFRPart=210).

[54] CFR – Code of Federal Regulations Title 21 – Title 21 -- Food and Drugs, Chapter 1 -- Food and Drug Administration, Department of Health and Human Services. Subchapter C – Drugs: General. Part 211 – Current Good Manufacturing Practice for Finished Pharmaceuticals. (Accessed Dec 30, 2024 at https://www.accessdata.fda.gov/scripts/cdrh/cfdocs/cfcfr/cfrsearch.cfm?cfrpart=211).

[55] Facts About the Current Good Manufacturing Practice (CGMP) 2024. Center for Drug Evaluation and Research, U. S. FDA. (Accessed Dec 30, 2024, at https://www.fda.gov/drugs/pharmaceutical-quality-resources/facts-about-current-good-manufacturing-practice-cgmp).

[56] Rathore AS. Setting specifications for a biotech therapeutic product in the quality by design paradigm. BioPharm International 2010; 23.

[57] U.S. Department of Health and Human Services, Food and Drug Administration, Center for Drug Evaluation and Research (CDER), Center for Biologics Evaluation and Research (CBER). Guidance for industry – Q11 Development and Manufacture of Drug Substances. U.S. Food and Drug Administration 2012.

[58] European Medicines Agency. EMA Assessment Report for Prevenar 20, EMA/66027/2024. Amsterdam, The Netherlands: European Medicines Agency, 2024.

[59] Grifols Therapeutics Inc. Gamunex-C [package insert]. 2020.

[60] Remington KM. Fundamental strategies for viral clearance – Part 1: exploring the regulatory implications. BioProcess International 2015; 13: 10–16.

[61] U.S. Dept. of Health and Human Services, Food and Drug Administration, Center for Drug Evaluation and Research (CDER), Center for Biologics Evaluation and Research (CBER), adopted from ICH. Q7 Good Manufacturing Practice Guidance for Active Pharmaceutical Ingredients – Guidance for Industry. U.S. Food and Drug Administration 2016.

[62] CMC Biotech Working Group. A–Mab: A Case Study in Bioprocess Development: ISPE | International Society for Pharmaceutical Engineering.

[63] Warikoo V, Godawat R, Brower K, et al. Integrated continuous production of recombinant therapeutic proteins. Biotechnology and Bioengineering 2012; 109: 3018–29.

[64] Hernandez R. Continuous manufacturing: a changing processing paradigm. Biopharm International 2015; 28.

[65] MOTIVE™ Inline Buffer Formulation Systems. Asahi Kasei Bioprocess publication UME1I01–1.0, Asahi Kasei Bioprocess America, Inc.

[66] Seiberling DA. Clean-In-Place for Biopharmaceutical Processes. Boca Raton: CRC Press, 2007.

[67] Sandle T. Sterility, Sterilisation and Sterility Assurance for Pharmaceuticals: Technology, Validation and Current Regulations. Oxford, UK: Woodhead Publishing Ltd., 2013.

[68] United States Pharmacopeia and National Formulary (USP 42-NF37 2S). United States Pharmacopeial Convention, 2018.

[69] American Society of Mechanical Engineers. Bioprocessing Equipment ASME BPE-2024 (Revision of ASME BPE-2022); An International Standard. New York, NY: ASME, American Society of Mechanical Engineers, 2024.

[70] ABEC Advances Single-Use Bioreactor Volumes To 6,000 Liters 2019. (Accessed Nov 9, 2020, at https://www.bioprocessonline.com/doc/abec-advances-single-use-bioreactor-volumes-to-liters-0001).

[71] Extractables and leachables subcommittee of the bio-process systems alliance. Recommendations for extractables and leachables testing, Part 1: introduction, regulatory issues, and risk assessment. BioProcess International 2007; 5: 36–49.

[72] Martin J. A brief history of single-use manufacturing. BioPharm International 2011; 5–6.

[73] Barnoon B, Bader B. Lifecycle cost analysis for single-use systems. BioPharm International 2008; 2008 Supplement.

[74] Jagschies G, Łącki K. Process Capability Requirements. In: Jagschies G, Lindskog E, Łącki K, Galliher P., eds. Biopharmaceutical Processing: Development, Design, and Implementation of Manufacturing Processes. Amsterdam, Netherlands: Elsevier, 2018: 73–94.

[75] Rader RA, Langer ES. 30 years of upstream productivity improvements. BioProcess International 2015; 2: 10–15.

[76] U.S. Department of Health and Human Services, Food and Drug Administration, Center for Drug Evaluation and Research (CDER), Center for Biologics Evaluation and Research (CBER), Center for Veterinary Medicine (CVM). Guidance for Industry – Process Validation: General Principles and Practices. 2011.

[77] Step 3: Clinical Research 2018. U.S. Dept. of Health and Human Services, Food and Drug Administration, Office of the Commissioner. (Accessed Nov 10, 2020, at https://www.fda.gov/pa tients/drug-development-process/step-3-clinical-research).

[78] Jiang C, Flansburg L, Ghose S, Jorjorian P, Shukla AA. Defining process design space for a hydrophobic interaction chromatography (HIC) purification step: Application of quality by design (QbD) principles. Biotechnology and Bioengineering 2010; 107: 985–97.

[79] DiMasi JA, Grabowski HG, Hansen RW. Innovation in the pharmaceutical industry: New estimates of R&D costs. Journal of Health Economics 2016; 47: 20–33.

[80] PDA Task Force on Technical Report No. 60. Technical Report No. 60 – Process Validation: A Lifecycle Approach: Parenteral Drug Association, Inc., 2013.

[81] Basics of process development for biotherapeutics - Definitions, activities, and things to consider when developing an upstream or downstream bioprocess. 2025. Cytiva. (Accessed Jan 05, 2025, at https://www.cytivalifesciences.com/en/us/solutions/emerging-biotech/knowledge-center/bio pharma-process-development-introduction?srsltid=AfmBOor9oVRaiR4a8UErxSoDRc_H-K2u2EcR9 ZuHW073aC2lvapF9yXq).

[82] Bioburden control in biopharmaceuticals: managing the risks. 2016. (Accessed Nov 10, 2020, at https://www.engineersireland.ie/Covid-19-information-base/bioburden-control-in-biopharmaceuticals-managing-the-risks).

[83] Von Wintzingerode F. Biologics production: impact of bioburden contaminations of non-sterile process intermediates on patient safety and product quality. American Pharmaceutical Review 2017; 20.

[84] Berk RA, Bozenhardt EH, Carabello-Oramas JA, et al. ISPE Baseline Guide: Volume 6 – Biopharmaceutical Manufacturing Facilities, Third. ISPE | International Society for Pharmaceutical Engineering, 2023.

[85] ISO/TC 209 Cleanrooms and associated controlled environments. ISO 14644-1:2015, Cleanrooms and associated controlled environments – Part 1: Classification of air cleanliness by particle concentration. 2015.

[86] ISO/TC 209 Cleanrooms and associated controlled environments. ISO 14644-4:2001, Cleanrooms and associated controlled environments – Part 4: Design, construction and start-up. 2001.

[87] Sun W. Cleanroom airlock performance and beyond. ASHRAE Journal 2018; 60.

[88] EudraLex – Volume 4 – EU Guidelines for Good Manufacturing Practice for Medicinal Products for Human and Veterinary Use, Annex 15: Qualification and Validation. 2015. (Accessed Nov 9, 2024, at https://ec.europa.eu/health/sites/health/files/files/eudralex/vol-4/2015-10_annex15.pdf).

[89] Biosafety in Microbiological and Biomedical Laboratories, 6th Edition. HHS Publication No. (CDC) 300859 2020.

[90] Department of Health and Human Services, National Institutes of Health. NIH Guidelines for Research Involving Recombinant or Synthetic Nucleic Acid Molecules (NIH Guidelines). 2024.

[91] Odum JN, Flickinger MC. Process Architecture in Biomanufacturing Facility Design. Hoboken, NJ: John Wiley & Sons, Hoboken, NJ:, 2018.

[92] Langer ES. Outsourcing trends in biopharmaceutical manufacturing. Pharmaceutical Outsourcing 2017; 18.

[93] McClung T. Biopharmaceutical manufacturing companies continue to expand their economic footprint across the United States. 2023. Pharmaceutical Research and Manufacturers of America®. (Accessed Mar 08, 2025, at https://phrma.org/blog/biopharmaceutical-manufacturing-companies-continue-to-expand-their-economic-footprint-across-the-united-states).

[94] Alonso W, Vandeberg P, Lang J, et al. Immune globulin subcutaneous, human 20% solution (Xembify®), a new high concentration immunoglobulin product for subcutaneous administration. Biologicals 2020; 64: 34–40.

[95] 21CFR211.25 – Code of Federal Regulations, Title 21, Vol. 4, Chapter I -- Food and Drug Administration, Department of Health and Human Services. Subchapter C – Drugs: General. Part 211 -- Current Good Manufacturing Practice for Finished Pharmaceuticals. Subpart B – Organization and Personnel. Sec. 211.25 Personnel qualifications. Current as of Aug 30, 2024. (Accessed Nov 9, 2024, at https://www.accessdata.fda.gov/scripts/cdrh/cfdocs/cfcfr/cfrsearch.cfm?fr=211.25).

[96] Joseph J. Facility Design and Process Utilities. In: Jagschies G, Lindskog E, Łącki K, Galliher P., eds. Biopharmaceutical Processing: Development, Design, and Implementation of Manufacturing Processes. Amsterdam, Netherlands: Elsevier, 2018: 933–86.

[97] Eudralex – Volume 4 – EU Guidelines for Good Manufacturing Practice for Medicinal Products for Human and Veterinary Use, Part I. Chapter 2: Personnel. 2014. (Accessed Nov, 9, 2024, at https://ec.europa.eu/health/documents/eudralex/vol-4_en).

[98] People in Cleanrooms: Understanding and Monitoring the Personnel Factor 2014. Sandle T (Accessed Jan 3, 2021, at http://www.ivtnetwork.com/article/people-cleanrooms-understanding-and-monitoring-personnel-factor).

[99] 21CFR211.28 – Code of Code of Federal Regulations, Title 21, Vol. 4, Chapter I -- Food and Drug Administration, Department of Health and Human Services. Subchapter C – Drugs: General. Part 211 -- Current Good Manufacturing Practice for Finished Pharmaceuticals. Subpart B – Organization and Personnel. Sec. 211.28 Personnel responsibilities. Current as of Aug 30, 2024. (Accessed Nov 9, 2024, at https://www.accessdata.fda.gov/scripts/cdrh/cfdocs/cfcfr/CFRSearch.cfm?fr=211.28).

[100] 21CFR211 – Code of Code of Federal Regulations, Title 21, Vol. 4, Chapter I -- Food and Drug Administration, Department of Health and Human Services. Subchapter C – Drugs: General. Part 211 -- Current Good Manufacturing Practice for Finished Pharmaceuticals. Subpart D – Equipment. Current as of Aug 30, 2024. (Accessed Nov 9, 2024, at https://www.accessdata.fda.gov/scripts/cdrh/cfdocs/cfcfr/CFRSearch.cfm?CFRPart=211&showFR=1&subpartNode=21:4.0.1.1.11.4).

[101] Good Manufacturing Practices (GMP) Guidelines for Active Pharmaceutical Ingredients (API) – (GUI-0104) 2022. Health Canada, Health Products and Food Branch Inspectorate. (Accessed Nov 9, 2024, at https://www.canada.ca/en/health-canada/services/drugs-health-products/compliance-enforcement/information-health-product/drugs/guidelines-active-pharmaceutical-ingredients-0104.html#s5.C.02.005).

[102] 21CFR211.68 – Code of Code of Federal Regulations, Title 21, Vol. 4, Chapter I -- Food and Drug Administration, Department of Health and Human Services. Subchapter C – Drugs: General. Part 211 -- Current Good Manufacturing Practice for Finished Pharmaceuticals. Subpart D – Equipment. Sec. 211.68 Automatic, mechanical, and electronic equipment. Current as of Aug 30, 2024. (Accessed Nov 9, 2024, at https://www.accessdata.fda.gov/scripts/cdrh/cfdocs/cfcfr/CFRSearch.cfm?fr=211.68).

[103] Eudralex - Volume 4 - EU Guidelines for Good Manufacturing Practice for Medicinal Products for Human and Veterinary Use, Part I. Chapter 4: Documentation. 2011. (Accessed Nov 9, 2024, at https://ec.europa.eu/health/sites/health/files/files/eudralex/vol-4/chapter4_01-2011_en.pdf).

[104] CFR – Code of Federal Regulations Title 21 –Title 21 -- Food and Drugs, Chapter 1 -- Food and Drug Administration, Department of Health and Human Services. Subchapter A – General. Part 11 – Electronic Records; Electronic Signatures. Current as of Aug 30, 2024. (Accessed Dec 30, 2024 at https://www.accessdata.fda.gov/scripts/cdrh/cfdocs/cfcfr/CFRSearch.cfm?CFRPart=11).

[105] 21CFR 211 – Code of Federal Regulations, Title 21 -- Food and Drugs, Chapter I -- Food and Drug Administration, Department of Health and Human Services. Subchapter C – Drugs: General. Part 211 -- Current Good Manufacturing Practice For Finished Pharmaceuticals. Subpart E – Control of Components and Drug Product Containers and Closures. Current as of Aug 30, 2024. (Accessed Nov 9, 2024, at https://www.accessdata.fda.gov/scripts/cdrh/cfdocs/cfcfr/CFRSearch.cfm?CFRPart= 211&showFR=1&subpartNode=21:4.0.1.1.11.5).

[106] 21CFR211.110 – Code of Federal Regulations, Title 21 -- Food And Drugs, Chapter I -- Food And Drug Administration, Department of Health And Human Services. Subchapter C – Drugs: General. Part 211 -- Current Good Manufacturing Practice For Finished Pharmaceuticals. Subpart F – Production and Process Controls. Sec. 211.110 Sampling and Testing of In-Process Materials and Drug Products. Current as of Aug 30, 2024. (Accessed Nov 9, 2024, at https://www.accessdata.fda. gov/scripts/cdrh/cfdocs/cfcfr/CFRSearch.cfm?fr=211.110).

[107] 21CFR211.165 – Code of Federal Regulations, Title 21 -- Food And Drugs Chapter I -- Food And Drug Administration, Department Of Health And Human Services. Subchapter C – Drugs: General. Part 211 -- Current Good Manufacturing Practice For Finished Pharmaceuticals. Subpart I – Laboratory Controls. Sec. 211.165 Testing and release for distribution. Current as of Aug 30, 2024. (Accessed Nov 9, 2024, at https://www.accessdata.fda.gov/scripts/cdrh/cfdocs/cfcfr/CFRSearch.cfm? fr=211.165).

[108] 21CFR820.3 – Code of Federal Regulations – Title 21 -- Food and Drugs Chapter I -- Food And Drug Administration, Department Of Health And Human Services. Subchapter H – Medical Devices Part 820 -- Quality System Regulation. Subpart A – General Provisions. Sec. 820.3 Definitions. Current as of Aug 30, 2024. (Accessed Dec 31, 2024, at https://www.accessdata.fda.gov/scripts/cdrh/ cfdocs/cfcfr/CFRSearch.cfm?fr=820.3).

[109] 21CFR211.192. – Code of Federal Regulations – Title 21 -- Food and Drugs Chapter I -- Food And Drug Administration, Department Of Health And Human Services. Subchapter C – Drugs: General. Part 211 -- Current Good Manufacturing Practice For Finished Pharmaceuticals. Subpart J – Records and Reports. Sec. 211.192 Production record review. Current as of Aug 30, 2024. (Accessed Nov 9, 2024, at https://www.accessdata.fda.gov/scripts/cdrh/cfdocs/cfcfr/cfrsearch.cfm?fr=211.192).

[110] Root Cause Analysis for Drugmakers. Washington Business Information Inc, 2013.

[111] What is Root Cause Analysis (RCA)? | ASQ. 2024. (Accessed Dec 31, 2024, at https://asq.org/quality-resources/root-cause-analysis).

[112] Eudralex - Volume 4 - EU Guidelines for Good Manufacturing Practice for Medicinal Products for Human and Veterinary Use, Part 1. Chapter 8: Complaints, Quality Defects and Product Recalls. 2014. (Accessed Nov 9, 2024, at https://ec.europa.eu/health//sites/health/files/files/eudralex/vol-4/ 2014-08_gmp_chap8.pdf).

[113] 21CFR211.198 – Code of Federal Regulations – Title 21 -- Food And Drugs. Chapter I -- Food And Drug Administration, Department Of Health And Human Services. Subchapter C – Drugs: General. Part 211 -- Current Good Manufacturing Practice For Finished Pharmaceuticals. Subpart J – Records and Reports. Sec. 211.198 Complaint files. Current as of Aug 30, 2024. (Accessed Nov 9, 2024, at https://www.accessdata.fda.gov/scripts/cdrh/cfdocs/cfcfr/CFRSearch.cfm?fr=211.198).

[114] 21CFR314.81 – Code of Federal Regulations – Title 21 -- Food And Drugs. Chapter I -- Food And Drug Administration, Department Of Health And Human Services. Subchapter D – Drugs for Human Use. Part 314 -- Applications For FDA Approval to Market a New Drug. Subpart B – Applications. Sec. 314.81 Other postmarketing reports. Current as of Aug 30, 2024. (Accessed Dec 31, 2024, at https://www.accessdata.fda.gov/scripts/cdrh/cfdocs/cfcfr/CFRSearch.cfm?fr=314.81).

[115] Blanch HW, Yee L. Recombinant protein expression in high cell density fed-batch cultures of Escherichia coli. Bio/Technology (New York, N.Y. 1983) 1992; 10: 1550–56.

[116] Porro D, Sauer M, Branduardi P, Mattanovich D. Recombinant protein production in yeasts. Molecular Biotechnology 2005; 31: 245–59.

[117] Gu MB, Todd P, Kompala DS. Foreign gene expression (β-galactosidase) during the cell cycle phases in recombinant CHO cells. Biotechnology and Bioengineering 1993; 42: 1113–23.

[118] Monod J. The growth of bacterial cultures. Annual Review of Microbiology 1949; 3: 371–94.

[119] Watanabe I, Okada S. Effects of temperature on growth rate of cultured mammalian cells (L5178Y). The Journal of Cell Biology 1967; 32: 309–23.

[120] Charoenchai C, Fleet GH, Henschke PA. Effects of temperature, pH, and sugar concentration on the growth rates and cell biomass of wine yeasts. American Journal of Enology and Viticulture 1998; 49: 283–88.

[121] Russell JB, Diez-Gonzalez F. The effects of fermentation acids on bacterial growth. Advances in Microbial Physiology 1998; 39: 205–34.

[122] Schmidt S. Drivers, opportunities, and limits of continuous processing. BioProcess International 2017; 15: 30–37.

[123] Steam In Place (Technical Report No. 61): Parenteral Drug Association, Inc., 2013.

[124] U.S. Dept. of Health and Human Services, Food and Drug Administration, Center for Biologics Evaluation and Research (CBER). Guidance for Industry – Characterization and Qualification of Cell Substrates and Other Biological Materials Used in the Production of Viral Vaccines for Infectious Disease Indications. 2010.

[125] ICH Topic Q 5 D – Quality of Biotechnological Products: Derivation and Characterisation of Cell Substrates Used for Production of Biotechnological/Biological Products. 1998.

[126] Gancedo C, Gancedo JM, Sols A. Glycerol metabolism in yeasts. Pathways of utilization and production. European Journal of Biochemistry 1968; 5: 165–72.

[127] Guo R, Gu J, Zong S, Wu M, Yang M. Structure and mechanism of mitochondrial electron transport chain. Biomedical Journal 2018; 41: 9–20.

[128] Boron WF. Regulation of intracellular pH. Advances in Physiology Education 2004; 28: 160–79.

[129] Orij R, Brul S, Smits GJ. Intracellular pH is a tightly controlled signal in yeast. Biochimica et Biophysica Acta. General Subjects 2011; 1810: 933–44.

[130] Booth IR. Regulation of cytoplasmic pH in bacteria. Microbiological Reviews 1985; 49: 359–78.

[131] Los DA, Murata N. Membrane fluidity and its roles in the perception of environmental signals. Biochimica et Biophysica Acta. Biomembranes 2004; 1666: 142–57.

[132] Beales N. Adaptation of microorganisms to cold temperatures, weak acid preservatives, low pH, and osmotic stress: a review. Comprehensive Reviews in Food Science and Food Safety 2004; 3: 1–20.

[133] Han Y, Liu X, Liu H, et al. Cultivation of recombinant Chinese hamster ovary cells grown as suspended aggregates in stirred vessels. Journal of Bioscience and Bioengineering 2006; 102: 430–35.

[134] Yeast Cell Architecture and Functions. In: Feldmann H, ed. Yeast: Molecular and Cell Biology. 2nd completely rev. and greatly enl. ed. Weinheim: John Wiley & Sons, Inc, 2012: 5–24.

[135] Orlean P. Architecture and biosynthesis of the saccharomyces cerevisiae cell wall. Genetics (Austin) 2012; 192: 775–818.

[136] Reshes G, Vanounou S, Fishov I, Feingold M. Cell shape dynamics in Escherichia coli. Biophysical Journal 2008; 94: 251–64.

[137] Components and Structure | Boundless Biology. 2024. Learneo, Inc. (Accessed Dec 31, 2024, at https://courses.lumenlearning.com/boundless-biology/chapter/components-and-structure/).

[138] Van Der Rest ME, Kamminga AH, Nakano A, Anraku Y, Poolman B, Konings WN. The plasma membrane of Saccharomyces cerevisiae: structure, function, and biogenesis. Microbiological Reviews 1995; 59: 304–22.

[139] Erickson HP. How bacterial cell division might cheat turgor pressure – a unified mechanism of septal division in gram-positive and gram-negative bacteria. Bioessays 2017; 39: 1700045.

[140] Petsch D, Anspach FB. Endotoxin removal from protein solutions. Journal of Biotechnology 2000; 76: 97–119.

[141] Singh SM, Panda AK. Solubilization and refolding of bacterial inclusion body proteins. Journal of Bioscience and Bioengineering 2005; 99: 303–10.

[142] Baneyx F, Mujacic M. Recombinant protein folding and misfolding in Escherichia coli. Nature Biotechnology 2004; 22: 1399–408.

[143] Yang Z, Zhang L, Zhang Y, et al. Highly efficient production of soluble proteins from insoluble inclusion bodies by a two-step-denaturing and refolding method. PLoS One 2011; 6: e22981.

[144] Jalalirad R. Selective and efficient extraction of recombinant proteins from the periplasm of Escherichia coli using low concentrations of chemicals. Journal of Industrial Microbiology & Biotechnology 2013; 40: 1117–29.

[145] Harcum SW, Lee KH. CHO cells can make more protein. Cell Systems 2016; 3: 412–13.

[146] Chen R. Bacterial expression systems for recombinant protein production: E. coli and beyond. Biotechnology Advances 2012; 30: 1102–07.

[147] Vandenberghe LH, Xiao R, Lock M, Lin J, Korn M, Wilson JM. Efficient serotype-dependent release of functional vector into the culture medium during adeno-associated virus manufacturing. Human Gene Therapy 2010; 21: 1251–57.

[148] Berman HM, Westbrook J, Feng Z, et al. The protein data bank. Nucleic Acids Research 2000; 28: 235–42.

[149] Salazar O, Asenjo JA. Enzymatic lysis of microbial cells. Biotechnology Letters 2007; 29: 985–94.

[150] Garcia FAP. Cell Wall Disruption. In: Flickinger M, ed. Encyclopedia of Industrial Biotechnology: Bioprocess, Bioseparation, and Cell Technology. John Wiley & Sons, Inc., 2009: 1–12.

[151] Kampen I, Kwade A. Cell Disruption, Micromechanical Properties. In: Flickinger M, ed. Downstream Industrial Biotechnology: Recovery and Purification. John Wiley & Sons, Inc., 2013: 49–64.

[152] Middleberg A. The Release of Intracellular Bioproducts. In: Bioseparation and Bioprocessing. Weinheim; New York: Wiley-VCH, 1998: 131–63.

[153] Hetherington P, Follows M, Dunnill P, Lilly M. Release of protein from baker's yeast (saccharomyces cerevisiae) by disruption in an industrial homogeniser. Transactions of the Institution of Chemical Engineers 1971; 49: 142–48.

[154] Kleinig AR, Mansell CJ, Nguyen QD, Badalyan A, Middelberg APJ. Influence of broth dilution on the disruption of Escherichia coli. Biotechnology 1995; 9: 759–62.

[155] GEA Ariete Series Homogenizers – Innovation designed to last. GEA Ariete Series Homogenizers Rev 2021, GEA Mechanical Equipment Italia S.p.A. 2021.

[156] Theory of Separation. Sweden: Alfa Laval, Industrial Separation Division.

[157] Richardson A, Walker J. Continuous solids-discharging centrifugation – a solution to the challenges of clarifying high–cell-density mammalian cell cultures. BioProcess International 2018; 16: 38–47.

[158] Kubitschek HE. Buoyant density variation during the cell cycle in microorganisms. Critical Reviews in Microbiology 1987; 14: 73–97.

[159] Axelsson H. Cell Separation, Centrifugation. In: Flickinger MC, ed. Downstream Industrial Biotechnology: Recovery and Purification. Hoboken, New Jersey: John Wiley & Sons, Inc, 2013: 27–48.

[160] Holdich RG, Rushton A. Centrifugal Separation. In: Rushton A, Ward AS, and Holdich RG, eds. Solid-Liquid Filtration and Separation Technology. Weinheim; New York: Wiley-VCH, 1996.

[161] Bell DJ, Hoare M, Dunnill P. The Formation of Protein Precipitates and their Centrifugal Recovery. In: Downstream Processing, Advances in Biochemical Engineering/ Biotechnology, Vol. 26. Berlin: Akademie-Verlag, 1983: 1–72.

[162] Harrison RG, Todd PW, Rudge SR, Petrides DP. Bioseparations Science and Engineering. New York: Oxford University Press, 2015.

[163] Wang DIC, Cooney CL, Demain AL, Dunnill P, Humphrey AE, Lilly MD. Fermentation and Enzyme Technology. New York: John Wiley & Sons, Inc, 1979.

[164] UniFuge® Pilot. CARR Biosystems. (Accessed Oct 17, 2024, at https://www.carrbiosystems.com/products/single-use-centrifuges/unifuge-single-use-centrifugation).

[165] U2k®. CARR Biosystems. (Accessed Oct 17, 2024, at https://www.carrbiosystems.com/products/single-use-centrifuges/u2k-single-use-centrifugation).

[166] Hebb MH, Smith FH. Encyclopedia of Chemical Technology, Vol. 3. In: Hebb MH, Smith FH, and Re K, eds. New York: Interscience Encyclopedia, Inc., 1948: 501–05.

[167] Ambler CM. The evaluation of centrifuge performance. Chemical Engineering Progress 1952; 48: 150–58.

[168] Ambler CM. The theory of scaling up laboratory data for the sedimentation type centrifuge. Journal of Biochemical and Microbiological Technology and Engineering 1959; 1: 185–205.

[169] Harrison RG, Todd PW, Rudge SR, Petrides DP. Bioseparations Science and Engineering (Topics in Chemical Engineering). New York: Oxford University Press, 2003.

[170] Aiba S, Kitai S, Heima H. Determination of equivalent size of microbial cells from their velocities in hindered settling. Journal of General and Applied Microbiology 1964; 10: 243–56.

[171] Shukla AA, Kandula JR. Harvest and recovery of monoclonal antibodies from large-scale mammalian cell culture. BioPharm International 2008; 21.

[172] Sullivan FE, Erikson RA. De Laval's "KQ value" spelled out. Design factors with the continuous disk centrifuge as a case in point. Industrial and Engineering Chemistry 1960; 53: 434–38.

[173] Baker RW. Membrane Technology and Applications. Oxford: Wiley-Blackwell, 2012.

[174] O'Brien TP, Brown LA, Battersby DG, Rudolph AS, Raman LP. Large-scale, single-use depth filtration systems for mammalian cell culture clarification. BioProcess International 2012; 10: 50–57.

[175] Quigley GT. Prefiltration in Biopharmaceutical Processes. In: Meltzer TH and Jornitz MW, eds. Filtration and Purification in the Biopharmaceutical Industry. Baton Rouge: CRC Press LLC, 2007: 1–21.

[176] Millistak+® HC Pro Fully synthetic depth filters for clarification and downstream filtration applications. Millipore Data Sheet No. MS_DS1240EN00 Ver. 5.0 48315 10/2023, Merck KGaA, Darmstadt, Germany and/or its affiliates. 2023.

[177] Provantage® Services – Millistak+® Pod Filter Holder qualification protocol and services. Merck Millipore publication Lit No. SDS1014EN00 Rev. C PS-SBU-12-06403, EMD Millipore Corporation, Billerica, MA. 2015.

[178] Millistak+® family and Clarisolve® depth filters at a glance. Millipore publication PB1900EN00 Ver. 3.0 EMD Millipore Corporation. 2017.

[179] Allegro™ MVP Single-use System. Cytiva publication CY40462-11Dec23-DF, Cytiva. 2023.

[180] Goldrick S, Joseph A, Mollet M, et al. Predicting performance of constant flow depth filtration using constant pressure filtration data. Journal of Membrane Science 2017; 531: 138–47.

[181] Lutz H, Chefer K, Felo M, et al. Robust depth filter sizing for centrate clarification. Biotechnology Progress 2015; 31: 1542–50.

[182] Frahm G, Clarke T, Johnston M. Evaluation of microflow digital imaging particle analysis for subvisible particles formulated with an opaque vaccine adjuvant. PLoS One 2016; 11.

[183] Prashad M, Tarrach K. Depth filtration: cell clarification of bioreactor offloads. Filtration & Separation 2006; 43: 28–30.

[184] Russell E, Wang A, Rathore AS. Harvest of a Therapeutic Protein Product from High Cell Density Fermentation Broths: Principles and Case Study. In: Shukla AA, Etzel MR, and Gadam S, eds. Process Scale Bioseparations for the Biopharmaceutical Industry. Boca Raton: CRC/Taylor & Francis, 2007: 1–58.

[185] Yavorsky D, Blanck R, Lambalot C, Brunkow R. The clarification of bioreactor cell cultures for biopharmaceuticals. Pharmaceutical Technology ((2003)) 2003; 27: 62–76.

[186] Pmax™/Tmax™ Constant Flow Rate Test for Depth Filter Sizing. Millipore Lit. No. MS_BR4506EN Ver. 2.0 33211, Merck KGaA, Darmstadt, Germany and/or its affiliates. 2020.

[187] Millistak+® Pod disposable depth filter system. Millipore Data Sheet Lit. No. DS0217EN00 Ver. 13.0 48314 10/2023, Merck KGaA, Darmstadt, Germany and/or its affiliates. 2023.

[188] Yigzaw Y, Piper R, Tran M, Shukla AA. Exploitation of the adsorptive properties of depth filters for host cell protein removal during monoclonal antibody purification. Biotechnology Progress 2006; 22: 288–96.

[189] Zhou JX, Solamo F, Hong T, Shearer M, Tressel T. Viral clearance using disposable systems in monoclonal antibody commercial downstream processing. Biotechnology and Bioengineering 2008; 100: 488–96.

[190] Lutz H. Ultrafiltration for Bioprocessing. Development and Implementation of Robust Processes. Waltham, MA, USA: Elsevier, 2015.

[191] Hughes DJ, Cui Z, Field RW, Tirlapur UK. In situ three-dimensional characterization of membrane fouling by protein suspensions using multiphoton microscopy. Langmuir 2006; 22: 6266–72.

[192] Chitale KC, Presz W, Ross-Johnsrud BP, Hyman M, Lamontagne M, Lipkens B Particle manipulation using macroscale angled ultrasonic standing waves. Proceedings of Meetings on Acoustics Jun 25, 2017; 30.

[193] Acoustic Cell Processing with ekko™ System. Millipore Application Note MS_AN8435EN Ver 1.0 37906 01/2022, Merck KGaA, Darmstadt, Germany and/or its affiliates. 2022.

[194] Scalable Cell Culture Perfusion, Applikon BioSep. LS3354-Applikon-BioSep-flyer-210923, Getinge AB, Delft, The Netherlands. 2024.

[195] Aunins JG, Wang DIC. Induced flocculation of animal cells in suspension culture. Biotechnology and Bioengineering 1989; 34: 629–38.

[196] pDADMAC flocculant reagent for use with Clarisolve® depth filters. Millipore Application Note AN33330000 Ver. 7.0 44054 09/2023, Merck KGaA, Darmstadt, Germany and/or its affiliates. 2023.

[197] Eichholz C, Silvestre M, Franzreb M, Nirschl H. Recovery of lysozyme from hen egg white by selective magnetic cake filtration. Engineering in Life Sciences 2011; 11: 75–83.

[198] Wang A, Lewus R, Rathore AS. Comparison of different options for harvest of a therapeutic protein product from high cell density yeast fermentation broth. Biotechnology and Bioengineering 2006; 94: 91–104.

[199] U.S. Dept. of Health and Human Services, Food and Drug Administration, Center for Drug Evaluation and Research (CDER), Center for Biologics Evaluation and Research (CBER). Guidance for Industry – Q6B Specifications: Test Procedures and Acceptance Criteria for Biotechnological/ Biological Products. 1999.

[200] Rosenberg AS. Effects of protein aggregates: an immunologic perspective. The AAPS Journal 2006; 8: E501–7.

[201] Vlasak J, Ionescu R. Fragmentation of monoclonal antibodies. Monoclonal Antibodies 2011; 3: 253–63.

[202] Mingozzi F, Maus M, Hui DJ, et al. CD8 + T-cell responses to adeno-associated virus capsid in humans. Nature Medicine 2007; 13: 419–22.

[203] Patel J, Kothari R, Tunga R, Ritter NM, Tunga BS. Stability considerations for biopharmaceuticals, Part 1 – overview of protein and peptide degradation pathways. BioProcess International 2011; 20–31.

[204] Annex 1 – Requirements for the use of animal cells as in vitro substrates for the production of biologicals (Requirements for Biological Substances No. 50): World Health Organization, 1998.

[205] Dept. of Health, Education, And Welfare, Public Health Service, Food And Drug Administration, *ORA/ORO/DEIO/IB*. FDA Inspection Technical Guide Number 40, Bacterial Endotoxins/ Pyrogens. 1985.

[206] U.S. Pharmacopeia (39th Revision) and National Formulary (34th Edition), USP 39 NF 34, (1132) Residual Host Cell Protein Measurement in Biopharmaceuticals. Rockville, MD: The United States Pharmacopeial Convention, 2016.

[207] Chon JH, Zarbis-Papastoitsis G. Advances in the production and downstream processing of antibodies. New Biotechnology 2011; 28: 458–63.

[208] Remington KM. Fundamental strategies for viral clearance, Part 2: technical approaches. BioProcess International 2015; 13: 10–17.

[209] Boström M, Tavares FW, Bratko D, Ninham BW. Specific ion effects in solutions of globular proteins: comparison between analytical models and simulation. The Journal of Physical Chemistry B 2005; 109: 24489–94.

[210] Carta G, Jungbauer A. Protein Chromatography: Process Development and Scale-Up. Hoboken: Wiley-VCH Imprint], Hoboken: John Wiley & Sons, Incorporated, 2020.

[211] Ion Exchange Chromatography – Principles and Methods. Cytiva handbook CY13983-08Feb21-HB, Cytiva. 2021.

[212] Process Resin Selection Guide. Bio-Rad Bulletin 6713 Ver C, Bio-Rad Laboratories, Inc. 2024.

[213] POROS™ CaptureSelect™ AAV Resins: AAV8, AAV9, AAVX User Guide. Thermo Scientific Pub. No. 100038399 Rev. E, Thermo Fisher Scientific Inc. 2017.

[214] Hydrophobic Interaction and Reversed Phase Chromatography – Principles and Methods. Cytiva handbook CY11248-28Sep22-HB, Cytiva. 2022.

[215] Sartobind® Membrane Chromatography. Sartorius Publication No.: SL-1533-e | Status: 11 | 22 | 2021, Sartorius Stedim Biotech GmbH, Goettingen, Germany. 2021.

[216] Natrix® Q Chromatography Membrane – For single-use, scalable purification. Millipore Data Sheet MS_DS2158EN Ver. 1.0 41362 07/2022. Merck KGaA, Darmstadt, Germany and/or its affiliates. 2022.

[217] Preparative Columns CIMmultus. 2024. Sartorius BIA Separations. (Accessed Dec 20, 2024, at https://www.biaseparations.com/consumables/purification/preparative-columns-cimmultus-line/).

[218] Subramanian G. Biopharmaceutical Production Technology. Weinheim: Wiley-VCH Verlag GmbH & Co. KGaA, 2012.

[219] Shukla AA, Hubbard B, Tressel T, Guhan S, Low D. Downstream processing of monoclonal antibodies – Application of platform approaches. Journal of Chromatography B, Analytical Technologies in the Biomedical and Life Sciences 2007; 848: 28–39.

[220] Eriksson K-O, Makonnel B. Hydrophobic Interaction Chromatography. In: Janson J-C, ed. Protein Purification: Principles, High Resolution Methods, and Applications. Hoboken, N.J.: John Wiley & Sons, 2011; 165–81.

[221] Karger BL, Snyder LR, Horvath C. An Introduction to Separation Science. New York: John Wiley & Sons, 1973.

[222] Van Deemter JJ, Zuiderweg FJ, Klinkenberg A. Longitudinal diffusion and resistance to mass transfer as causes of nonideality in chromatography. Chemical Engineering Science 1956; 5: 271–89.

[223] Boi C, Malavasi A, Carbonell RG, Gilleskie G. A direct comparison between membrane adsorber and packed column chromatography performance. Journal of Chromatography. A 2020; 1612: 460629.

[224] Goyon A, Excoffier M, Janin-Bussat M, et al. Determination of isoelectric points and relative charge variants of 23 therapeutic monoclonal antibodies. Journal of Chromatography.B, Analytical Technologies in the Biomedical and Life Sciences 2017; 1065–1066: 119–28.

[225] Ghose S, Jin M, Jia L, et al. Integrated Polishing Steps for Monoclonal Antibody Purification. In: Gottschalk U., ed. Process Scale Purification of Antibodies. Hoboken, NJ: John Wiley & Sons, Inc., 2017; 303–23.

[226] Bemberis I, Noyes A, Natarajan V. Column packing for process-scale chromatography: guidelines for reproducibility. Biopharm International 2003; 2003 Supplement.

[227] GE Healthcare Application note 28-9372-07 AA – Column efficiency testing. 28-9372-07 AA 01/2010. General Electric Company. 2010.

[228] ÄKTA ready™ systems – Single-use Chromatography Systems. Cytiva publication CY10671-07Jul23-DF, Cytiva. 2023.

[229] Resolute® Flowdrive SU – Easy to Use, Flexible Single-Use Chromatography Systems. Sartorius Product Data Sheet DIR: 2578687-000-01 Status: 05 | 2024. Sartorius Stedim Biotech GmbH. Goettingen, Germany. 2024.

[230] Mobius® FlexReady Solution with Smart Flexware™ Assemblies for Chromatography and TFF – For Purification of Clinical and Process Scale Biologics. Millipore Data Sheet DS2251EN00 Ver. 8.0 43310 09/2022, Merck KGaA, Darmstadt, Germany and/or its affiliates. 2022.

[231] Mirasol F. Single-use for downstream chromatography: benefit or hindrance? BioPharm International 2019; 32: 34–37.

[232] OPUS® Pre-packed Chromatography Columns – Leader of the Pack from process validation to GMP manufacturing. Data Sheet. Repligen Corporation. 2021.

[233] Fahrner RL, Knudsen HL, Basey CD, et al. Industrial purification of pharmaceutical antibodies: development, operation, and validation of chromatography processes. Biotechnology & Genetic Engineering Reviews 2001; 18: 301–27.

[234] Hanke AT, Ottens M. Purifying biopharmaceuticals: knowledge-based chromatographic process development. Trends in Biotechnology (Regular ed.) 2014; 32: 210–20.

[235] Shukla AA, Han XS. Screening of Chromatographic Stationary Phases. In: Shukla AA, Etzel MR, and Gadam S, eds. Process Scale Bioseparations for the Biopharmaceutical Industry. Boca Raton: CRC/Taylor & Francis, 2007: 227–44.

[236] PreDictor 96-well filter plates and Assist software – High-Throughput Process Development. Cytiva publication CY13663-29Jul20-DF, Cytiva. 2020.

[237] OPUS® RoboColumn® Pre-packed Chromatography Columns User Guide. 01-08R-E03, Repligen Corporation. 2021.

[238] Welsh JP, Petroff MG, Rowicki P, et al. A practical strategy for using miniature chromatography columns in a standardized high-throughput workflow for purification development of monoclonal antibodies. Biotechnology Progress 2014; 30: 626–35.

[239] Jensen OE, Kidal S. Using volumetric flow to scaleup chromatographic processes. BioPharm International 2006; 19.

[240] Bracewell DG, Pathak M, Ma G, Rathore AS. Re-use of protein A resin: fouling and economics. BioPharm International 2015; 28.

[241] Zhang M, Miesegaes GR, Lee M, et al. Quality by design approach for viral clearance by protein A chromatography. Biotechnology and Bioengineering 2014; 111: 95–103.

[242] Strauss DM, Cano T, Cai N, et al. Strategies for developing design spaces for viral clearance by anion exchange chromatography during monoclonal antibody production. Biotechnology Progress 2010; 26: 750–55.

[243] U.S. Department of Health and Human Services, Food and Drug Administration, Center for Drug Evaluation and Research (CDER), Center for Biologics Evaluation and Research (CBER), adopted from ICH. Guidance for Industry – Q5A(R2) Viral Safety Evaluation of Biotechnology Products Derived From Cell Lines of Human or Animal Origin. 2024.

[244] MilliporeSigma – Protein Concentration and Diafiltration by Tangential Flow Filtration. Millipore Technical Brief Lit. No. TB032 Rev. C 06/03 03-117, Millipore Corporation, Billerica, MA. 2003.

[245] Challener CA. Evolving UF/DF capabilities. BioPharm International 2018; 31: 24–27.

[246] Sartoflow® 4500 Single-use Tangential Flow Filtration System. Sartorius Product Datasheet DIR: 2722133-000-01, Sartorius Stedim Biotech GmbH, Goettingen, Germany. 2023.

[247] Mobius® FlexReady Solution for TFF. 2024. MilliporeSigma. (Accessed Nov 29, 2024, at https://www.emdmillipore.com/US/en/product/Mobius-FlexReady-Solution-for-TFF,MM_NF-C100747#overview).

[248] KrosFlo® KTF TFF Systems. 2024. Repligen Corporation. (Accessed Nov 29, 2024, at https://www.repligen.com/products/downstream-filtration/krosflo-tff-systems/krosflo-ktf).

[249] Latulippe DR, Molek JR, Zydney AL. Importance of biopolymer molecular flexibility in ultrafiltration processes. Industrial & Engineering Chemistry Research 2009; 48: 2395–403.

[250] Van Reis R, Zydney A. Bioprocess membrane technology. Journal of Membrane Science 2007; 297: 16–50.

[251] Find a Spectrum Hollow Fiber Filter. 2024. Repligen Corporation. (Accessed Nov 29, 2024, at https://login.repligen.com/resources/configurators/selection-tools/find-hollow-fiber-filter).

[252] Membrane separations-Selecting a cross flow cartridge. Cytiva publication CY16044-05Jan21-SG, Cytiva. 2020.

[253] Pellicon®3 Cassettes with Ultracel® Membrane. Millipore Data Sheet Lit. No. DS1209EN00 Ver. 10.0 2016 – 00188 03/2018, Merck KGaA, Darmstadt, Germany and/or its affiliates. 2018.

[254] Yee Lau S, Pattnaik P, Raghunath B. Integrity testing of ultrafiltration systems for biopharmaceutical applications. BioProcess International 2012; 10: 52–66.

[255] ProCell™ large volume process-scale cartridges. Cytiva data file CY14299-11Sep20-DF, Cytiva. 2020.

[256] MaxCell™ process-scale hollow fiber cartridges. Cytiva data file CY13411-22Jul20-DF DF, Cytiva. 2020.

[257] Ultracel® Membranes – The membrane of choice for ultra-low protein binding and robust process performance. Millipore Data Sheet Lit No. PF1401EN00 Ver. 5.0 42711 11/2022. Merck KGaA, Darmstadt, Germany and/or its affiliates. 2022.

[258] Pellicon® Capsules with Ultracel® Membrane. Millipore Data Sheet Lit. No. MS_DS1285EN Ver. 7.0 50218 09/2023. Merck KGaA, Darmstadt, Germany and/or its affiliates. 2023.

[259] Biomax® Membranes – The membrane of choice for fast processing and exceptional chemical resistance. Millipore Data Sheet Lit No. PF1402EN00 Ver. 4.0 35173 11/2022. Merck KGaA, Darmstadt, Germany and/or its affiliates. 2022.

[260] Prostak™ Microfiltration Modules – For Convenient and Economical Perfusion and Clarification/Concentration Applications. Millipore Data Sheet DS4775EN00 Ver. 4.0 2016 – 00991, Merck KGaA, Darmstadt, Germany and/or its affiliates. 2018.

[261] Microfiltration Cassettes with Supor® Membrane. Pall publication USD 3253, GN17.10044, Pall Life Sciences, Port Washington, NY. 2017.

[262] T-Series TFF Cassettes with Omega™ Membrane. Pall Data File USD 2322d 11/20 GN20.10354, Pall Corporation. Port Washington, NY. 2020.

[263] T-Series cassette with Delta membrane. Cytiva publication CY41932-29Feb24-DF, Cytiva. 2024.

[264] TangenX® Flat Sheet Membranes. 2024. Repligen Corporation. (Accessed Nov 29, 2024, at https://www.repligen.com/products/downstream-filtration/tangenx-cassettes/tangenx-membranes).

[265] Tangential Flow Filtration Cassettes. 2024. Sartorius AG. (Accessed Nov 29, 2024, at https://www.sartorius.com/en/products/process-filtration/tangential-flow-filtration/tff-cassettes).

[266] Pellicon® 2 Cassettes Installation, User, and Maintenance Guide. MK_UG2996EN Rev 3 01/2022. Merck KGaA, Darmstadt, Germany and/or its affiliates. 2022.

[267] Korson L, Drost-Hansen W, Millero FJ. Viscosity of water at various temperatures. The Journal of Physical Chemistry 1969; 73: 34–39.

[268] Pellicon® Capsules for Single-Use Tangential Flow Filtration – Enhancing ease-of-use, batch turnaround, and process flexibility. Millipore Application Note Lit. No. MS_AN2608EN Ver. 2.0 2017-08175 09/2020, Merck KGaA, Darmstadt, Germany and/or its affiliates. 2020.

[269] Vilker VL, Colton CK, Smith KA, Green DL. The osmotic pressure of concentrated protein and lipoprotein solutions and its significance to ultrafiltration. Journal of Membrane Science 1984; 20: 63–77.

[270] Binabaji E, Rao S, Zydney AL. The osmotic pressure of highly concentrated monoclonal antibody solutions: Effect of solution conditions. Biotechnology and Bioengineering 2014; 111: 529–36.

[271] Van Reis R, Goodrich EM, Yson CL, Frautschy LN, Dzengeleski S, Lutz H. Linear scale ultrafiltration. Biotechnology and Bioengineering 1997; 55: 737–46.

[272] Parenteral Drug Association Technical Report No. 15 (Revised 2009): Validation of Tangential Flow Filtration in Biopharmaceutical Applications. 2009. (Accessed Feb 28, 2021, at https://store.pda.org/TableOfContents/TR1509_TOC.pdf).

[273] De Los Reyes G, Mir L. Method and apparatus for the filtration of biological solutions. 2008; 11/615 (028): 1–56.

[274] Cadence™ Systems Employ New Single-Pass TFF Technology to Simplify Processes and Lower Costs – Understanding the New Technology. Pall Life Sciences publication Application Note GN11.4802 USD 2789. Pall Corporation. 2011.

[275] BioContinuum™ Ultrafiltration Platform – Pellicon® Single-Pass Tangential Flow Filtration. Millipore Data Sheet Lit. No. MS_DS2735EN Ver. 2.0, Merck KGaA, Darmstadt, Germany and/or its affiliates. 2019.

[276] Cadence™ Single-Pass TFF Modular Kit. USD3284 GN18.07131. Pall Biotech. Port Washington, NY, 2018.

[277] Cadence™ Inline Diafiltration Module. USD3248 GN17.0912. Pall Life Sciences. Port Washington, NY, 2017.

[278] Langer ES, Rader RA. Biopharmaceutical manufacturing is shifting to single-use systems. Are the dinosaurs, the large stainless steel facilities, becoming extinct? American Pharmaceutical Review 2018; 21.

[279] Whitford W, Nelson D. From interest in intensification to a factory of the future. BioProcess International 2019; 17: 24–30.

[280] U.S. Department of Health and Human Services Food and Drug Administration Center for Drug Evaluation and Research (CDER) Center for Veterinary Medicine (CVM) Office of Regulatory Affairs (ORA). Guidance for Industry: PAT – A Framework for Innovative Pharmaceutical Development, Manufacturing, and Quality Assurance. 2004.

[281] Jones S. BioPhorum operations group technology roadmapping, Part 2 – efficiency, modularity, and flexibility as hallmarks for future key technologies. BioProcess International 2017; 15: 14–19.

[282] Jiang M, Severson KA, Love JC, et al. Opportunities and challenges of real-time release testing in biopharmaceutical manufacturing. Biotechnology and Bioengineering 2017; 114: 2445–56.

[283] Glossary – EudraLex – Volume 4 – Good Manufacturing Practice (GMP) guidelines. 2004. (Accessed Jan 02, 2025, at https://ec.europa.eu/health/sites/health/files/files/eudralex/vol-4/pdfs-en/glos4en200408_en.pdf).

[284] ISO/TC 209 Cleanrooms and associated controlled environments. ISO 14644-2:2015 Cleanrooms and associated controlled environments – Part 2: Monitoring to provide evidence of cleanroom performance related to air cleanliness by particle concentration. 2015.

[285] Good Manufacturing Practice (GMP) Resources. 2024. International Society for Pharmaceutical Engineering. (Accessed Jan 02, 2025, at https://ispe.org/initiatives/regulatory-resources/gmp).

[286] U.S. Food & Drug Administration, Office of Regulatory Affairs Center for Drug Evaluation and Research Center for Biologics Evaluation and Research. Guidance Document, Sterile Drug Products Produced by Aseptic Processing – Current Good Manufacturing Practice, Guidance for Industry.

[287] 100 Years of Insulin. U.S. Food and Drug Administration. (Accessed Dec 30, 2024 at https://www.fda.gov/about-fda/fda-history-exhibits/100-years-insulin).

[288] Novo Nordisk Inc. Ozempic [package insert]. 2020.

[289] EudraLex – Volume 4 – EU Guidelines for Good Manufacturing Practice for Medicinal Products for Human and Veterinary Use, Annex 1: Manufacture of Sterile Medicinal Products. 2022. (Accessed November 3, 2024, at https://health.ec.europa.eu/document/download/e05af55b-38e9-42bf-8495-194bbf0b9262_en?filename=20220825_gmp-an1_en_0.pdf).

[290] Łącki K. Introduction to Preparative Protein Chromatography. In: Jagschies G, Lindskog E, Łącki K, Galliher P., eds. Biopharmaceutical Processing: Development, Design, and Implementation of Manufacturing Processes. Amsterdam, Netherlands: Elsevier, 2018: 319–66.

[291] BPG columns Bioprocess Chromatography Column. Cytiva Publication CY978-19Jan24-DF, Cytiva. 2024.

[292] Resolute® BioSMB 80 and BioSMB 350 Single-Use Continuous Chromatography Systems for Perfusion and Batch Bioreactor Processes. Sartorius Product Datasheet Status: 09 | 2024, Sartorius Stedim Biotech GmbH, Goettingen, Germany. 2024.

[293] U.S. Dept. of Health and Human Services, Food and Drug Administration, Center for Drug Evaluation and Research (CDER), Center for Biologics Evaluation and Research (CBER), adopted from ICH. Guidance for Industry – Q10 Pharmaceutical Quality System. U.S. Food and Drug Administration 2009.

Index

https://doi.org/10.1515/9783111112459-013